Lecture Notes in Computer Science 14235

The series Lecture Notes in Computer Science (LNCS), including its subseries Lecture Notes in Artificial Intelligence (LNAI) and Lecture Notes in Bioinformatics (LNBI), has established itself as a medium for the publication of new developments in computer science and information technology research, teaching, and education.

LNCS enjoys close cooperation with the computer science R & D community, the series counts many renowned academics among its volume editors and paper authors, and collaborates with prestigious societies. Its mission is to serve this international community by providing an invaluable service, mainly focused on the publication of conference and workshop proceedings and postproceedings. LNCS commenced publication in 1973.

Olivier Bournez · Enrico Formenti · Igor Potapov
Editors

Reachability Problems

17th International Conference, RP 2023
Nice, France, October 11–13, 2023
Proceedings

Editors
Olivier Bournez
École Polytechnique
Paris, France

Enrico Formenti
Université Côte d'Azur
Nice, France

Igor Potapov
University of Liverpool
Liverpool, UK

ISSN 0302-9743 ISSN 1611-3349 (electronic)
Lecture Notes in Computer Science
ISBN 978-3-031-45285-7 ISBN 978-3-031-45286-4 (eBook)
https://doi.org/10.1007/978-3-031-45286-4

This Springer imprint is published by the registered company Springer Nature Switzerland AG
The registered company address is: Gewerbestrasse 11, 6330 Cham, Switzerland

Paper in this product is recyclable.

Preface

This volume contains the papers presented at the 17th International Conference on Reachability Problems (RP 2023), organized by the Laboratoire d'Informatique, Signaux et Systèmes de Sophia Antipolis (I3S) of the Université Côte d'Azur (France) and the Laboratoire d'Informatique de l'École Polytechnique (LIX), Paris-Saclay (France). Previous events in the series were located at the University of Kaiserslautern, Germany (2022); the University of Liverpool, UK (2021); Université Paris Cité, France (2020); Université Libre de Bruxelles, Belgium (2019); Aix-Marseille University, France (2018); Royal Holloway, University of London, UK (2017); Aalborg University, Denmark (2016); the University of Warsaw, Poland (2015); the University of Oxford, UK (2014); Uppsala University, Sweden (2013); the University of Bordeaux, France (2012); the University of Genoa, Italy (2011); Masaryk University, Czech Republic (2010); École Polytechnique, France (2009); the University of Liverpool, UK (2008); and Turku University, Finland (2007).

The aim of the conference is to bring together scholars from diverse fields with a shared interest in reachability problems, and to promote the exploration of new approaches for the modeling and analysis of computational processes by combining mathematical, algorithmic, and computational techniques. Topics of interest include (but are not limited to) reachability for infinite state systems; rewriting systems; reachability analysis in counter/timed/cellular/communicating automata; Petri nets; computational game theory; computational aspects of semigroups, groups, and rings; reachability in dynamical and hybrid systems; frontiers between decidable and undecidable reachability problems; complexity and decidability aspects; predictability in iterative maps; and new computational paradigms.

We are very grateful to our invited speakers, who gave the following talks:

Nathalie Aubrun (CNRS, Paris-Saclay, France)
"*The Domino problem extended to groups*"
Jarkko Kari (University of Turku, Finland)
"*Low complexity colorings of the two-dimensional grid*"
Bruno Martin (Université Côte d'Azur, France)
"*Randomness quality and trade-offs for random number generators*"
Shinnosuke Seki (University of Electro-Communications, Japan)
"*How complex shapes can RNA fold into?*"
Dmitry Zaitsev (Odessa State Environmental University, Ukraine)
"*Sleptsov net computing resolves modern supercomputing problem*"

The conference received 33 submissions (19 regular, one invited paper and 13 presentation-only submissions) from which two regular papers were withdrawn by their authors. Each submission was carefully single reviewed by at least three Program Committee (PC) members. Based on these reviews, the PC decided to accept 13 regular papers and 13 presentation-only submissions, in addition to the five invited speakers' contributions. The members of the PC and the list of external reviewers can be found at the end of this preface. We are grateful for the high-quality work produced by the PC and the external reviewers. Overall this volume contains 13 contributed papers and two abstracts from invited speakers which cover their talks.

The conference also provided the opportunity to other young and established researchers to present work in progress or work already published elsewhere. This year in addition to the thirteen regular submissions, the PC accepted 13 high-quality informal presentations on various reachability aspects in theoretical computer science. A list of accepted presentation-only submissions is given later in this front matter. So overall, the conference program consisted of five invited talks, 13 presentations of contributed papers, and 13 informal presentations in the area of reachability problems, stretching from results on fundamental questions in mathematics and computer science up to efficient solutions of practical problems.

It is a pleasure to thank the team behind the EasyChair system and the Lecture Notes in Computer Science team at Springer, who together made the production of this volume possible in time for the conference. Finally, we thank all the authors and invited speakers for their high-quality contributions, and the participants for making RP 2023 a success. We are also very grateful to Springer for their financial sponsorship.

October 2023

Olivier Bournez
Enrico Formenti
Igor Potapov

Organization

Program Committee Chairs

Olivier Bournez	École Polytechnique, LIX, IPP, France
Enrico Formenti	Université Côte d'Azur, France

Steering Committee

Parosh Aziz Abdulla	Uppsala University, Sweden
Olivier Bournez	École Polytechnique, LIX, IPP, France
Vesa Halava	University of Turku, Finland
Alain Finkel	École normale supérieure Paris-Saclay, France
Oscar Ibarra	University of California Santa Barbara, USA
Juhani Karhumaki	University of Turku, Finland
Jérôme Leroux	Université de Bordeaux, France
Joël Ouaknine	Max Planck Institute for Software Systems, Germany
Igor Potapov	University of Liverpool, UK
James Worrell	University of Oxford, UK

Program Committee

Parosh Aziz Abdulla	Uppsala University, Sweden
Udi Boker	Reichman University, Israel
Christel Baier	Technical University of Dresden, Germany
Veronica Becher	Universidad de Buenos Aires, Argentina
Benedikt Bollig	ENS Paris-Saclay, France
Olivier Bournez	École Polytechnique, LIX, IPP, France
Emilie Charlier	Université de Liège, Belgium
Stéphane Demri	ENS Paris-Saclay, France
Javier Esparza	Technical University of Munich, Germany
Enrico Formenti	Université Côte d'Azur, France
Gilles Geeraerts	Université libre de Bruxelles, Belgium
Roberto Giacobazzi	University of Verona, Italy
Daniel Graça	University of Algarve, Portugal
Christoph Haase	University of Oxford, UK

Peter Habermehl	Université Paris Cité, France
Vesa Halava	University of Turku, Finland
Mika Hirvensalo	University of Turku, Finland
Akitoshi Kawamura	Kyoto University, Japan
Dietrich Kuske	Technical University of Ilmenau, Germany
Martin Kutrib	University of Giessen, Germany
Jérôme Leroux	Université de Bordeaux, France
Igor Potapov	University of Liverpool, UK
Tali (Nathalie) Sznajder	Sorbonne Université, France

Additional Reviewers

Thomas Brihaye
Tero Harju
Lucie Guillou
Roland Guttenberg
Andreas Malcher
Pablo Terlisky
Takao Yuyama

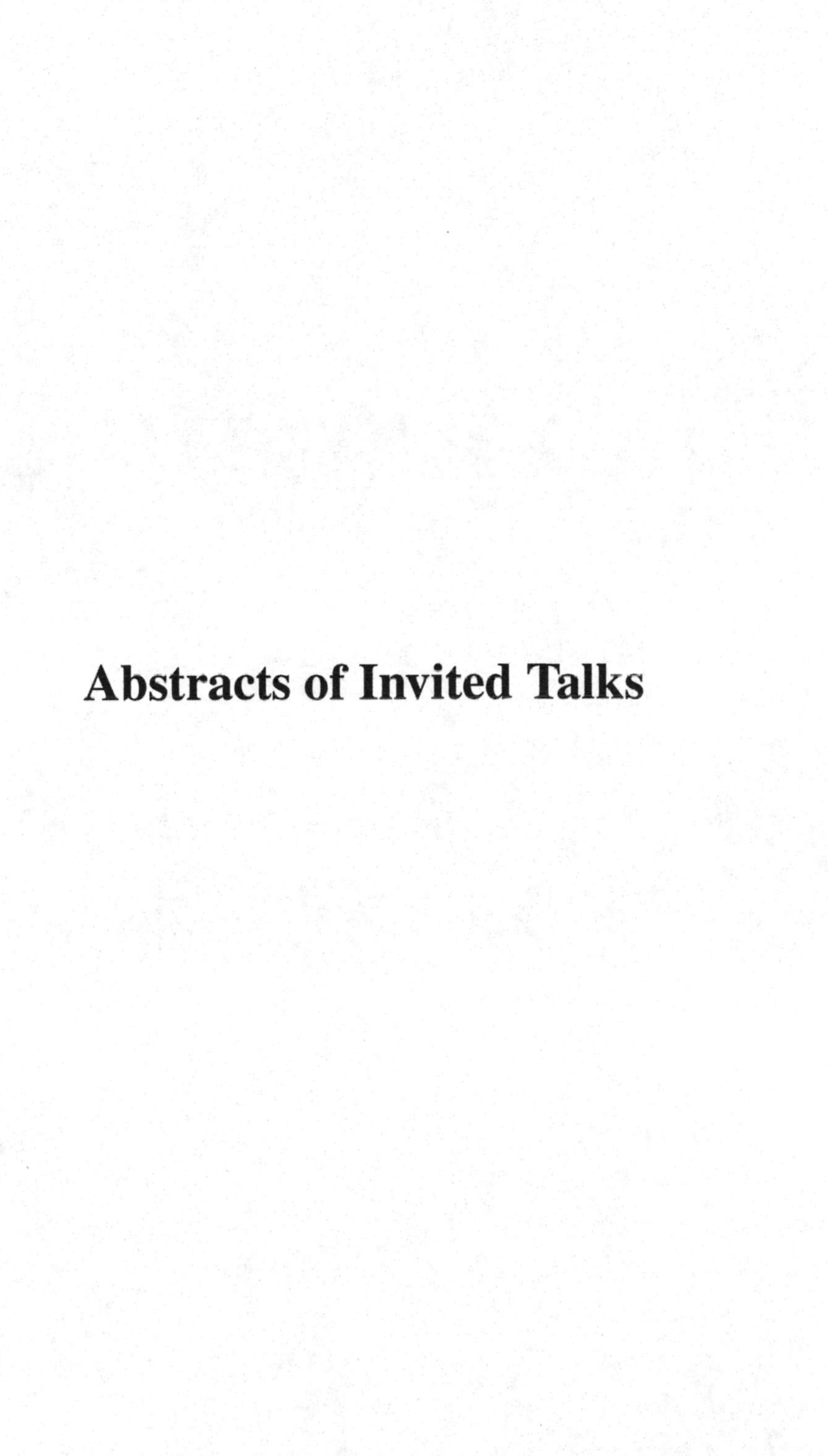

Abstracts of Invited Talks

Low Complexity Colorings of the Two-Dimensional Grid

Jarkko Kari

Department of Mathematics and Statistics, University of Turku, Finland
jkari@utu.fi

Abstract. A two-dimensional configuration is a coloring of the infinite grid $\mathbb{Z}^2$ using a finite number of colors. For a finite subset D of Z^2, the D-patterns of a configuration are the patterns of shape D that appear in the configuration. The number of distinct D-patterns of a configuration is a natural measure of its complexity. We consider low-complexity configurations where the number of distinct D-patterns is at most $|D|$, the size of the shape. We use algebraic tools to study periodicity of such configurations [3]. We show, for an arbitrary shape D, that a low-complexity configuration must be periodic if it comes from the well-known Ledrappier subshift, or from a wide family of other similar algebraic subshifts [1]. We also discuss connections to the well-known Nivat's conjecture: In the case D is a rectangle – or in fact any convex shape – we establish that a uniformly recurrent configuration that has low-complexity with respect to shape D must be periodic [2]. This implies an algorithm to determine if a given collection of mn rectangular patterns of size $m \times n$ admit a configuration containing only these patterns. Without the complexity bound the question is the well-known undecidable domino problem.

References

1. Kari, J., Moutot, E.: Nivat's conjecture and pattern complexity in algebraic subshifts. Theor. Comput. Sci. **777**, 379–386 (2019)
2. Kari, J., Moutot, E.: Decidability and periodicity of low complexity tilings. In: Paul, C., Bläser, M. (eds.) STACS 2020, 10–13 March 2020, Montpellier, France. LIPIcs, vol. 154, pp. 14:1–14:12. Schloss Dagstuhl - Leibniz-Zentrum für Informatik (2020)
3. Kari, J., Szabados, M.: An algebraic geometric approach to Nivat's conjecture. Inf. Comput. **271**, 104481 (2020)

How Complex Shapes Can RNA Fold Into?

Shinnosuke Seki

University of Electro-Communications, 1-5-1 Chofugaoka, Chofu, Tokyo, 1828585 Japan
s.seki@uec.ac.jp

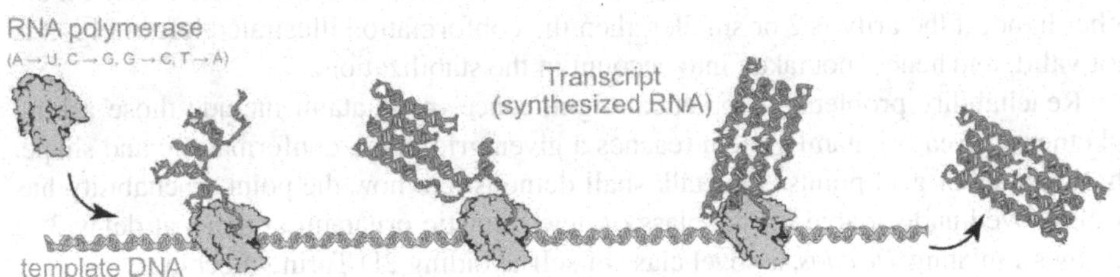

Fig. 1. RNA co-transcriptional folding.

Transcription is a process in which an RNA sequence of nucleotides (A, C, G, and U), colored in blue in Fig. 1, is synthesized from its DNA template sequence (gray) by a molecular Xerox called polymerase (orange), which scans the template uni-directionally and maps it nucleotide by nucleotide according to the rule A → U, C → G, G → C, and T → A, While being thus synthesized (*transcribed*), the product sequence, or *transcript*, folds upon itself into a structure by stabilizing via hydrogen bonds primarily of the types A−U, C−G, and G−U, but also by having RNA helices be stacked coaxially. This phenomenon termed *co-transcriptional folding* has proven programmable *in-vitro* for assembling an RNA rectangular tile structure; indeed, Geary, Rothemund, and Andersen demonstrated how to design a DNA template sequence such that the corresponding RNA transcript co-transcriptionally folds into the tile structure highly probably [2].

Oritatami is a model of computation by RNA co-transcriptional folding [1]. In this model, a system folds a sequence of abstract molecules, called *beads*, of finitely many types co-transcriptionally over the 2-dimensional triangular grid into a non-self-crossing directed path. Let Σ be a finite alphabet of bead types. A configuration in oritatami, or *conformation*, is a triple $(w, P, \mathcal{H})$ of a sequence w over Σ, a self-avoiding directed path over the triangular grid that is as long as w, and a set $\mathcal{H}$ of pairs of integers (see Fig. 2); it is to be interpreted that w is folded along P, that is, the i-th bead of w is at the i-th vertex of P, and $(j, k) \in \mathcal{H}$ means that $k \geq j{+}2$, $P[j]$ is adjacent to $P[k]$, and a hydrogen bond forms between the beads at these adjacent points, that is, $w[j]$ and $w[k]$. A conformation is of *arity* α if each of its beads forms at most α hydrogen bonds;

His work is supported in part by the JSPS KAKENHI Grant-in-Aids for Scientific Research (B) No. 20H04141 and (C) No. 20K11672.

for instance, the conformation in Fig. 2 is of arity 3. An oritatami system is a tuple $(\Sigma, \delta, \alpha, \sigma, w, \heartsuit)$ of two integer parameters δ and α called *delay* and arity, respectively, an initial conformation, or *seed*, σ of arity α, a transcript w over the alphabet Σ, and a symmetric affinity relation $\heartsuit \subseteq \Sigma \times \Sigma$, which determines what types of beads are allowed to bind, being placed at unit distance. It has the transcript w elongate at the end of the seed one bead per step (oritatami is a discrete-time model), and fold co-transcriptionally in the sense that only the most nascent δ beads are allowed to move, and the oldest among these δ beads is stabilized according to the conformations that form the largest number of hydrogen bonds. For instance, in Fig. 2, the bead `H7` is going to stabilize at delay $\delta = 3$. If either the arity of the system is set to 3 or it is set to 4 or larger but there is no other way to fold the suffix `H7-H8-H9` so as to form 4 or more new bonds, then `H7` can be stabilized along with its bond with `H4` as illustrated there. On the other hand, if the arity is 2 or smaller, then the conformation illustrated left in Fig. 2 is not valid, and hence, not taken into account in the stabilization.

Reachability problems of practical significance in oritatami include those asking whether a given oritatami system reaches a given grid point, conformation, and shape, that is, a set of grid points. This talk shall demonstrate how the point reachability has been proved undecidable for the class of deterministic oritatami systems at delay 3 in [3] by simulating *Turedos*, a novel class of self-avoiding 2D Turing machines.

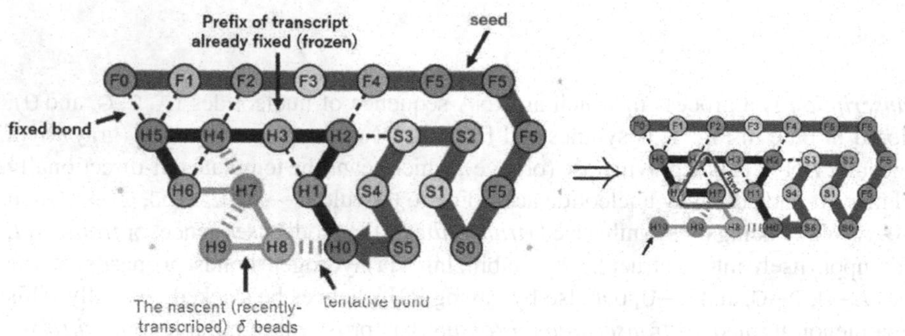

Fig. 2. A conformation and stabilization of a bead (`H7` here) in oritatami

References

1. Geary, C., Meunier, P., Schweabanel, N., Seki, S.: Programming biomolecules that fold greedily during transcription. In: Proceedings of the MFCS 2016. LIPIcs, vol. 58, pp. 43:1–43:14 (2016)
2. Geary, C., Rothemund, P.W.K., Andersen, E.S.: A single-stranded architecture for cotranscriptional folding of RNA structures. Science **345**(6198), 799–804 (2014)
3. Pchelina, D., Schabanel, N., Seki, S., Theyssier, G.: Oritatami systems assemble shapes no less complex than tile assembly model (aTAM). In: Proceedings of the STACS 2022. LIPIcs, vol. 219, pp. 51:1–51:23 (2022)

Sleptsov Net Computing Resolves Modern Supercomputing Problems

Dmitry Zaitsev

Odessa State Environmental University, St. Lvivska 15, Odessa, 65016 Ukraine
daze@acm.org
http://daze.ho.ua

Modern HPC Problems. In his Turing Award Lecture, Jack Dongarra revealed a drastic problem of modern HPC: low efficiency on real-life task mixtures, 0.8% for the best supercomputer Frontier [1]. For dozens of years, HPC architecture design was influenced by LINPACK benchmarks based on dense computations, filling-in cache perfectly and reusing its content with matrix arithmetic. Real-life tasks are often sparse, decreasing computer efficiency considerably.

Sleptsov Net. It was proven that a Sleptsov net (SN), which fires a transition in a multiple number of copies at a step, runs exponentially faster than a Petri net [2]. SNs are Turing-complete [3]. That opens the prospect of SNs application as a general-purpose concurrent programming language. An example of the SN program is shown in Fig. 1.

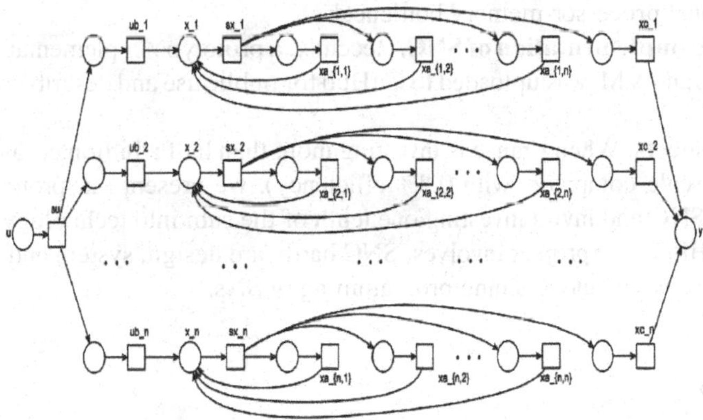

Fig. 1. SN program for linear control in 2 time cycles capable of hypersonic object control.

Computability vs Reachability. Reachability problems are transformed into computability problems when the system in question becomes Turing-complete. In spite of Rice's theorem, the formal verification of programs succeeds in proving some program properties, including the program function, which, in essence, represents a symbolic specification of a reachable set of states. Concurrent programs induce additional

Partially supported by Fulbright for the OLF talk on October 10, 2017 at Stony Brook University, New York, USA.

aspects of reachability by reaching a terminal state, usually the only one, invariant with respect to the transitions' firing sequence, which represents a definite implementation of concurrency.

Verifying SN Programs. The reachability of a symbolic terminal state and its invariance with respect to the transitions' firing sequence become basic criteria for conventional formal verification of programs. Conservativeness means the absence of overflow. Liveness, and liveness enforcement, are central concepts for debugging concurrent programs.

Drawing SN Programs. Graphical programming was a serious alternative to conventional textual programming in a series of real-life projects. R-technology of programming is a convincing example. UML uses a graphical approach and Petri net notation. There are approaches that combine a Petri net graph with a programming language to load a graph's elements, for example, CPN Tools. SN Computing (SNC) represents a purely graphical approach that uses text as comments only; the hierarchical structure of a program is implemented via transition substitution by a subnet. SNC's homogeneous structure simplifies concurrent programming and provides fine granulation of parallel processes starting from the arithmetic and logic operations level. We can use either control or data flow, or a combined approach. For compatibility, traditional operators of programming languages, such as branching and loops, are easily implemented. Transitions are concurrent initially; a graph restricts this concurrency only when the application domain requires it.

Running SN Programs on Dedicated Hardware. SNC hardware is foreseen as a re-configurable sparse multidimensional matrix of computing memory with primordially concurrent transitions and conflicts resolved by locks and arbiters. It allows us to get rid of the traditional processor-memory bottleneck.

Prototype Implementation of SNC. Recently, a prototype implementation of SNC, including IDE and VM, was uploaded to GitHub for public use and described in a journal paper [4].

SNC Prospects. When France is investing more than half a billion euros into a conventional exascale computer (with 0.8% efficiency), we present our prototype implementation of SNC and invite investing one tenth of the sum into technology that brings efficiency to HPC. The project involves: SNC hardware design, system and application SNC software design, and teaching programming in SNs.

References

1. Zaitsev, D.: Sleptsov net computing resolves problems of modern supercomputing revealed by Jack Dongarra in his turing award talk in November 2022. Int. J. Parallel, Emergent Distrib. Syst. **38**(4), 275–279 (2023)
2. Zaitsev, D.: Sleptsov nets run fast, IEEE Trans. Syst. Man Cybern. Syst. **46**(5), 682–693 (2016)
3. Zaitsev, D.: Strong Sleptsov nets are turing complete. Inf. Sci. **621**, 172–182 (2023)
4. Zaitsev, D., Shmeleva, T.R., Zhang, Q., Zhao, H.: Virtual machine and integrated developer environment for Sleptsov net computing. Parallel Process. Lett. (2023)

Presentation-Only Submissions

Sleptsov Nets Are Turing-Complete

Dmitry Zaitsev and Bernard Berthomieu

Abstract. The present paper proves that a Sleptsov net (SN) is Turing-complete, which considerably improves, with a brief construct, the previous result that a strong SN is Turing-complete. Recall that, unlike Petri nets, an SN always fires enabled transitions at their maximal firing multiplicity, as a single step, leaving a nondeterministic choice of which fireable transitions to fire. A strong SN restricts nondeterministic choice to firing only the transitions having the highest firing multiplicity.

On Computing Optimal Temporal Branchings

Daniela Bubboloni, Costanza Catalano, Andrea Marino, and Ana Silva

Abstract. A spanning out-branching (in-branching), also called directed spanning tree, is the smallest subgraph of a digraph that makes every node reachable from a root (the root reachable from any node). The computation of spanning branchings is a central problem in theoretical computer science due to its application in reliable network design. This concept can be extended to temporal graphs, which are digraphs where arcs are available only at prescribed times and paths make sense only if the availability of the arcs they traverse is non-decreasing. In this context, the paths of the out-branching from the root to the spanned vertices must be valid temporal paths. While the literature has focused only on minimum weight temporal out-branchings or the ones realizing the earliest arrival times to the vertices, the problem is still open for other optimization criteria. In this work we define four different types of optimal temporal out-branchings (TOB) based on the optimization of the traveling time (ST-TOB), of the travel duration (FT-TOB), of the number of transfers (MT-TOB) or of the departure time (LD-TOB). For d in ST,MT,LD, we provide necessary and sufficient conditions for the existence of spanning d-TOBs; when those do not exist, we characterize the maximum vertex set that a d-TOB can span. Moreover, we provide a log linear algorithm for computing such d-TOBs. Oppositely, we show that deciding the existence of an FT-TOB spanning all the vertices is NP-complete. This is quite surprising, as all the above distances, including FT, can be computed in polynomial time, meaning that computing temporal distances is inherently different from computing d-TOBs. Finally, we show that the same results hold for optimal temporal in-branchings.

Positivity Problems for Reversible Linear Recurrence Sequences

George Kenison, Joris Nieuwveld, Joel Ouaknine, and James Worrell

Abstract. It is a longstanding open problem whether there is an algorithm to decide the Positivity Problem for linear recurrence sequences (LRS) over the integers, namely whether given such a sequence, all its terms are non-negative. Decidability is known for LRS of order 5 or less, i.e., for those sequences in which every new term depends linearly on the previous five (or fewer) terms. For *simple* LRS (i.e., those whose characteristic polynomial has no repeated roots), decidability of Positivity is known up to order 9.

In this paper, we focus on the important subclass of reversible LRS, i.e., those integer LRS $\langle u_n \rangle_{n=0}^{\infty}$ whose bi-infinite completion $\langle u_n \rangle_{n=-\infty}^{\infty}$ also takes exclusively integer values; a typical example is the classical Fibonacci (bi-)sequence $\langle \ldots, 5, -3, 2, -1, 1, 0, 1, 1, 2, 3, 5, \ldots \rangle$. Our main results are that Positivity is decidable for reversible LRS of order 11 or less, and for simple reversible LRS of order 17 or less.

Discontinuous IVPs with Unique Solutions

Riccardo Gozzi and Olivier Bournez

Abstract. Motivated by discussing the hardness of reachability questions for dynamical systems, we develop a theory for initial value problems (involving ordinary differential equations) with a discontinuous right-hand side, in the case one knows that the solution is unique. We first discuss our results and our obtained theory. We will then review some consequences on the hardness of reachability questions, in particular, related to the questions of measuring hardness by a rank similar to the Denjoy hierarchy or Kechris and Woodin's rank for differentiability.

The study of ordinary differential equations (ODEs) and initial value problems (IVPs) with discontinuous right-hand terms has many applications to a wide range of problems in mechanics, electrical engineering and theory of authomatic control. Broadly speaking, discontinuous ODEs of the form $y'(t) = f(t, y)$ can be divided into two main categories [5]: one in which f is continuous in y for almost all t and one in which f is discontinuous on an arbitrary subset of its domain. In the first case, existence and unicity for solutions of the IVPs can be discussed under specific requirements on f, such as the Carathéodory conditions [1]. In the second case, the most common approach is to study the dynamic using differential inclusions of the form $y'(t) \in F(t, y)$ by identifying the correct definition of F on the set of discontinuity points. In both cases, the solution, when unique, is an absolutely continuous function y such that $y'(t) = f(t, y)$ is defined almost everywhere in an interval I.

We choose to analyze a different scenario: discontinuous IVPs for which the solution is necessarily unique and the equation $y'(t) = f(t, y)$ is defined everywhere on I. In other words, we assume existence and unicity and we focus on finding an analytical procedure to obtain such a solution from f and the initial condition. In this sense, the point of view is similar to the one in [2], where it is shown that when y is unique, then it is computable if the IVP is. Nonetheless, unicity when f is discontinuous might imply noncomputability of y even when the set of discontinuity points is trivial.

We first demonstrate the difficulty by means of an example: we construct a bidimensional IVP such that, despite having computable initial condition and f computable everywhere except a straight line, has a solution that assumes a noncomputable value at a fixed integer time. This demonstrates the capability of these dynamical systems of generating highly complicated solutions even when the structure of the discontinuity points in the domain is particularly simple. Therefore, our goal is to characterize the definability of y, generalizing the result obtained in [2] for computability to a wider class of IVPs with unique solutions.

Our approach is close in spirit to that of A. Denjoy, who provided with his totalization method an extension to the Lebesgue integral in order to generalize the operation of antidifferentiation to a wider class of derivatives [3]. This perspective fits within a wider research field that explores set theoretical descriptions of the complexity of operations such as differentiation and integration. For example, [12] makes use of the notion of differentiable rank from [7] to present a precise lightface classification of differentiable functions based on how complex their derivative can be. In [4, 11] instead the authors conduct similar treatments for antidifferentiation.

More formally, the dynamical systems we are considering are IVPs of the form: given an interval $[a, b]$, a closed domain $E \subset \mathbb{R}^r$ for some $r \in \mathbb{N}$, a point $y_0 \in E$ and a function $f : E \to E$ we have:

$$\begin{cases} y'(t) = f(y(t)) \\ y(a) = y0 \end{cases} \tag{1}$$

for some $y : [a, b] \to \mathbb{R}^r$ with $y([a, b]) \subset E$. Under the assumption of unicity of y we show that two conditions on the right-hand term f are sufficient for obtaining the solution via transfinite recursion up to a countable ordinal. In order to do so we make use of a construction inspired by the Cantor-Bendixson analysis where the set derivative operator is replaced by the action of excluding the set of discontinuity points of f. More precisely, we define:

Definition 1. *Consider a closed domain $E \subset \mathbb{R}^r$ for some $r \in \mathbb{N}$ and a function $f : E \to \mathbb{R}^r$. Let $\{E_\alpha\}_{\alpha<\omega_1}$ and $\{f_\alpha\}_{\alpha<\omega_1}$ be transfinite sequences such that $f_\alpha = f\restriction_{E_\alpha} : E_\alpha \to \mathbb{R}^r$ defined as following: let $E_0 = E$; for all $\alpha = \beta + 1$ let $E_\alpha = \{x \in E_\beta : f_\beta \text{ is discontinuous in } x\}$; for all α limit ordinal let E_α be $E_\alpha = \cap_\beta E_\beta$ with $\beta < \alpha$. We call $\{E_\alpha\}_{\alpha<\omega_1}$ the sequence of f-removed sets on E.*

Studying the structure of the sequence of f-removed sets on E allows us to prove that the two following conditions on f are sufficient for obtaining the solution. The conditions are: 1) f is a function of class Baire one 2) For all closed $K \subseteq E$ the set of discontinuity points of $f\restriction_K$ is a closed set. This result is obtained with our main theorem:

Theorem 1. *Consider a closed interval, a closed domain $E \subset \mathbb{R}^r$ for some $r \in \mathbb{N}$ and a function $f : E \to E$ such that, given an initial condition, the IVP of the form of Equation 1 with right-hand term f has a unique solution on the interval. If f is a function of class Baire one such that for every closed $K \subseteq E$ the set of discontinuity points of $f\restriction_K$ is closed, then we can obtain the solution analytically via transfinite recursion up to an ordinal α such that $\alpha < \omega_1$.*

This result expresses the analogy between this context and the totalization method in [3], leaving open the possibility of defining a rank for the IVPs related to constructible ordinals in order to populate a hierarchy similar to the one presented in [12] for differentiable functions.

The process of associating with each of these discontinuous IVPs a correspective rank is deeply connected with evaluating an upper bound for the complexity of the reachability problem within each domain of definition. Indeed, for a generic dynamical system in a given initial state, the reachability problem is defined as the problem of determining which sets of states that dynamical system can reach. In the context of bounded IVPs with a unique solution, which are continuous dynamical systems, the reachability problem is then expressed as the problem of verifying whether the solution has reached a specific target area in the domain. Hence, reachability analysis of continuous systems of ODEs involves the study of attractors, fixed points, and periodic trajectories and has had huge relevance in the last century. To cite just one example, it is enough to mention the investigation over the existence of a strange attractor for the Lorenz system [8], included by Smale in his list of problems of the century in [9], and later answered affirmatively in [10]. As long as the solution is always unique, the method we introduced for proving the above theorem satisfies the property of being uniform with respect to the choice of the initial condition within E. Indeed, it performs a careful, controlled exploration of the whole search space E by subdividing it into a transfinite number of adequate rational boxes. This implies that the number of maximum transfinite steps required to obtain the solution from any initial state can be interpreted as a good upper bound for the complexity of the reachability problem for the whole domain E.

For the case of computable, unbounded IVPs involving ODEs with unique solutions, it has been proved in [6] that the reachability problem is not computable, since the time required for the solution to reach its maximal interval of definition can be a noncomputable real number. As mentioned above, we have constructed an example that demonstrates that, once we allow simple discontinuities in the right-hand term f, the solution can be noncomputable also for the case of bounded domains with the maximum interval of definition being a closed interval. Indeed, in our example, the noncomputable real is no longer expressed by the time variable but instead by the actual value assumed by the solution at a given integer time. Consequently, this result establishes a lower bound for the complexity of the reachability problem of these types of continuous dynamical systems.

References

1. Carathéodory, C.: Vorlesungen über reelle Funktionen. BG Teubner (1918)
2. Collins, P., Graça, D.: Effective computability of solutions of differential inclusions-the ten thousand monkeys approach. J. Univ. Comput. Sci. (15), 1162–1185 (2009)
3. Denjoy, A.: Une extension de l'intégrale de m. lebesgue. CR Acad. Sci. Paris **154**, 859–862 (1912)
4. Dougherty, R., Kechris, A.S.: The complexity of antidifferentiation. Adv. Math. **88**(2), 145–169 (1991)
5. Filippov, A.F.: Differential equations with discontinuous right-hand side. Matematicheskii sbornik **93**(1), 99–128 (1960)
6. Graça, D.S., Zhong, N., Buescu, J.: Computability, noncomputability and undecidability of maximal intervals of IVPs. Trans. Am. Math. Soc. **361**(6), 2913–2927 (2009)
7. Kechris, A.S., Hugh Woodin, W.: Ranks of differentiable functions. Mathematika **33**(2), 252–278 (1986)
8. Lorenz, E.N.: Deterministic non-periodic flow. J. Atmos. Sci. **20**, 130–141 (1963)
9. Smale,S.: Mathematical problems for the next century. Math. Intell. **20**, 7–15 (1998). https://doi.org/10.1007/BF03025291
10. Tucker, W.: A rigorous ODE solver and Smale's 14th problem. Found. Comput. Mathe. **2**(1), 53–117 (2002)

11. Westrick, L.: An effective analysis of the Denjoy rank (2020)
12. Westrick, L.B.: A lightface analysis of the differentiability rank. J. Symb. Logic **79**(1), 240–265 (2014)

Geometry of Reachability Sets of Vector Addition Systems

Roland Guttenberg, Michael Raskin, and Javier Esparza

Abstract. Vector Addition Systems (VAS), aka Petri nets, are a popular model of concurrency. The reachability set of a VAS is the set of configurations reachable from the initial configuration. Leroux has studied the geometric properties of VAS reachability sets, and used them to derive decision procedures for important analysis problems. In this paper we continue the geometric study of reachability sets. We show that every reachability set admits a finite decomposition into disjoint almost hybridlinear sets enjoying nice geometric properties. Further, we prove that the decomposition of the reachability set of a given VAS is effectively computable. As a corollary, we derive a new proof of Hauschildt's 1990 result showing the decidability of the question whether the reachability set of a given VAS is semilinear. As a second corollary, we prove that the complement of a reachability set, if it is infinite, always contains an infinite linear set.

Semënov Arithmetic, Affine VASS, and String Constraints

Andrei Draghici, Christoph Haase, and Florin Manea

Abstract. We study extensions of Semënov arithmetic, the first-order theory of the structure $\langle \mathbb{N}, +, 2^x \rangle$. It is well-known that this theory becomes undecidable when extended with regular predicates over tuples of number strings, such as the Büchi V_2-predicate. We therefore restrict ourselves to the existential theory of Semënov arithmetic and show that this theory is decidable in EXPSPACE when extended with arbitrary regular predicates over tuples of number strings. Our approach relies on a reduction to the language emptiness problem for a restricted class of affine vector addition systems with states, which we show decidable in EXPSPACE. As an application of our result, we settle an open problem from the literature and show decidability of a class of string constraints involving length constraints.

Multiplicity Problems on Algebraic Series and Context-Free Grammars

Nikhil Balaji, Lorenzo Clemente, Klara Nosan, Mahsa Shirmohammadi, and James Worrell

Abstract. In this paper we obtain complexity bounds for computational problems on algebraic power series over several commuting variables. The power series are specified by systems of polynomial equations: a formalism closely related to weighted context-free grammars. We focus on three problems—decide whether a given algebraic series is identically zero, determine whether all but finitely many coefficients are zero, and compute the coefficient of a specific monomial. We relate these questions to well-known computational problems on arithmetic circuits and thereby show that all three problems lie in the counting hierarchy. Our main result improves the best known complexity bound on deciding zeroness of an algebraic series. This problem is known to lie in PSPACE by reduction to the decision problem for the existential fragment of the theory of real closed fields. Here we show that the problem lies in the counting hierarchy by reduction to the problem of computing the degree of a polynomial given by an arithmetic circuit. As a corollary we obtain new complexity bounds on multiplicity equivalence of context-free grammars restricted to a bounded language, language inclusion of a non-deterministic finite automaton in an unambiguous context-free grammar, and language inclusion of a non-deterministic context-free grammar in an unambiguous finite automaton.

Linear Loop Synthesis for Polynomial Invariants

George Kenison, Laura Kovács, and Anton Varonka

Abstract. A loop invariant is a formal property specifying a relationship between variables that holds before and after every iteration of a program loop. Invariants provide inductive arguments that are key in automating the verification of loops. In this line of work, we advocate an alternative solution to invariant generation. Rather than inferring invariants from loops, we generate loops satisfying a given set of invariants. As such, the correctness of synthesised loops is guaranteed by construction. From the reachability perspective, the objective of loop synthesis is to generate non-trivial behaviours that never reach a "bad state". We show that already loops with linear updates (or linear dynamical systems) exhibit behaviours specified by arbitrary polynomial invariants from a broad class: e.g., quadratic equations or conjunctions of pure difference binomial equalities. We introduce an algorithmic approach that constructs linear loops from such polynomial invariants, by generating linear recurrence sequences that have given algebraic relations amongst their terms. This work extends the paper presented at ISSAC'23.

Higher-Dimensional Automata Theory

Uli Fahrenberg

Abstract. I will give a gentle introduction to higher-dimensional automata (HDAs) and their language theory. HDAs have been introduced some 30 years ago as a model for non-interleaving concurrency which generalizes, for example, Petri nets while retaining some automata-theoretic intuition. They have been studied mostly for their operational and geometric aspects and are one of the original motivations for directed algebraic topology. In a series of papers we have recently started to work on the language theory of HDAs: we have introduced languages of HDAs as weak sets of interval pomsets with interfaces [1, 2] and shown that they satisfy Kleene and Myhill-Nerode type theorems [3, 4]. Further, HDAs are not generally determinable nor complementable, but language inclusion is decidable [4, 5]. The picture which emerges is that, even though things can sometimes get hairy in proofs, HDAs have a rather pleasant language theory, a fact which should be useful in the theory of non-interleaving concurrency and its applications. Joint work with Amazigh Amrane, Hugo Bazille, Christian Johansen, Georg Struth, and Krzysztof Ziemiański.

References

1. Fahrenberg, U., Johansen, C., Struth, G., Ziemiański, K.: Languages of higher-dimensional automata. Math. Struct. Comput. Sci. **31**(5) (2021)
2. Fahrenberg, U., Johansen, C., Struth, G., Ziemiański, K.: Posets with interfaces as a model for concurrency. Inf. Comput. **285**(2) (2022)
3. Fahrenberg, U., Johansen, C., Struth, G., Ziemiański, K.: A Kleene Theorem for Higher-Dimensional Automata. CONCUR (2022)
4. Fahrenberg, U., Ziemiański, K.: A Myhill-Nerode theorem for higher-dimensional automata. In: Gomes, L., Lorenz, R. (eds.) PETRI NETS 2023. LNCS, vol 13929, pp 167–188. Springer, Cham (2023). https://doi.org/10.1007/978-3-031-33620-1_9
5. Amrane, A., Bazille, H., Fahrenberg, U., Ziemiański, K.: Developments in Higher-Dimensional Automata Theory. CoRR abs/2305.02873 (2023)

Universality and Forall-Exactness of Cost Register Automata with Few Registers

Laure Daviaud and Andrew Ryzhikov

Abstract. The universality problem asks whether a given finite state automaton accepts all the input words. For quantitative models of automata, where input words are mapped to real values, this is naturally extended to ask whether all the words are mapped to values above (or below) a given threshold. This is known to be undecidable for commonly studied examples such as weighted automata over the positive rational (plus-times) or the integer tropical (min-plus) semirings, or equivalently cost register automata (CRAs) over these semirings. In this paper, we prove that when restricted to CRAs with only three registers, the universality problem is still undecidable, even with additional restrictions for the CRAs to be copyless linear with resets.

In contrast, we show that, assuming the unary encoding of updates, the ∀-exact problem (does the CRA output zero on all the words?) for integer min-plus linear CRAs can be decided in polynomial time if the number of registers is constant. Without the restriction on the number of registers this problem is known to be PSPACE-complete.

This paper was published at MFCS 2023.

History-Determinism vs. Simulation

Karoliina Lehtinen

Abstract. Language inclusion between automata is a key problem in verification: given an automaton representing a program and another one representing a specification, language inclusion of the former in the latter captures precisely whether all executions of the program satisfy the specification. Unfortunately, in the presence of nondeterminism, inclusion is algorithmically hard. (Fair) Simulation, which implies language inclusion—but is, in general, strictly stronger—is easier to check. Therefore, automata for which language inclusion and simulation coincide are particularly well-suited for model-checking. We call such automata guidable, after a similar notion used previously by Colcombet and Löding for tree automata.

Guidability, however, is not an easy property to decide, since it is contingent on an automaton simulating a potentially infinite number of language-included automata. Hence we would like to have, whenever possible, a characterisation that is more amenable to algorithmic detection.

Deterministic automata are of course always guidable, and so are history-deterministic automata. These are mildly nondeterministic automata, in which nondeterministic choices are permitted, but they must only depend on the word read so far, rather than the future of the word. Such automata are guidable. In fact, the connection between guidability and history-determinism seems profound: for several classes of automata, including the class of all labelled transition systems, guidability and history-determinism coincide. In these cases, history-determinism, which can in many cases be efficiently decided, is a convenient characterisation of guidability. However, the two notions do not always coincide.

In this joint work in progress with Udi Boker, Tom Henzinger and Aditya Prakash, we study under what conditions history-determinism and guidability coincide.

Energy Büchi Problems

Sven Dziadek, Uli Fahrenberg and Philipp Schlehuber

Abstract. We show how to efficiently solve energy Büchi problems in finite weighted automata and in one-clock weighted timed automata. Solving the former problem is our main contribution and is handled by a modified version of Bellman-Ford interleaved with Couvreur's algorithm. The latter problem is handled via a reduction to the former relying on the corner-point abstraction. All our algorithms are freely available and implemented in a tool based on the open-source platforms TChecker and Spot.

In a recent extension to the FM paper we also investigate the extraction of an actual witness for the energy feasibility. We discuss why this is a non-trivial task and how it can be solved efficiently.

Reentrable Colored Petri Net Model of Ebola Virus Dynamics

Tatiana Shmeleva

Abstract. Early developed technique for modeling cellular automata by colored Petri nets has been further refined with regard to the model transformation into the reenterable form. Reenterable model contains each component in a single copy that makes it invariant to the actual connection of components, definite topology of modeled system. Cellular automaton of Burkhead and Hawkins for modeling Ebola virus dynamics possesses a regular topology given by a square plain lattice. Within reenterable model, each token is supplied with topology tag that specifies the token location within the lattice using a pair of indexes. Check of neighbors regarding to a definite neighborhood, Moore for the automaton in question, is implemented via check of tokens having the corresponding value of indexes with respect to the index difference specified by set $\{-1, 1\}$. Reenterable model is rather convenient for simulation and model-driven development of treatment prescriptions because it does not require graphical editing, supposing modification of the model parameters only, represented by constants of the model declarations section. This especially concerns the lattice size which can be rather big corresponding to the number of cells in deceased organ. A series of simulations acknowledges that the model behavior closely corresponds to the behavior of either source cellular automata or conventional colored Petri net model with respect to statistical error of 2–3%. Thus, the reenterable format is recommended as a reference one for modeling viruses by Petri nets.

Contents

Invited Papers

Randomness Quality and Trade-Offs for CA Random String Generators

Bruno Martin(✉)

Université Côte d'Azur, CNRS, I3S, Sophia Antipolis, France
Bruno.Martin@univ-cotedazur.fr

Abstract. We present classical theories for randomness, starting from mathematical ones to others focusing on randomness testing, more useful in computer science. Those characterisations are made by bounding the computational resources required for testing. Next, we present some suitable practical randomness testing suites designed to measure the quality of random strings that can be efficiently generated. Finally, random string generation by binary uniform cellular automata of increasing quality illustrates the improvements of the randomness testing suites.

Keywords: Randomness · pseudo-randomness · random number generation · Boolean functions

1 Introduction

We present the theories of randomness that are suitable in computer science together with the practice of random number generators although, as von Neumann said: *Anyone who considers arithmetical methods of producing random numbers is, of course, in a state of sin.*

We begin with the three theories of randomness that were developed in the last half of the XX-th Century.

The first, initiated by Shannon comes from the theory of probability and considers distributions that are not perfectly random. Shannon's information theory characterises perfect randomness as the extreme case where the information content is maximised.

The second theory due to Solomonov, Kolmogorov [9], Chaitin and Martin-Löf is rooted in computability theory. Intuitively, it combines randomness with incompressibility. Unfortunately, Kolmogorov's approach is not computable and limits its use in generating randomness.

To get a more effective view of randomness, we turn our attention to the theory of *pseudo-randomness* which comes from complexity theory. This notion is due to Blum, Goldwasser, Micali and Yao [1,6]. This approach aims at providing a theory that allows the generation of perfect random strings. Perhaps one of the most important consequences of this theory is an effective construction of pseudo-random generators from cryptographic one-way functions [5]. But it generates the pseudo-random sequence bit after bit at the cost of some computational cost

O. Bournez et al. (Eds.): RP 2023, LNCS 14235, pp. 3–12, 2023.
https://doi.org/10.1007/978-3-031-45286-4_1

(the computation of the image of a cryptographic one-way function is required to generate a single bit).

In practice, random number generators took a different route. The history of random number generators begins with von Neumann in 1946. To do this, one can use a random set of numbers, although if the initial seed value is the same, it will produce an identical sequence. However, this has been improved, as have methods for measuring randomness. In the light of the above theories, we aim to present various random cellular automata string generators with their advantages and with their disadvantages depending on their use. This string generation by cellular automata was initiated by Wolfram [23] and further investigated in [14]. This last approach combines cellular automata with the Boolean functions which also yields interesting random number generators. Recent works [16] add an evolutionary approach, initiated in [17], to improve the quality of randomness.

Section 2 presents the definitions and notation that are used in the paper, including the definition of cellular automata and some basic properties of Boolean functions. In Sect. 3, we introduce Martin-Löf randomness, the pseudorandomness introduced by Blum, Golwasser, Micali and Yao. Both characterisations of randomness strongly use the notion of testing but are not suitable to efficiently generate long random strings. To that end, we recall some classing testing suites of increasing quality. Finally, Sect. 4 illustrates the improvements in the generation of random strings by uniform binary cellular automata in the light of the successive testing suites.

2 Definition and Notation

We denote by Σ a finite alphabet and by $\Sigma^\star$ the free monoid generated by Σ whose elements are called *strings*. Σ^ω denotes the set of infinite strings over Σ.

A *probability distribution* is a mathematical function that gives the probabilities of occurrence of different possible outcomes for an experiment. As a special case, the discrete *uniform distribution* is a symmetric probability distribution wherein a finite number of values are equally likely to be observed. The *Bernoulli distribution* is the discrete probability distribution of a binary random variable which takes the value 1 with probability p and the value 0 with probability $1 - p$.

When dealing with Boolean functions, we will use the finite field $\mathbb{F}_2$ as a special alphabet with the classical field operations.

2.1 Cellular Automata

One-dimensional binary cellular automata (CA for short) consist of a (finite for practical purposes) line of cells taking their states among binary values. A CA has *periodic boundary conditions* if the cells are arranged in a ring and *null boundary conditions* when both extreme cells are continuously fixed to zero. All the cells are finite state machines with an updating function which gives the new state of the cell according to its current state and the current state of its nearest neighbors. For a presentation of CAs, see [7].

Binary CAs with l cells ($l = 2N + 1$ for $N \in \mathbb{N}$) are considered. For a CA, the values of the cells at time $t \geq 0$ are updated synchronously by a Boolean function f with $n = r_1 + r_2 + 1$ variables by the rule $x_i(t+1) = f(x_{i-r_1}(t), \ldots, x_i(t), \ldots, x_{i+r_2}(t))$. Elementary CAs are such that $r_1 = r_2 = 1$. For a fixed t, the sequence of the values $x_i(t)$ for $1 \leq i \leq 2N + 1$, is the *configuration* at time t. It is a mapping $c : [\![1, l]\!] \rightarrow \mathbb{F}_2$ which assigns a Boolean state to each cell. The initial configuration (at $t = 0$) $x_1(0), \ldots, x_l(0)$ is the *seed*, the sequence $(x_N(t))_t$ is the *output sequence* and, when $r_1 = r_2 = r$, the number r is the *radius* of the rule. The *Wolfram numbering* associates a rule number to any one of the 256 elementary CA; it takes the binary expansion of a rule number as the truth table of a 3-variable Boolean function in ascending numerical order.

2.2 Boolean Functions

A Boolean function is a mapping from $\mathbb{F}_2^n$ into $\mathbb{F}_2$. In the sequel, additions in $\mathbb{Z}$ (resp. $\mathbb{F}_2$) will be denoted by $+$ and Σ (resp. $\oplus$ and $\bigoplus$), products by $\times$ and $\prod$ (resp. $\cdot$ and $\prod$). When there is no ambiguity, $+$ will denote the addition of binary vectors. If x and y are binary vectors, their inner product is $x \cdot y = \sum_{i=1}^{n} x_i y_i$. The classical representation for a Boolean function is the *algebraic normal form*:

Definition 1 (ANF). *A Boolean function f with n variables is represented by a unique binary polynomial in n variables, called* algebraic normal form*: $f(x) = \bigoplus_{u \in \mathbb{F}_2^n} a_u (\prod_{i=1}^{n} x_i^{u_i})$ $a_u \in \mathbb{F}_2$, u_i is the i-th projection of u.*

The *degree of the ANF* or *algebraic degree* of f corresponds to the number of variables in the longest term $x_1^{u_1} \ldots x_n^{u_n}$ in its ANF. The *Hamming weight* $w_H(f)$ of f counts the $x \in \mathbb{F}_2^n$ such that $f(x)=1$. The *Hamming weight* $w_H(x)$ of $x \in \mathbb{F}_2^n$ counts the number of 1-valued coordinates in x. f is *balanced* if $w_H(f) = w_H(1 \oplus f) = 2^{n-1}$.

Definition 2. *f and g Boolean functions in n variables are* equivalent *iff*

$$f(x) = g\left((x \cdot A) \oplus a\right) \oplus \left(x \cdot B^T\right) \oplus b, \quad \forall x \in \mathbb{F}_2^n \tag{1}$$

where A is a non-singular binary $n \times n$ matrix, b a binary constant, a and $B \in \mathbb{F}_2^n$.

An important tool in the study of Boolean functions is the *Fourier-Hadamard transform*, a linear mapping which maps a Boolean function f to the real-valued function $\widehat{f}(u) = \sum_{x \in \mathbb{F}_2^n} f(x)(-1)^{u \cdot x}$, which describes the *spectrum* of the latter. When applied to the *sign function* $\mathrm{f}_\chi(x) = (-1)^{f(x)}$, the Fourier-Hadamard transform is the *Walsh transform*: $\widehat{\mathrm{f}_\chi}(u) = \sum_{x \in \mathbb{F}_2^n} (-1)^{f(x) \oplus u \cdot x}$. Since $\mathrm{f}_\chi(u) = 1 - 2f(u)$, the Fourier-Hadamard transform is:

$$\widehat{f}(u) = \frac{1}{2} \sum_{x \in \mathbb{F}_2^n} (-1)^{u \cdot x} - \frac{1}{2} \widehat{\mathrm{f}_\chi}(u) \quad , \tag{2}$$

Using Eq. (2), we obtain that $\widehat{f_\chi}(u) = 2^n\delta_0 - 2\widehat{f}(u)$, where δ_0 denotes the *Dirac symbol* defined by $\delta_0(u) = 1$ if u is the null vector and $\delta_0(u) = 0$ otherwise [2].

If f and g are two equivalent Boolean functions in n variables, it holds that:

$$\widehat{f_\chi}(u) = (-1)^{a \cdot A^{-1}(u^t + B^T) + b}\, \widehat{g_\chi}((u \oplus B)(A^{-1})^T) \ . \tag{3}$$

The Walsh transform allows to study the *correlation-immunity* of a function.

Definition 3. *A Boolean function f in n variables is k-*correlation-immune *$(0 < k < n)$ if, given any n independent and identically distributed binary random variables $x_1, \cdots, x_n$ according to a uniform Bernoulli distribution, then the random variable $Z = f(x_1, \ldots, x_n)$ is independent from any random vector $(x_{i_1}, x_{i_2}, \ldots, x_{i_k})$, $1 \leq i_1 < \cdots < i_k < n$. When f is k-correlation immune and balanced, it is k-*resilient.

In [24], a spectral characterization of resilient functions was given and it concerns both transforms (refer to [2] for further details):

Theorem 1. *A Boolean function f in n variables is k-resilient iff it is balanced and $\widehat{f}(u) = 0$ for all $u \in \mathbb{F}_2^n$ s.t. $0 < w_H(u) \leq k$. Equivalently, f is k-resilient iff $\widehat{f_\chi}(u) = 0$ for all $u \in \mathbb{F}_2^n$ s.t. $w_H(u) \leq k$.*

Theorem 2 (Siegenthaler Bound). *For a k-resilient $(0 \leq k < n-1)$ Boolean function in n variables, there is an upper bound for its algebraic degree d: $d \leq n - k - 1$ if $k < n - 1$ and $d = 1$ if $k = n - 1$.*

3 Theories of Randomness

In this paper, we focus on theories of randomness that are suitable for computer science and, more specifically, that make use of testing. But let us start with a first theory, initiated by Shannon (cf. [19]) in the second half of the *XX*-th Century, which is rooted in probability theory and focused on distributions that are not perfectly random. Shannon's information theory characterizes perfect randomness as the extreme case in which the information content is maximized (and there is no redundancy at all). Thus, perfect randomness is associated with the uniform distribution. And, by definition, it is not possible to generate such perfect random strings from shorter ones. This approach is not suitable for computer science.

Almost at the same time than Shannon's information theory, Solomonov, Kolmogorov and Chaitin [10] proposed another way to characterise randomness. Their work is strongly connected with computability theory and, more specifically with the existence of a universal machine. It measures the complexity of objects in terms of the shortest machine (given a fixed universal machine) that prints out the object on its standard output. Chaitin-Kolmogorov complexity is quantitative, and perfect random objects appear as an extreme case. Intuitively, it expresses that a string is random if it is uncompressible. Interestingly, one may say that a single object, rather than a distribution over objects, is perfectly random. Still, Chaitin-Kolmogorov's approach is inherently uncomputable. As a consequence, one cannot generate strings of high Chaitin-Kolmogorov complexity from short random strings.

3.1 Martin-Löf Randomness

The Martin-Löf tests [15] give a statistical interpretation of the Chaitin-Kolmogorov theoretic notion of randomness. He defined a set of sequences of measure one, called Martin-Löf random sequences, which satisfies all of the probability laws. These sequences are defined as those satisfying all Martin-Löf tests, which can be described as follows ([3]).

Definition 4 (Martin-Löf test (ML test)). *Let ϕ be an algorithm which generates a sequence of sets O_m, $m \in \mathbb{N}$. Each O_i is a computably enumerable set of binary strings x interpreted as the dyadic interval $[0.x, 0.x+2^{-|x|})$. When ϕ receives as an input (m, k) it returns the k-th interval (binary string) in the enumeration of O_m. Also assume that $\mu(O_i) < 2^{-i}$ and $O_i \supset O_{i+1}$.*

Definition 4 defines a decreasing sequence of sets O_i of intervals such that the Lebesgue measure upper-bounds O_i by 2^{-i}. If one wants to test the randomness of the infinite string $x \in \Sigma^\omega$, it can be either rejected as non-random or accepted as being random. x is rejected at order n if $x \in O_n$ for some $n \geq 1$. We say that x is ML-random if it passes all the ML tests.

It has been proved that ML randomness corresponds to Chaitin-Kolmogorov's as stated in Theorem 3, due to Schnorr.

Theorem 3. *A real number is Kolmogorov-Chaitin random if and only if it is Martin-Löf random.*

Martin-Löf approach introduced the notion of testing to characterise randomness. This notion is of great theoretical interest but needs to be adapted to be useful in practice. In some sense (at least in the writer's opinion), this is what has been done by restricting the resources used by the model (switching from computability to complexity). This is basically the notion of pseudo-randomness described in the next section.

3.2 Pseudo-randomness

The approach of pseudo-randomness (see [4]) aimed at providing a theory of perfect randomness that nevertheless allows the efficient generation of perfect random strings from shorter random strings. It strongly relies on the notion of indistinguishability that claims that two strings are equal if they cannot be distinguished as stated in Definition 5

Definition 5 (Computational Indistinguishability). *Two probability distributions, $\{X_n\}_{n\in\mathbb{N}}$ and $\{Y_n\}_{n\in\mathbb{N}}$ are called* indistinguishable *if for any probabilistic polynomial time algorithm A, any polynomial p and all sufficiently large n,*

$$|Pr_{x\sim X_n}[A(x) = 1] - Pr_{y\sim Y_n}[A(y) = 1]| < \frac{1}{p(n)}$$

The probability is taken over X_n (resp. Y_n) as wall as over the coin tosses of algorithm A.

In Definition 5, the probabilistic algorithm A is called a distinguisher. In our case, the probabilistic distinguisher tries to determine if its input follows the distribution X_n or the distribution Y_n. If its output is close to one half, it means intuitively that it cannot decide (its outcome is close to tossing a fair coin to get the answer).

Technically, no fixed string can be said to be "pseudo-random". Rather, pseudo-randomness actually refers to a distribution on strings, and when we say that a distribution D over strings of length l is pseudo-random this means that D is indistinguishable from the uniform distribution over strings of length l.

Yao [25] provides a definition of pseudo-random number generator which is based on computational complexity, and proposes a definition of "perfect"-in current terminology, "pseudo-random"-probability distribution. (A distribution is perfect if it cannot be distinguished from a truly random distribution in the sense of Definition 5). Yao relates his notion of pseudo-randomness to the idea of a statistical test, a notion already used in the study of pseudo-random number generators, and shows that one particular test, known as the next-bit test, is adequate for characterizing pseudo-randomness. Having defined perfect distributions, Yao then defined a pseudo-random number generator as an efficient probabilistic algorithm which uses a limited number of truly random bits in order to output a sample from a perfect distribution whose size is polynomial in the number of random bits used.

If we continue to decrease the resources given to define randomness, we obtain the following series of tests presented in decreasing quality in the next three sections.

3.3 FIPS-140-2 and 140-3 Tests

The National Institute of Standards and Technology (NIST) issued the FIPS 140 Publication Series to coordinate the requirements and standards for cryptography modules that include both hardware and software components. Federal agencies and departments can validate that the module in use is covered by an existing FIPS 140 certificate that specifies the exact module name, hardware, software, firmware, and/or applet version numbers. Its Annex C provides a list of approved random number generators as well as a `linux` utility named `rngtest` which implements the series of statistical tests to conduct against a random number generator.

3.4 Marsaglia's Tests

Marsaglia from the Florida State University has proposed in 1985 a lower battery of tests packaged in the Diehard test suite, a widely used tool. It consists of 17 different tests which have become something which could be considered as a "benchmarking tool" for random sequences generators (see [11]). It is meant to evaluate if a stream of numbers is a good generator. We will not explain how

Diehard really works and we refer the reader to [12] for further details. Basically, Diehard uses Kolmogorov-Smirnov normality test to quantify the distance between the distribution of a given data set and the uniform distribution; and as the documentation says:

> Each Diehard test is able to provide probability values (p-value) which should be uniformly distributed on $[0, 1)$ if the sequence is made of truly independent bits. Those p-values are obtained by $p = F(X)$ where F is the assumed distribution of the sample random variable X–often normal. But that assumed F is just an asymptotic approximation, for which the fit will be worse in the tail of the distribution. Thus, we should not be surprised with occasional p-values close to 0 or 1. When a stream really fails, one gets p-values of 0 or 1 to six or more places. Otherwise, for each test, its p-value should lie in the interval $(0.025, 0.975)$.

3.5 Knuth's Tests

In order to evaluate whether pseudo-random numbers are independent and unpredictable, Knuth [8] proposed 11 randomness test methods in 1968 such as frequency test, run-length test, poker test, etc., and became the pioneer of systematic randomness testing. The Knuth test suite was one of the statistical randomness suites, but the suite is mainly used for real number sequences, and the test parameters are not explicitly given.

4 Pseudo-random Strings Generation

4.1 Radius One CA Rules

In 1985, Wolfram [22] proposed to use CA rule 30 as random strings generator still in use in Mathematica^{TM}. He justified the quality of the pseudo-random sequence by the use of the tests of Knuth (see Sect. 3.5).

But if we consider the CA rule as a Boolean function, by Theorem 1 and an exhaustive search of 3-variable Boolean update function [13], we can state that:

Theorem 4. *There is no non-linear correlation-immune elementary CA.*

The same result can be obtained by applying the Siegenthaler bound (Theorem 2) with $n = 3$ variables and testing for $k = 1$-resiliency. It tells that the algebraic degree is $d \leq n - k - 1 = 1$. Thus, only linear functions can be resilient but providing rules that are not interesting for generating randomness.

Despite this, CA may be used for generating random strings by increasing the number of variables in the Boolean function which is used as a local CA rule. In the next section, we present a way to gather radius 2 CA rules for generating better random strings.

4.2 Radius Two CA Rules

In [14], we considered radius two CA rules as five variable Boolean functions and select CA rules that satisfy the criteria of Theorem 1. We used those rules as a random string generator as in [18].

We set up two rings of cells. Although Wolfram used a ring of 127 cells and Preneel (1993) suggested a ring of 1024 cells to ensure a better quality (both used a slightly different mechanisms for random bit extraction), we use perimeters 64 and 65 as done in [18]. The initial configuration of these rings is of Hamming weight 1. We let the CA iterate about 2 million times. Then, from each configuration obtained, we extract two 32-bits words: the "even" (resp. "odd"), word is built with the state of the first 32 "even" (resp. "odd") cells. The sequences of these "even" (resp. "odd") words constitute two different sequences of 16 Mbytes.

Then, we use Diehard test suite to produce p-values for each test. We were able to find some CAs (like the one with rule `0x69999999`). This suggests that it may be possible to obtain a good random string generator from such a CA.

With such "good rules" selected, we tried to extend them. We consider those rules as five variable Boolean functions, and extend them to Boolean functions in 9 variables (or radius 4 CA rules) simply by making two iterations of the local CA rule. We proved that 1-resiliency is preserved upon iteration only when we negate the truth table of the Boolean function or when we take the mirror image of its truth table. This means that, in general, resiliency is not preserved upon iteration. And so, we obtain CA rules to generate random strings that pass the Marsaglia's tests.

In a more recent paper, Wang et al. [20] use a novel particle swarm cellular automata (PSCA). They apply PSCA to generate pseudo-random strings that pass all tests of diehard and FIPS 140-2.

Conclusion

From the theory to the practice, testing is used to characterise or to measure randomness, depending on the resource we allow to a model of computation. Theories come from the mid of the XX-th Century and testing suites started in the last 30 years of the same Century. The testing suites are still under strong improvements using the theory and the random string generators follow this evolution.

We have given an illustration with the generation of random strings with cellular automata which follows the evolution of testing suites. And as long as we give more resources to testing, we tend to the definition of pseudo-random strings as defined by Yao.

Today's best current testing suite (given by FIPS 140-2) is not an end and already faces controversy (see [21] for instance) and also needs improvements that will probably be achieved soon. This will lead to new advances in random sequence generation.

References

1. Blum, M., Micali, S.: How to generate cryptographically strong sequences of pseudo-random bits. SIAM J. Comput. **13**, 850–864 (1984)
2. Carlet, C.: Boolean functions for cryptography and error-correcting codes. Technical report, University of Paris 8 (2011)
3. Davie, G.: Characterising the Martin-Löf random sequences using computably enumerable sets of measure one. Inf. Process. Lett. **92**(3), 157–160 (2004)
4. Goldreich, O.: Pseudorandomness. Not. AMS **46**(10), 1209–1216 (1999)
5. Goldreich, O., Levin, L.A.: A hard core predicate for any one way function. In: 21st STOC, pp. 25–32 (1989)
6. Goldwasser, S., Micali, S.: Probabilistic encryption. JCSS **28**(2), 270–299 (1984)
7. Gruska, J.: Foundations of Computing. International Thomson Publishing, London (1997)
8. Knuth, D.: Seminumerical Algorithms. Addison Wesley, Boston (1969)
9. Kolmogorov, A.: Three approaches to the quantitative definition of information. Problemy Pederachi Informatsii **1**, 3–11 (1965)
10. Li, M., Vitányi, P.: An Introduction to Kolmogorov Complexity and Its Applications. TCS, Springer, New York (2008). https://doi.org/10.1007/978-0-387-49820-1
11. Marsaglia, G.: A current view of random number generators. In: Computer Sciences and Statistics, pp. 3–10 (1985)
12. Marsaglia, G.: Diehard (1995). http://www.stat.fsu.edu/pub/diehard/
13. Martin, B.: A Walsh exploration of elementary CA rules. J. Cell. Autom. **3**(2), 145–156 (2008)
14. Formenti, E., Imai, K., Martin, B., Yunès, J.-B.: Advances on random sequence generation by uniform cellular automata. In: Calude, C.S., Freivalds, R., Kazuo, I. (eds.) Computing with New Resources. LNCS, vol. 8808, pp. 56–70. Springer, Cham (2014). https://doi.org/10.1007/978-3-319-13350-8_5
15. Martin-Löf, P.: The definition of random sequences. Inf. Control **9**, 602–619 (1966)
16. Ryan, C., Kshirsagar, M., Vaidya, G., Cunningham, A., Sivaraman, R.: Design of a cryptographically secure pseudo random number generator with grammatical evolution. Sci. Rep. **12**(1), 8602 (2022). https://doi.org/10.1038/s41598-022-11613-x
17. Seredynski, F., Bouvry, P., Zomaya, A.Y.: Cellular automata computations and secret key cryptography. Parallel Comput. **30**(5–6), 753–766 (2004)
18. Shackleford, B., Tanaka, M., Carter, R.J., Snider, G.: FPGA implementation of neighborhood-of-four cellular automata random number generators. In: Proceedings of the 2002 ACM/SIGDA Tenth International Symposium on Field-programmable Gate Arrays, pp. 106–112. FPGA 2002, ACM (2002)
19. Shannon, C., Weaver, W.: The mathematical theory of communication. University of Illinois Press (1964)
20. Wang, Q., Yu, S., Ding, W., Leng, M.: Generating high-quality random numbers by cellular automata with PSO. In: 2008 Fourth International Conference on Natural Computation, vol. 7, pp. 430–433 (2008)
21. Wertheimer, M.: The mathematics community and the NSA. Not. Am. Math. Soc. **62**(2), 165–167 (2015)
22. Wolfram, S.: Cryptography with cellular automata. In: Williams, H.C. (ed.) CRYPTO 1985. LNCS, vol. 218, pp. 429–432. Springer, Heidelberg (1986). https://doi.org/10.1007/3-540-39799-X_32

23. Wolfram, S.: Random sequence generation by cellular automata. Adv. Appl. Math. **7**, 123–169 (1986)
24. Xiao, G.Z., Massey, J.L.: A spectral characterization of correlation-immune combining functions. IEEE Trans. Inf. Theory **34**(3), 569–571 (1988)
25. Yao, A.: Theory and application of trapdoor functions. In: 23d Symposium on Foundations of Computer Science (1982)

Regular Papers

Complexity of Reachability Problems in Neural Networks

Adrian Wurm(✉)

BTU Cottbus-Senftenberg, Lehrstuhl Theoretische Informatik, Platz der Deutschen Einheit 1, 03046 Cottbus, Germany
wurm@b-tu.de
https://www.b-tu.de/

Abstract. In this paper we investigate formal verification problems for Neural Network computations. Various reachability problems will be in the focus, such as: Given symbolic specifications of allowed inputs and outputs in form of Linear Programming instances, one question is whether valid inputs exist such that the given network computes a valid output? Does this property hold for all valid inputs? The former question's complexity has been investigated recently in [20] by Sälzer and Lange for nets using the Rectified Linear Unit and the identity function as their activation functions. We complement their achievements by showing that the problem is NP-complete for piecewise linear functions with rational coefficients that are not linear, NP-hard for almost all suitable activation functions including non-linear ones that are continuous on an interval, complete for the Existential Theory of the Reals $\exists\mathbb{R}$ for every non-linear polynomial and $\exists\mathbb{R}$-hard for the exponential function and various sigmoidal functions. For the completeness results, linking the verification tasks with the theory of Constraint Satisfaction Problems turns out helpful.

1 Introduction

Given the huge success of utilizing Neural Networks, NN for short, in the last decade, such nets are nowadays widely used in all kind of data processing, including tasks of varying difficulty. There is a wide range of applications, the following exemplary references (mostly taken from [20]) just collect non-exclusively some areas for further reading: Image recognition [15], natural language processing [9], autonomous driving [8], applications in medicine [16], and prediction of stock markets [7], just to mention a few. Khan et al. [14] provide a survey of such applications, a mathematically oriented textbook concerning structural issues related to Deep Neural Networks is provided by [4]. Among the many different aspects of areas where the use of Neural Networks seems appropriate, some also involve safety-critical systems like autonomous driving or power grid management. In such a setting, when security issues become important, aspects of certification come into play [10].

In the present paper we are interested in studying certain verification problems for NNs in form of particular reachability problems. Starting point is the work by Sälzer and Lange [20] being based on [13,18]. The authors of these

O. Bournez et al. (Eds.): RP 2023, LNCS 14235, pp. 15–27, 2023.
https://doi.org/10.1007/978-3-031-45286-4_2

papers analyze the computational complexity of one particular such verification task. It deals with a reachability problem in networks using the Rectified Linear Unit together with the identity function as its activation function. In such a net, specifications describe the set of valid inputs and outputs in form of two Linear Programming instances. The question then is to decide whether a valid input exists such that the network's result on that input is a valid output, i.e., whether the set of valid outputs is reachable from that of valid inputs. In the above references the problem is shown to be NP-complete, even for one hidden layer and output dimension one, or some restricted set of weights being used [20]. Note that the network in principle is allowed to compute with real numbers, so the valid inputs we are looking for belong to some space $\mathbb{R}^n$, but the network itself is specified by its discrete structure and rational weights defining the linear combinations computed by its single neurons.

Obviously, one can consider a huge variety of networks created by changing the underlying activation functions. There are of course many activations frequently used in NN frameworks, and in addition we could extend reachability questions to nets using all kinds of activation. One issue to be discussed is the computational model in which one argues. If, for example, the typical sigmoid activation $f(x) = 1/(1+e^{-x})$ is used, it has to be specified in which sense it is computed by the net: For example exactly or approximately, and at which costs these operations are being performed.

In the present work we study the reachability problem for commonly used activation functions and show that for most of them it will be complete either in (classical) NP or in the presumably larger class $\exists\mathbb{R}$, which captures the so-called existential theory of the reals. Our main results are as follows: The reachability problem is in P for linear activations, in NP for semilinear activations, NP-hard for all non-linear activations that are continuous on an interval, and ETR-hard for several commonly used activations such as arctan and the exponential function. These results imply, for example, NP-completeness for frequently used activations such as (Leaky) ReLU, Heaviside and Signum.

A most helpful tool for establishing these results is linking the problems under consideration to the area of constraint satisfaction problem CSP and known complexity results for special instances of the latter. This connection will provide us with a classification of a vast set of activation functions in the complexity classes between P and $\exists\mathbb{R}$. We also consider a variant of the reachability problem asking whether for all valid inputs the computed output is necessarily valid and establish several complexity results as well.

The paper is organized as follows: In Sect. 2 we collect basic notions, recall the definition of feedforward neural nets as used in this paper as well as useful facts about Constraint Satisfaction Problems. Section 3 studies various activation functions and their impact on the complexity of reachability problems. We show that the reachability problem is basically the same as the CSP containing the graphs of the activation functions together with relations necessary to express linear programming instances. We show that adding the identity as activation does not change the complexity of the reachability problem in several cases, for example when either ReLU is used as activation or if a network connection is allowed to skip a layer. We show that the problem is NP-hard for every sensible non-linear activation and finally discuss problems that are hard or complete for $\exists\mathbb{R}$. The paper ends with some open questions. Lacking proofs are given in the full version.

2 Preliminaries and Network Reachability Problems

We start by defining the problems we are interested in; here, we follow the definitions and notions of [20] for everything related to neural networks. The networks considered are exclusively feedforward. In their most general form, they can process real numbers and contain rational weights.

Definition 1. *A (feedforward) neural network N is a layered graph that represents a function of format $\mathbb{R}^n \to \mathbb{R}^m$, for some $n, m \in \mathbb{N}$. The first layer with label $\ell = 0$ is called the* input layer *and consists of n nodes called* input nodes. *The input value x_i of the i-th node is also taken as its output $y_{0i} := x_i$. A layer $1 \leq \ell \leq L-2$ is called* hidden *and consists of $k(\ell)$ nodes called* computation nodes. *The i-th node of layer ℓ computes the output $y_{\ell i} = \sigma_{\ell i}(\sum_j c_{ji}^{(\ell-1)} y_{(\ell-1)j} + b_{\ell i})$. Here, the $\sigma_{\ell i}$ are (typically nonlinear)* activation *functions (to be specified later on) and the sum runs over all output neurons of the previous layer. The $c_{ji}^{(\ell-1)}$ are real constants which are called* weights, *and $b_{\ell i}$ is a real constant called* bias. *The outputs of all nodes of layer ℓ combined gives the output $(y_{\ell 0}, \ldots, y_{\ell(k-1)})$ of the hidden layer. The final layer $L-1$ is called* output layer *and consists of m nodes called* output nodes. *The i-th node computes an output $y_{(L-1)i}$ in the same way as a node in a hidden layer. The output $(y_{(L-1)0}, \ldots, y_{(L-1)(m-1)})$ of the output layer is considered the output $N(x)$ of the network N.*

Note that above, as in [20], we allow several different activation functions in a single network. This basically is because for some results technically the identity is necessary as a second activation function beside the 'main' activation function used. We next recall from [20] the definition of the reachability problem NNREACH. Since we want to study its complexity in the Turing model, we restrict all weights and biases in a NN to the rational numbers. The problem involves two Linear Programming LP instances in a decision version, recall that such an instance consists of a system of (componentwise) linear inequalities $A \cdot x \leq b$ for a rational matrix A and vector b of suitable dimensions. The decision problem asks for the existence of a real solution vector x.

Definition 2. *a) Let F be a set of activation functions from $\mathbb{R}$ to $\mathbb{R}$. An instance of the* reachability problem for neural networks NNREACH(F) *consists of an $n \in \mathbb{N}$, a (feedforward) neural network N with n inputs and all its activation functions belonging to F, rational data as weights and biases, and two instances of LP in decision version with rational data, one with the input variables of N as variables, and the other with the output variables of N as variables. These* *instances are also called* input *and* output specification, *respectively. The problem is to decide if there exists an $x \in \mathbb{R}^n$ that satisfies the input specification such that the output $N(x)$ satisfies the output specification.*

b) The problem verification of interval property VIP*(F) consists of the same instances, except for the output specification being the open polyhedron, meaning the interior of the solution space. This is due to technical reasons that will later on simplify the reductions. The question is whether for all $x \in \mathbb{R}^n$ satisfying the input specification, $N(x)$ will satisfy the output specification (cf. [10]).*

As for NNReach*, we denote by (A, B, N) such an instance, assuming n is obvious from the context.*

c) Let $F = \{f_1, ..., f_n\}$ be a set of activation functions. Then the Network Equivalence problem NE(F) *is the decision problem whether two F-networks describe the same function or not.*

d) The size of an instance is given as the sum of the (usual) bit-sizes of the two LP instances and $T \cdot L$; here, T denotes the number of neurons in the net N and L is the maximal bit-size of any of the weights and biases.

As usual for neural networks, we consider different choices for the activation functions used. Typical activation functions are $ReLU(x) = max\{0, x\}$, the Heaviside function or sigmoidal functions like $\sigma(x) = \frac{1}{1+e^{-x}}$. By technical reasons, in some situations the identity function $\sigma(x) = x$ is also allowed, Sälzer and Lange [20], for example, examined NNReach$(id, ReLU)$. We name nodes according to their internal activation function, so we call nodes with activation function $\sigma(x) = x$ identity nodes and nodes with activation function $\sigma(x) = ReLU(x)$ ReLU-nodes etc. Note that the terminology of the LP-specifications has its origin in software verification.

2.1 Basics on Constraint Satisfaction Problems CSP

As we shall see, analyzing the complexity of the above reachability problems is closely related to suitable questions in the framework of Constraint Satisfaction Problems CSP. This is a well established area in complexity, see for example the survey [6]. Here, we collect the basic notions and results necessary for our purposes.

Informally, a CSP deals with the question whether values from a set A can be assigned to a set of variables so that given conditions (constraints) hold. These conditions are taken from a set of relations over A that, together with the set A, define the CSP. This can be formalized as follows:

Definition 3. *A (relational)* signature *is a pair $\tau = (\mathbf{R}, a)$, where $\mathbf{R}$ is a finite set of* relation symbols *and $a \colon \mathbf{R} \to \mathbb{N}$ is a function called the* arity.

A (relational) τ-structure *is a tuple $\mathcal{A} = (A, \mathbf{R}^{\mathcal{A}})$, where A is a set called the* domain *and $\mathbf{R}^{\mathcal{A}}$ is a set containing precisely one relation $R^{\mathcal{A}} \subseteq A^{a(R)}$ for each relation symbol $R \in \mathbf{R}$.*

An instance of a CSP over a given τ-structure is a conjunction of constraints, where a single constraint restricts a variable tuple to belong to a particular relation of the structure under a suitable assignment of values from the domain to the variables. For the entire instance one then asks for the existence of an assignment satisfying all its constraints.

Definition 4. *Let τ be a signature and $\mathcal{A}$ a τ-structure with domain A and relations $\mathbf{R}$. We always assume equality to be among the structure's relations. Let $X = \{x_1, x_2, \ldots\}$ be a countable set of variables.*

a) A constraint *for $\mathcal{A}$ is an expression $R(y_1, \ldots, y_{a(R)})$, where $R \in \mathbf{R}$, $a(R)$ its arity and all $y_i \in X$. For $z \in A^{a(R)}$ we say that $R(z)$ is true over $\mathcal{A}$ iff $z \in R^{\mathcal{A}}$.*

b) A formula ψ is called primitive positive *if it is of the form*

$$\exists x_{n+1}, \ldots, \exists x_t : \psi_1 \wedge \ldots \wedge \psi_k,$$

where each ψ_i either is a constraint, $\top$ (true), or $\bot$ (false).

A formula with no free variables is a sentence.

c) The decision problem CSP($\mathcal{A}$) *is the following: Let $n \in \mathbb{N}$ and a primitive positive τ-sentence over X, i.e., a finite collection of constraints involving variables $\{x_1, \ldots, x_n\} \subset X$ be given. The question then is, whether there exists an assignment $f : \{x_1, \ldots, x_n\} \to A$ for the variables such that each given constraint is true under the assignment f, i.e., all $R(f(y_1), \ldots, f(y_{a(R)}))$ are true.*

The size of an instance is $n+m$, where m denotes the number of constraints.

Example 1 (folklore). Consider as domain the real numbers $\mathbb{R}$, together with the binary order relation $\leq$, the ternary relation R_+ defined via $R_+(x, y, z) \Leftrightarrow x + y = z$, and the unary relation $R_{=1}(x) \Leftrightarrow x = 1$. Then CSP($\mathbb{R}; \leq, R_+, R_{=1}$) is polynomial time equivalent to the Linear Programming problem in feasibility form with rational input data. Reducing the former to the latter is obvious, for the reverse direction first multiply all inequalities with a sufficiently large natural number to obtain integer coefficients only. Now observe that any natural number n can be expressed as (one component of) a solution of a set of constraints involving $a = 1$ and doubling a number via $c = b + b$. This way, the binary expansion of n can be constructed with $O(\log n)$ constraints. Apply this construction similarly to a variable x of an instance of LP to obtain the term $n \cdot x$; now adding as constraint the equation $nx = 1$ similarly allows to express rational numbers as coefficients. Clearly the size of the resulting CSP instance is polynomially bounded in the (bit-)size of the given LP instance. Note that due to the theory of Linear Programming an instance with rational data has the same answer, independently of whether the considered domain is $\mathbb{R}$ or $\mathbb{Q}$.

Definition 5. *A relation R is called* primitive positive definable *(pp-definable) over $\mathcal{A}$, iff it can be defined by a primitive positive formula ψ over $\mathcal{A}$, i.e.,*

$$R(x_1, \ldots, x_n) \Leftrightarrow \exists x_{n+1}, \ldots, \exists x_t : \psi(x_1, \ldots, x_t).$$

It was shown by Jeavons, Bulatov and Krokhin [3] that CSP($\mathcal{A}$) and CSP($\mathcal{A}'$), where the latter structure arises from the former by attaching finitely many relations being pp-definable over $\mathcal{A}$, are linear-time equivalent. The obvious idea of replacing every occurrence of the new relation suffices to prove the statement. This argument will be used below once in a while.

Definition 6. *a) A set $R \subseteq \mathbb{R}^n$ is called* semilinear, *iff it is a boolean combination of half-spaces*[1].

b) A set $R \subseteq \mathbb{R}^n$ is called essentially convex, *iff for any two points $x, y \in \mathbb{R}^n$ the intersection of the line segment $[x, y]$ contains only finitely many points that are not in R. If $R \subseteq \mathbb{R}^n$ is not essentially convex, any two points for which the property fails are called* witnesses *for the set not being essentially convex.*

This gives us access to the following results of Bodirsky, Jonsson and von Oertzen:

Theorem 1 ([2]). *a) Let $R_1, ..., R_n$ be semilinear relations. Then* $\mathrm{CSP}(\mathbb{Q}; \leq, R_+, R_{=1}, R_1, ..., R_n)$ *is in P if $R_1, ..., R_n$ are essentially convex and is NP-complete otherwise.*

b) Let $R_1, ..., R_n$ be relations such that at least one of them is not essentially convex witnessed by two rational points. Then $\mathrm{CSP}(\mathbb{R}; \leq, R_+, R_{=1}, R_1, ..., R_n)$ *is NP-hard.*[2]

3 Complexity Results for Reachability

We shall now study the complexity of the reachability problem for various sets of activation functions used by the neural network under consideration. Starting point will be the result from [13,20] that NNREACH(*id*, *ReLU*) is NP-complete. We analyze the problem for a larger repertoire of activation functions. To do so, in a first step it will be very helpful to relate these problems to instances of certain CSP problems which can be attached to a network canonically. This relation is made precise in the following theorem. The fact that input and output specifications are LP instances causes, that the structures below naturally contain the relations $R_{=1}, R_+$, and $\leq$. Further relations then will be determined by the activation functions used.

Theorem 2. *For any set of unary real functions $F = \{f_1, ..., f_s\}$, interpreted as relations via their graphs,* $\mathrm{CSP}(\mathbb{R}; \leq, R_+, R_{=1}, f_1, ..., f_s)$ *and* NNREACH$(id, f_1, ..., f_s)$ *are linear-time equivalent.*

Proof. We prove both directions explicitly for the case $s = 1$, then the conclusion for $s > 1$ is immediate. For reducing NNREACH(id, f) to $\mathrm{CSP}(\mathbb{R}; \leq, R_+, R_{=1}, f)$, let N be a network using *id* and f as activation functions. The weights and biases of N are assumed to be rational numbers. The variable set of the CSP we construct contains one variable for each input and output node of N. For each node v in a hidden layer we introduce two variables v_{sum} and v_f. Note that according to Example 1 any linear inequality with rational coefficients can be expressed as an instance of $\mathrm{CSP}(\mathbb{R}; \leq, R_+, R_{=1})$ of linear size. Thus, the

[1] i.e., finite unions, intersections and complements of sets of the form $Ax \leq b$.

[2] Note that we can not switch between the domains $\mathbb{Q}$ and $\mathbb{R}$ at will any more after dropping semilinearity with rational coefficients, for it could in this case change solvability.

input and output specifications of N can be expressed as a set of constraints in CSP($\mathbb{R}; \leq, R_+, R_{=1}$) using the corresponding variables, which is of linear size with respect to the size of those specifications. For the nodes in the hidden layers, we proceed similarly. If node v receives a linear sum $\sum_{i=1}^{k} c_i \cdot u_i + b$ as its input, where c_i, b are the rational weights and bias and the u_i represent the outputs of the previous layer, then as in Example 1 we add the constraint $v_{sum} = \sum_{i=1}^{k} c_i \cdot u_{i,f} + b$ to the constructed instance. In case v has f as activation, we add the constraint $v_f = f(v_{sum})$ and if v was an *id*-node we add the constraint $v_f = v_{sum}$. Obviously, the size of the CSP instance is linearly bounded in that of the given net. Moreover, NNReach(*id*, f) is solvable for N if and only if the above CSP has a solution.

For the reverse direction, we translate an instance of CSP($\mathbb{R}; \leq, R_+, R_{=1}, f$) into an instance of NNReach(*id*, f) with only one hidden layer. For each variable in the instance we introduce a node in the input layer and encode all constraints of the form $\leq, R_+$ and $R_{=1}$ into the input specification. For every constraint $y = f(x)$ we introduce a new f-node $\bar{x}$ in the hidden layer connected only to x with bias 0 and weight 1. Next, we allocate an identity-node $\bar{y}$ in the hidden layer connected only to y also with bias 0 and weight 1 for the connections. Finally, we require both nodes $\bar{x}$ and $\bar{y}$ to be equal by adding the equation $\bar{x} = \bar{y}$ to the output-specification.

It is obvious that both reductions can be performed inductively for all the functions in F, so the statement holds for the entire set. ■

Note that the proof does not depend on formalizing the specifications as LP instances. It would similarly hold if the specifications would be given by (in-) equality systems involving polynomials and adding a relation for multiplication on the CSP-side. However, in this case checking feasibility of the specifications is already difficult, see below.

Before studying NNReach for different activations, we briefly discuss a more technical issue, namely the necessity of adding *id* as activation.

The above proof implies that using an injective activation allows to omit *id*, if we drop the condition that the network has to be layered, meaning that a connection can skip layers:

Lemma 1. *For f injective,* NNReach(*id*, f) *and* NNReach(f) *are linear-time equivalent.*

Proof. Given the proof of Theorem 2 it only remains to avoid *id*-nodes when reducing an instance of CSP($\mathbb{R}; \leq, R_+, R_{=1}, f$) to one of NNReach($f$). Identity nodes were used to propagate the value of a node y in order to include a constraint $y = f(x)$ in the output specification. Instead, if f is injective one can use a network node for $f(x)$ and one for y and connect them with biases 0 and weights 1 and -1, respectively, to an f-activation node computing $f(f(x) - y)$. Use another f-node to compute $f(0)$. This is possible by demanding a further input node to have value 0. Finally, in the output specification we add the equality between these two nodes with values $f(f(x)-y)$ and $f(0)$; injectivity provides the equivalence of this condition with $f(x)=y$. ■

For example, for nets using sigmoidal activation functions identity activations are not necessary.

Sälzer and Lange [20] asked whether for the NP-completeness result involving *id* and *ReLU* as activations one can avoid *id* as activation. Though *ReLU* is not injective, this in fact holds as well, even when we do not allow connections to skip layers:

Proposition 1. *The following problems are linear-time equivalent:*
i) $\text{CSP}(\mathbb{R}; \leq, R_+, R_{=1}, ReLU)$,
ii) NNREACH$(id, ReLU)$ *and*
iii) NNREACH$(ReLU)$ *with only one hidden layer.*

As consequence, all three are NP-complete.

Proof. Given the NP-completeness of NNREACH$(id, ReLU)$ and Theorem 2 above, it remains to reduce NNREACH$(id, ReLU)$ to NNREACH$(ReLU)$ in linear time. Towards this goal, we show that an identity node can be replaced by two ReLU-nodes in the following way: In the neural net to be constructed use two copies of the identity node and let both have the same incoming and outgoing connections as the original node. Replace the identity map by the ReLU map in both and invert all incoming and outgoing weights as well as the bias in the second one. Delete the initial identity node. This does not change the computed function of the network, because

$$\begin{aligned}
\sum_{i=1}^{n} a_i x_i + b &= max\{0, \sum_{i=1}^{n} a_i x_i + b\} + min\{0, \sum_{i=1}^{n} a_i x_i + b\} \\
&= max\{0, \sum_{i=1}^{n} a_i x_i + b\} - max\{0, -(\sum_{i=1}^{n} a_i x_i + b)\} \\
&= ReLU(\sum_{i=1}^{n} a_i x_i + b) - ReLU(-(\sum_{i=1}^{n} a_i x_i + b)) \\
&= ReLU(\sum_{i=1}^{n} a_i x_i + b) - ReLU(\sum_{i=1}^{n} (-a_i) x_i - b)
\end{aligned}$$

Applying this to every node gives us at most twice as many nodes with at most four times as many connections, thus the construction runs in linear time. ■

Theorem 2 enables us treating network reachability complexity questions by using the rich fund of complexity results for CSP problems of various types.

As an easy warm up, convince yourself that NNREACH(id) is by the previous theorem equivalent to $\text{CSP}(\mathbb{R}; \leq, R_+, R_{=1}, id)$ which in turn is equivalent to LP by Example 1, a problem well known to belong to P in Turing model complexity [12].

In [20] it was shown that NNREACH($id, ReLU$) is NP-complete, the link to CSPs however provides us with a much shorter proof. We will make use of Theorem 1 and apply Theorem 2:

Corollary 1. *Let $f_1, ..., f_s$ be unary real functions. If their graphs $g_1, ..., g_s$ are semilinear with rational coefficients, then* NNREACH($id, f_1, ..., f_s$) *is in P if and only if $g_1, ..., g_s$, interpreted as binary relations, are essentially convex, and NP-complete otherwise. If at least one of the graphs $g_1, ..., g_s$ is not essentially convex witnessed by two rational points, then* NNREACH($id, f_1, ..., f_s$) *is NP-hard.*

Though Corollary 1 certainly is not that surprising from the CSP point of view, it gives an elegant way to argue about the complexity of reachability problems for neural networks. Of interest now is to investigate the latter for many more activation functions. NP-hardness is for example the case for $f(x) = n^x, n \in \mathbb{N}, n > 1$, the witnesses are $f(0) = 1$ and $f(1) = n$, but not necessarily for $f(x) = e^x$ for it lacks a second rational point. Other activation functions that immediately give us NP-hardness this way are all non-linear rational functions (including polynomials) with rational coefficients, the square-root, all other rational roots, and the binary logarithm. Reasonable non-linear functions that lack rational points can be compressed to do so, such as $f(x) = sin(\frac{x}{\pi})$ instead of $f(x) = sin(x)$.

Moreover, if we use the strong version that requires semilinearity, we get that NNREACH(id, f) is NP-complete for f being either the absolute value, the floor or ceiling function on a bounded domain, the sign or the Heaviside function, piecewise linear functions such as $f_\alpha(x) = \begin{cases} -1 & if\ x \leq -\alpha \\ \frac{x}{\alpha} & if\ -\alpha < x < \alpha \\ 1 & if\ x \geq \alpha \end{cases}$, the ReLU function, Leaky ReLU given via $R_\alpha(x) = \begin{cases} x & if\ x \geq 0 \\ \alpha x & if\ x < 0 \end{cases}, \alpha \in \mathbb{Q}$ as well as all their scalar generalizations and any other non-trivial rational step function such as the indicator function on an interval.

We will now see a much more powerful version of the NP-hardness part of Corollary 1:

Theorem 3. NNREACH(id, f) *is NP-hard for any non-linear function $f : \mathbb{R} \to \mathbb{R}$ that is continuous on an interval $[a, b] \subseteq \mathbb{R}$, including $f(x) = \frac{x}{1+e^{-x}}$ and all sigmoidal functions such as $f(x) = \frac{1}{1+e^{-x}}$, $f(x) = \frac{x}{\sqrt{1+x^2}}$ and $f(x) = tanh(x)$ the hyperbolic tangent.*

Proof. By the previous Corollary, it suffices to show that with such a function f we can pp-define a new function $\bar{f}$ that excludes an interval witnessed by two rational points. The construction of $\bar{f}$ proceeds in several steps.
First, by density of $\mathbb{Q}$ there exist $\bar{a}, \bar{b} \in [a, b] \cap \mathbb{Q}, \bar{a} < \bar{b}$ so that $f|_{[\bar{a},\bar{b}]}$ is still non-linear and continuous, we apply a linear transformation on the argument of f so that we can assume $\bar{a} = 0$ and $\bar{b} = 1$. This is pp-definable for rational $\bar{a}, \bar{b}$ and neither changes non-linearity nor continuity. There must exist $c, d \in [0, 1] \cap \mathbb{Q}$, such that

$$f\left(\frac{c+d}{2}\right) \neq \frac{f(c) + f(d)}{2},$$

for f would otherwise fulfill Cauchy's functional equation and therefore be linear.

Apply again a linear transformation on the argument that maps c to 0 and d to 1 and call the resulting function $\hat{f}$. By construction,

$$\hat{f}\left(\frac{1}{2}\right) \neq \frac{\hat{f}(0)+\hat{f}(1)}{2}$$

Define the function $\bar{f} : \mathbb{R} \to \mathbb{R}$ by $\bar{f}(x) = \hat{f}(x) + \hat{f}(1-x) - \hat{f}(0) - \hat{f}(1)$, this is primitive positive, for addition, affine transformation and the constants 0 and 1 are pp-definable in $(\mathbb{R}; \leq, R_+, R_{=1})$. It also matches the requirements for Corollary 1 part 2: The rational points are $\bar{f}(0) = 0$ and $\bar{f}(1) = 0$ and the function must exclude an interval because

$$\bar{f}\left(\frac{1}{2}\right) = \hat{f}\left(\frac{1}{2}\right) + \hat{f}\left(\frac{1}{2}\right) - \hat{f}(0) - \hat{f}(1) = 2(\hat{f}\left(\frac{1}{2}\right) - \frac{\hat{f}(0)+\hat{f}(1)}{2}) \neq 0$$

and by continuity. ∎

Note that any sensible/common activation function is either linear or of this type. We have shown that in the latter case, together with the identity as activation function, we have an NP-hard reachability problem. Theorem 3 however only states NP-hardness for this vast set of reachability problems. One might wonder about membership in NP. Our next main result will show that membership in NP and thus NP-completeness is unlikely for many of the activations, because reachability becomes complete for a complexity class conjectured to be much larger than NP.

Definition 7 (cf. [19]**).** *The problem of deciding whether a system of polynomial equations with integer coefficients is solvable over $\mathbb{R}$ is called the* Existential Theory of the Reals *ETR. The complexity class $\exists\mathbb{R}$ is defined as the set of all decision problems that reduce to ETR in polynomial time. A problem is called $\exists\mathbb{R}$-complete if it is in $\exists\mathbb{R}$ and ETR reduces to the problem in polynomial time.*

It was shown by Canny [5] that ETR is in PSPACE and it is easily seen that ETR is NP-hard. These are currently the best known bounds and it is widely believed that NP$\subsetneq \exists\mathbb{R} \subsetneq$PSPACE.

ETR can be formulated as CSP$(\mathbb{R}, E)$, where E is the set of all polynomial relations $R_p := \{x \in \mathbb{R}^n \mid p(x)\nabla 0, \nabla \in \{=, <, \leq\}, p \in \mathbb{Z}[x_1, ..., x_n]\}$ and inequalities. Note that E does not have finite signature any more, this issue has to be resolved before talking about algorithms and complexities of the CSP, for the encoding into Turing Machines is only possible for finite sets of relations. However, E is a first-order reduct of $(\mathbb{R}; 1, +, \cdot)$, any integer polynomial can be described by the integers, addition, multiplication and logical combinations of these like we did for LP in Example 1. It thus suffices to represent the relations by their first-order definition. The integer coefficients can be assumed to be encoded in binary by the same ongoing that we used in Example 1.

The following Theorem will provide us with NNREACH problems that are hard or even complete for $\exists\mathbb{R}$.

Theorem 4. *a)* NNREACH(id, f) *is* $\exists\mathbb{R}$*-complete for* f *any non-linear polynomial and* $\exists\mathbb{R}$*-hard for* f *any function that coincides with a non-linear polynomial on an interval.*
b) NNREACH(id, f) *is* $\exists\mathbb{R}$*-hard in the following cases:*

i) for any function f *that allows to pp-define a function that coincides with the exponential function on an interval, especially the exponential function itself,* $ELU(x) := \begin{cases} x & \text{if } x \geq 0 \\ \lambda(e^x - 1) & \text{if } x \leq 0 \end{cases}$ *where* $\lambda \in \mathbb{Q}$*, and* $f(x) = e^{-|x|}$

ii) for the Gaussian function $f(x) = e^{-x^2}$

iii) for the arctan function $f(x) = arctan(x)$ *and*

iv) for the Gallant-White cosine-smasher $f(x) = \begin{cases} 0 & x \leq -\frac{\pi}{2} \\ \frac{1+cos(\frac{3}{2}\pi)}{2} & x \in [-\frac{\pi}{2}, \frac{\pi}{2}] \\ 0 & x \geq \frac{\pi}{2} \end{cases}$.

Remark 1. Note that NNREACH$(id, e^{(\cdot)})$ is equivalent to CSP$(\mathbb{R}; R_+, R_., 1, e^{(\cdot)})$, also known as Tarski's exponential function problem. It is not even known whether this problem is decidable or not (cf. [17]). Similar approaches have recently been made in [11].

4 Network Equivalence and Verification of Interval Property

In this section, we study the complexity of the remaining decision problems VIP and NE introduced in Definition 2. We will see that in a lot of cases these problems are essentially the same.

Theorem 5. *Let* F *be a set of activation functions such that sign, id* $\in F$.

a) NE(F) *one-one reduces to the complement of* NNREACH(F) *in linear time. Consequently,* NE*(id) is in P.*

b) NNREACH(F) *one-one reduces to the complement of* NE(F) *in linear time. Consequently,* NE*(ReLU) is co-NP-complete and* NE*(f) is co-NP-hard for any non-linear f that is continuous on an interval.*

c) NE *truth-table reduces to* NE *with just one output dimension in linear time independent of the set of activation functions.*

The same holds for Heaviside or a similar step function instead of sign, id can independently be replaced by ReLU.

Theorem 6. *Let* F *be a set of activation functions.*

a) VIP(F) *truth-table reduces to the complement of* NNREACH(F) *in linear time. Consequently,* VIP*(id) is in P and* VIP*(ReLU) is in co-NP.*

b) Let F *contain at least one among the functions* H *(the Heaviside function), sign or ReLU. Then* NNREACH(F) *one-one reduces to the complement of* VIP(F) *in linear time.*

c) VIP(F) *truth-table reduces to* VIP(F) *with just one output condition in linear time, meaning the output constraint is just one strict inequality.*

Theorem 7. *Let F be a set of activation functions containing id or ReLU and H or sign.*

a) NE(F) *one-one reduces to* VIP(F) *in linear time.*

b) VIP(F) *one-one reduces to* NE(F) *in linear time.*

5 Conclusion and Further Questions

We examined the computational complexity of the reachability problem for neural networks in dependence of the activation functions used. We provided conditions for includedness and hardness for NP, translated a dichotomy result for certain CSPs into the language of reachability problems, and showed ETR-hardness of the reachability problems for several typical activation functions. We also showed that NE and VIP are in many cases the same problem which is essentially "co-NNREACH". Further open questions are:

1.) Do there exist activation functions that are not semilinear but still lead to a reachability problem in NP?

2.) Can the reachability problem for the sigmoid function $f(x) = 1/(1 + e^{-x})$, one oft the most frequently used activations, be classified any better than just as NP-hard? Can it be related to ETR?

3.) Can the discussed ETR-hard problems be classified with respect to (potentially) larger complexity classes such as PSPACE, EXP-Time, decidable,...?

4.) Can reductions between VIP, NE and NNREACH be found that do not rely on certain functions to be included in the set of activations?

5.) How do these problems behave when the deciding algorithm is considered as a computation model with reals as entities? For example of which complexity are the respective problems in a model of real computations like the Blum-Shub-Smale model [1]? What if the underlying neural net may have any real weights and biases instead of just rational ones?

Acknowledgment. I want to thank Klaus Meer for helpful discussion and the anonymous referees for several hints improving the writing.

References

1. Blum, L., Cucker, F., Shub, M., Smale, S.: Complexity and Real Computation. Springer, New York (1998). https://doi.org/10.1007/978-1-4612-0701-6
2. Bodirsky, M., Jonsson, P., von Oertzen, T.: Essential convexity and complexity of semi-algebraic constraints. Log. Methods Comput. Sci. **8**, 1–25 (2012)
3. Bulatov, A.A., Jeavons, P., Krokhin, A.: Classifying the complexity FO constraints using finite algebras. SIAM J. Comput. **34**, 720–742 (2003)
4. Calin, O.: Deep Learning Architectures - A Mathematical Approach. Springer, Cham (2020). https://doi.org/10.1007/978-3-030-36721-3
5. Canny, J.: Some algebraic and geometric computations in PSPACE. Computer Science Division, University of California, Berkeley, Technical report (1988)

6. Carbonnel, C., Cooper, M.C.: Tractability in constraint satisfaction problems: a survey. Assoc. Comput. Mach. **21**, 115–144 (2016)
7. Dixon, M., Klabjan, D., Bang, J.H.: Classification-based financial markets prediction using deep neural networks. Alg. Financ. **6**, 67–77 (2017)
8. Grigorescu, S., Trasnea, B., Cocias, T., Macesanu, G.: A survey of deep learning techniques for autonomous driving. J. Field Robot. **37**, 362–386 (2019)
9. Hinton, G., et al.: Deep neural networks for acoustic modeling in speech recognition: the shared views of four research groups. IEEE Signal Process **29**, 82–97 (2012)
10. Huang, X., et al.: A survey of safety and trustworthiness of deep neural networks: verification, testing, adversarial attack and defence, and interpretability. Comput. Sci. Rev. **37**, 100270 (2020)
11. Isac, O., Zohar, Y., Barrett, C., Katz, G.: DNN Verification, Reachability, and the Exponential Function Problem (2023)
12. Karmarkar, N.: A new polynomial-time algorithm for linear programming. Combinatorica **4**, 373–396 (1984)
13. Katz, G., Barrett, C., Dill, D., Julian, K., Kochenderfer, M.: Reluplex: an efficient SMT solver for verifying deep neural networks. Comput. Aided Verif. **10426**, 97–117 (2017)
14. Khan, A., Sohail, A., Zahoora, U., Qureshi, A.S.: A survey of the recent architectures of deep convolutional neural networks. Artif. Intell. Rev. **53**(8), 5455–5516 (2020). https://doi.org/10.1007/s10462-020-09825-6
15. Krizhevsky, A., Sutskever, I., Hinton, G.E.: Imagenet classification with deep convolutional neural networks. Assoc. Comput. Mach. **25**, 1–9 (2017)
16. Litjens, G., Kooi, T., Bejnordi, B.E., Setio, A.A.A., Ciompi, F., Ghafoorian, M., van der Laak, J.A., van Ginneken, B., Sánchez, C.I.: A survey on deep learning in medical image analysis. Med. Image Anal. **42**, 60–88 (2017)
17. Macintyre, A., Wilkie, A.: On the Decidability of the Real Exponential Field. Kreiseliana: About and Around Georg Kreisel, pp. 451–477 (1996)
18. Ruan, W., Huanga, X., Kwiatkowska, M.: Reachability analysis of deep neural networks with provable guarantees. In: Proceedings of the Twenty-Seventh International Joint Conference on Artificial Intelligence, IJCAI, pp. 2651–2659 (2018)
19. Schaefer, M., Štefankovič, D.: Fixed points, Nash equilibria, and the existential theory of the reals. Theory Comput. Syst. **60**, 172–193 (2012)
20. Sälzer, M., Lange, M.: Reachability is NP-complete even for the simplest neural networks. Int. Conf. Reachabi. Probl. **13035**, 149–164 (2021)

Weakly Synchronous Systems with Three Machines Are Turing Powerful

Cinzia Di Giusto[1(✉)], Davide Ferre'[2], Etienne Lozes[1], and Nicolas Nisse[2]

[1] Université Côte d'Azur, CNRS, I3S, Nice, France
cinzia.di-giusto@univ-cotedazur.fr
[2] Université Côte d'Azur, Inria, CNRS, I3S, Nice, France

Abstract. Communicating finite-state machines (CFMs) are a Turing powerful model of asynchronous message-passing distributed systems. In weakly synchronous systems, processes communicate through phases in which messages are first sent and then received, for each process. Such systems enjoy a limited form of synchronization, and for some communication models, this restriction is enough to make the reachability problem decidable. In particular, we explore the intriguing case of p2p (FIFO) communication, for which the reachability problem is known to be undecidable for four processes, but decidable for two. We show that the configuration reachability problem for weakly synchronous systems of three processes is undecidable. This result is heavily inspired by our study on the treewidth of the Message Sequence Charts (MSCs) that might be generated by such systems. In this sense, the main contribution of this work is a weakly synchronous system with three processes that generates MSCs of arbitrarily large treewidth.

Keywords: Distributed Systems · Message-Passing · Treewidth

1 Introduction

Systems of communicating finite-state machines (CFMs) are a simple, yet expressive, model of asynchronous message-passing distributed systems. In this model, each machine performs a sequence of send and receive actions, where a send action can be matched by a receive action of another machine. For instance, the system in Fig. 1 (left), models a protocol between three processes a, b, and r.

A computation of such a system can be represented graphically by a Message Sequence Chart (MSC), a simplified version of the ITU recommendation [17]. Each machine of the system has its own "timeline" on the MSC, where actions are listed in the order in which they are executed, and message arrows link a send action to its matching receive action. For instance, the MSC of Fig. 1 (right) represents one of the many computations of the system in Fig. 1 (left). The set of all MSCs that the system may generate is determined both by the machines, since the sequence of actions of each timeline must be a sequence of action in the corresponding CFM, and by the "transport layer" or "communication model" employed by the machines. Roughly speaking, a communication model is a class

O. Bournez et al. (Eds.): RP 2023, LNCS 14235, pp. 28–41, 2023.
https://doi.org/10.1007/978-3-031-45286-4_3

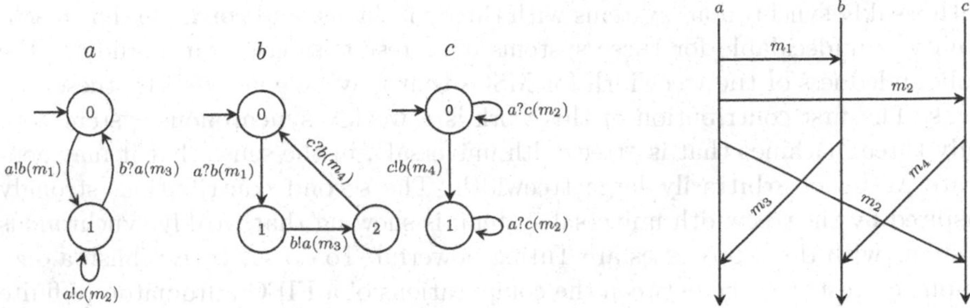

Fig. 1. Example of a system of 3 CFMs (left) and of an MSC generated by it (right). $a!b(m_1)$ (resp., $b?a(m_1)$) denotes the sending (reception) of message m_1 from (by) process a to (from) process b.

of MSCs that are considered "realizable" within that model of communications. For instance, for rendezvous synchronization, an MSC is considered to be realizable with synchronous communication if the only path between a sending and its matching receipt is the direct one through the message arrow that relates them. Among the various communication models that have been considered, we can cite p2p (or FIFO) model, where each ordered pair of machines defines a dedicated FIFO queue; causal ordering (CO), where a message cannot overtake the messages that were sent causally before it; the mailbox model, where each machine holds a unique FIFO queue for all incoming messages; the bag (or simply asynchronous) model, where a message can overtake any other message (see [3,9,10] for various presentations of these communication models).

The configuration reachability problem for a system of CFMs consists in checking whether a control state, together with a given content of the queues, is reachable from the initial state. This problem is decidable for synchronous communication, as the state space of the system is finite, and also for bag communication, by reduction to Petri nets [19]. For other communication models, as soon as two machines are allowed to exchange messages through two FIFO queues, reachability becomes undecidable [8]. Due to this strong limitation, there has been a wealth of work that tried to recover decidability of reachability by considering systems of CFMs that are "almost synchronous".

In weakly synchronous systems, processes communicate through phases in which messages are first sent and then received, for each process; graphically, the MSCs of such systems are the concatenation of smaller, independent MSCs, within which no send happens after a receive. For instance, the MSC in Fig. 1 (right) is weakly synchronous, as it is the concatenation of three "blocks" (namely $\{m_1\}$, $\{m_2\}$, and $\{m_2, m_3, m_4\}$), within which all sends of a given machine happen before all the receives of this same machine. It is known that reachability is decidable for mailbox weakly synchronous systems [6], whereas it is undecidable for either p2p or CO weakly synchronous systems with at least four machines. On the other hand, reachability is decidable for two machines (since any p2p MSC with two machines is also mailbox). In this work, we deal with weakly synchronous systems with three machines, and conclude that reachability is undecidable for these systems. Our result is based on a study of the unboundedness of the treewidth for MSCs that may be generated by these systems. The first contribution of this work is a weakly synchronous system with only three machines that is "treewidth universal", in the sense that it may generate MSCs of arbitrarily large treewidth. The second contribution, strongly inspired by the treewidth universal system, is showing that weakly synchronous systems with three processes are Turing powerful. To do so, we establish a one-to-one correspondence between the computations of a FIFO automaton (a finite state machine that may push and pop from a FIFO queue, which is known to be a Turing powerful computational model) on the one hand, and a subset of the MSCs of the treewidth universal system on the other hand.

Related Work. Beyond weakly synchronous systems, several similar notions have been considered to try to capture the intuition of an "almost synchronous" system. Reachability of existentially bounded systems [14,18] is decidable for FIFO, CO, p2p, or bag communications. Synchronizable systems [2] were an attempt to define a class of systems with good decidability properties, however reachability for such systems with FIFO communications is undecidable [12]. The status of reachability for k-stable systems [1] is unknown. Finally, reachability for k-synchronous systems [7] is decidable for FIFO, CO, p2p, or bag communications.

Another form of under-approximation of the full behaviour of a system of CFMs is the bounded context-switch reachability problem, which is known to be decidable for systems of CFMs, even with a controlled form of function call [15, 20].

Finally, weak synchronisability share some similarities with reversal-bounded counter machines [13,16]: in the context of bag communications, a send is a counter increment, a receive a decrement, and weak synchronisability is a form of bounding the number of reversals of increment and decrement phases.

Outline. Section 2 introduces the necessary terminology. Section 3 presents the weakly synchronous system with three machines that may generate MSCs of arbitrarily large treewidth. Then, Sect. 4 discusses the undecidability of the configuration reachability problem for weakly-synchronous systems with three machines. Finally, Sect. 5 concludes with some final remarks. Proofs and additional material can be found in [11].

2 MSCs and Communicating Automata

We recall here concepts and definitions related to MSCs and communicating automata. Assume a finite set of processes $\mathbb{P}$ and a finite set of messages $\mathbb{M}$. A send action is of the form $p!q(m)$ where $p, q \in \mathbb{P}$ and $m \in \mathbb{M}$; it is executed by p and sends message m to process q. The corresponding receive action, executed by q, is $p?q(m)$. Let $\mathit{Send}(p, q, \_) = \{p!q(m) \mid m \in \mathbb{M}\}$ and $\mathit{Rec}(p, q, \_) = \{p?q(m) \mid m \in \mathbb{M}\}$. For $p \in \mathbb{P}$, we set $\mathit{Send}(p, \_, \_) = \{p!q(m) \mid q \in \mathbb{P} \setminus \{p\} \text{ and } m \in \mathbb{M}\}$, etc. Moreover, $\Sigma_p = Send(p,\_,\_) \cup Rec(\_,p,\_) \cup \{\varepsilon\}$ will denote the set of all actions that are executed by p. Finally, $\Sigma = \bigcup_{p\in\mathbb{P}} \Sigma_p$ is the set of all the actions.

Definition 1 (p2p MSC). *A (p2p) MSC M over $\mathbb{P}$ and $\mathbb{M}$ is a tuple $M = (\mathcal{E}, \rightarrow, \lhd, \lambda)$ where $\mathcal{E}$ is a finite (possibly empty) set of* events *and $\lambda : \mathcal{E} \rightarrow \Sigma$ is a labeling function. For $p \in \mathbb{P}$, let $\mathcal{E}_p = \{e \in \mathcal{E} \mid \lambda(e) \in \Sigma_p\}$ be the set of events that are executed by p. $\rightarrow$ (the* process relation*) is the disjoint union $\bigcup_{p\in\mathbb{P}} \rightarrow_p$ of relations $\rightarrow_p \subseteq \mathcal{E}_p \times \mathcal{E}_p$ such that $\rightarrow_p$ is the direct successor relation of a total order on $\mathcal{E}_p$. $\lhd \subseteq \mathcal{E} \times \mathcal{E}$ (the* message relation*) satisfies the following:*

(1) for every pair $(s,r) \in \lhd$, there is a send action $p!q(m) \in \Sigma$ such that $\lambda(s) = p!q(m)$, $\lambda(r) = p?q(m)$, and $p \neq q$;
(2) for all $r \in \mathcal{E}$ with $\lambda(r) = p?q(m)$, there is a unique $s \in \mathcal{E}$ such that $s \lhd r$;
(3) letting $\leq_M = (\rightarrow \cup \lhd)^*$, we require that $\leq_M$ is a partial order;
(4) for every $s_1 \in \mathcal{E}$ and pair $(s_2, r_2) \in \lhd$ with $\lambda(s_1) = p!q(m_1)$ and $\lambda(s_2) = p!q(m_2)$, if $s_1 \rightarrow_p^+ s_2$, then there exists r_1 such that $(s_1, r_1) \in \lhd$ and $r_1 \rightarrow_q^+ r_2$.

Condition (1) above ensures that message arrows relate a send event to a receive event on a distinct machine. By Condition (2), every receive event has a matching send event. Note that, however, there may be unmatched send events in an MSC. An MSC is called *orphan free* if all send events are matched. Condition (3) ensures that there exists at least one scheduling of all events such that each receive event happens after its matching send event. Condition (4) captures the p2p communication model: a message cannot overtake another message that has the same sender and same receiver as itself.

Let $M = (\mathcal{E}, \rightarrow, \lhd, \lambda)$ be an MSC, then $SendEv(M) = \{e \in \mathcal{E} \mid \lambda(e)$ is a send action$\}$, $RecEv(M) = \{e \in \mathcal{E} \mid \lambda(e)$ is a receive action$\}$, $Matched(M) = \{e \in \mathcal{E} \mid$ there is $f \in \mathcal{E}$ such that $e \lhd f\}$, and $Unm(M) = \{e \in \mathcal{E} \mid \lambda(e)$ is a send action and there is no $f \in \mathcal{E}$ such that $e \lhd f\}$. We do not distinguish isomorphic MSCs. Let $E \subseteq \mathcal{E}$ such that E is $\leq_M$*-downward-closed*, i.e., for all $(e, f) \in \leq_M$ such that $f \in E$, we also have $e \in E$. Then the MSC $M' = (E, \rightarrow, \lhd, \lambda)$ obtained by restriction to E is called a *prefix* of M. If $M_1 = (\mathcal{E}_1, \rightarrow_1, \lhd_1, \lambda_1)$ and $M_2 = (\mathcal{E}_2, \rightarrow_2, \lhd_2, \lambda_2)$ are two MSCs, their *concatenation* $M_1 \cdot M_2 = (\mathcal{E}, \rightarrow, \lhd, \lambda)$ is as expected: $\mathcal{E}$ is the disjoint union of $\mathcal{E}_1$ and $\mathcal{E}_2$, $\lhd = \lhd_1 \cup \lhd_2$, λ is the "union" of λ_1 and λ_2, and $\rightarrow = \rightarrow_1 \cup \rightarrow_2 \cup R$. Here, R contains, for all $p \in \mathbb{P}$ such that $(\mathcal{E}_1)_p$ and $(\mathcal{E}_2)_p$ are non-empty, the pair (e_1, e_2), where e_1 and e_2 are the last and the first event executed by p in M_1 and M_2, respectively. Due to condition (4), concatenation is a partially defined operation: $M_1 \cdot M_2$ is defined if for all $s_1 \in Unm(M_1)$ and $s_2 \in SendEv(M_2)$ that have the same sender and destination ($\lambda(s_1) \in Send(p,q,\_)$ and $\lambda(s_2) \in Send(p,q,\_)$), we have $s_2 \in Unm(M_2)$. In particular, $M_1 \cdot M_2$ is defined when M_1 is orphan-free. Concatenation is associative.

We recall from [5] the definition of weakly synchronous MSC. We say that an MSC is weakly synchronous if it can be broken into phases where all sends are scheduled before all receives.

Definition 2 (weakly synchronous). *We say that $M \in$ MSC is* weakly synchronous *if it is of the form $M = M_1 \cdot M_2 \cdots M_n$ such that for every $M_i = (\mathcal{E}, \rightarrow, \lhd, \lambda)$ $SendEv(M_i)$ is a $\leq_{M_i}$-downward-closed set.*

We now recall the definition of communicating system, which consists of finite-state machines A_p (one per process $p \in \mathbb{P}$) that can exchange messages.

Definition 3 (communicating system). *A (communicating) system over* $\mathbb{P}$ *and* $\mathbb{M}$ *is a tuple* $\mathcal{S} = ((A_p)_{p\in\mathbb{P}})$. *For each* $p \in \mathbb{P}$, $A_p = (Loc_p, \delta_p, \ell_p^0, \ell_p^{acc})$ *is a finite transition system where:* Loc_p *is the finite set of (local) states of* p, $\delta_p \subseteq Loc_p \times \Sigma_p \times Loc_p$ *(also denoted* $\ell \xrightarrow[A_p]{a} \ell'$*) is the transition relation of* p, $\ell_p^{acc} \in Loc_p$ *is the final state of* p.

Given $p \in \mathbb{P}$ and a transition $t = (\ell, a, \ell') \in \delta_p$, we let $source(t) = \ell$, $target(t) = \ell'$, $action(t) = a$, and $msg(t) = m$ if $a \in Send(\_,\_,m) \cup Rec(\_,\_,m)$.

An *accepting run* of $\mathcal{S}$ on an MSC M is a mapping $\rho : \mathcal{E} \to \bigcup_{p\in\mathbb{P}} \delta_p$ that assigns to every event e the transition $\rho(e)$ that is executed at e by A_p. Thus, we require that (i) for all $e \in \mathcal{E}$, we have $action(\rho(e)) = \lambda(e)$, (ii) for all $(e, f) \in \to$, $target(\rho(e)) \xrightarrow[A_p]{\varepsilon}{}^{*} source(\rho(f))$, (iii) for all $(e, f) \in \lhd$, $msg(\rho(e)) = msg(\rho(f))$, (iv) for all $p \in \mathbb{P}$ and $e \in \mathcal{E}_p$ such that there is no $f \in \mathcal{E}$ with $f \to e$, we have $source(\rho(e)) = \ell_p^0$, (v) for all $p \in \mathbb{P}$ and $e \in \mathcal{E}_p$ such that there is no $f \in \mathcal{E}$ with $e \to f$, we have $target(\rho(e)) = \ell_p^{acc}$ and, (vi) $Unm(M) = \emptyset$.

Essentially, in an accepting run of $\mathcal{S}$ every A_p takes a sequence of transitions that lead to its final state ℓ_p^{acc}, and such that each send action will have a matching receive action (i.e., there are no unmatched messages). The *language* of $\mathcal{S}$ is $L(\mathcal{S}) = \{M \in \mathsf{MSC} \mid \text{there is an accepting run of } \mathcal{S} \text{ on } M\}$. We say that $\mathcal{S}$ is weakly synchronous if for all $M \in L(\mathcal{S})$, M is weakly synchronous.

The *emptiness problem* is the decision problem that takes as input a system $\mathcal{S}$ and addresses the question "is $L(\mathcal{S})$ empty?". This problem is a configuration reachability problem, and under several circumstances, its decidability is closely related to the one of the control state reachability problem. In this work, we will study the emptiness problem with the additional hypothesis that $\mathcal{S}$ is a weakly synchronous system with three machines only.

Finally, we recall the less known notion of "FIFO automaton", a finite state machine that can push into and pop from a FIFO queue. This is a system of communicating machines with just one machine, whose semantics is a set of MSCs with a single timeline, for which we exceptionally relax condition (1) of Definition 1, so to allow a send event and its matching receive event to occur on the same machine. The following result is proved in [12, Lemma 4].

Lemma 1 ([12]). *The emptiness problem for FIFO automata is undecidable.*

3 Treewidth of Weakly Synchronous p2p MSCs

There is a strong correlation between MSCs and graphs. An MSC is a directed graph (digraph in the following) where the nodes are the events of the MSC and the arcs are represented by the $\to$ and the $\lhd$ relations. We are, therefore, able to use some tools and techniques from graph theory to possibly derive some interesting results about MSCs. A graph parameter which is particularly important in this context is the *treewidth* [4] mostly due to Courcelle's theorem that, roughly, states that many properties can be checked in classes of MSCs with bounded treewidth[1]. For instance, in [5], it is shown that the class of weakly synchronous mailbox MSCs has bounded treewidth. Interestingly enough, it is also shown that the bigger class of weakly synchronous **p2p** MSCs has unbounded treewidth, by means of a reduction to the Post correspondence problem. Here we give a more direct proof, for all weakly synchronous systems that have at least three processes. We begin with some terminology:

Definition 4. To contract *an arc (u, v) in a (di)graph G means replacing u and v by a single vertex whose neighborhood is the union of the neighborhoods of u and v. A (di)graph H is a* minor *of a (di)graph G if H can be obtained from a subgraph of G by contracting some edges/arcs.*

Next, we show how to build a family of weakly synchronous MSCs with three processes (a, b and c) and unbounded treewidth. We want to find a class of MSCs that admit grids of unbounded size as a minor. The idea is illustrated in Fig. 2, and it consists in bouncing groups of messages between processes so to obtain the depicted shape. The class of MSCs is indexed by two non-zero natural numbers: h and ℓ. Intuitively, h represents the number of consecutive events in a group, and ℓ is the number of groups per process, divided by 2. The graph depicted on the top left of Fig. 2 is not an MSC, because it is undirected and there are multiple actions associated to the same event. Nonetheless, the connection with MSCs is quite intuitive, and formalized in Lemma 2.

We, now, specify how to build a digraph $G_{h,\ell} = (V(G_{h,\ell}), E(G_{h,\ell}))$, from which our MSC $G^*_{h,\ell}$ will be obtained. The set of vertices $V(G_{h,\ell}) = \mathcal{A} \cup \mathcal{B} \cup \mathcal{C}$ contains all the events of each process: $\mathcal{A} = \{s_a^{i,j}, r_a^{i,j} \mid 1 \le i \le h, 1 \le j \le \ell\}$, $\mathcal{B} = \{s_b^{i,j}, r_b^{i,j} \mid 1 \le i \le h, 1 \le j \le \ell\}$, and $\mathcal{C} = \{s_c^{i,j}, r_c^{i,j} \mid 1 \le i \le h, 1 \le j \le \ell\}$.

For $x \in \{a, b, c\}$ and $y \in \{r, s\}$, we add the following arcs to $E(G_{h,\ell})$, which will represent the "timelines" connecting events of each process:

1. for each group of h events/messages and $1 \le j \le \ell$, $Col_{x,y,j} = \{(y_x^{i,j}, y_x^{i+1,j}) \mid 1 \le i < h\}$;
2. then, to link groups together $\{(y_x^{h,j}, y_x^{1,j+1}) \mid 1 \le j < \ell\}$;
3. and finally, to link the phase of sendings with the one of receptions: $(s_x^{h,\ell}, r_x^{1,1})$.

It remains to add the arcs that correspond to the messages exchanged by the processes. Intuitively, each vertex $s_x^{i,j}$ corresponds to two messages sent by process x to the other two processes (except for $j = 1$ and $x = a$, in which case it will correspond to a single message), and each vertex $r_x^{i,j}$ will correspond to two messages received by process x from the other two processes (except for $j = \ell$ and $x = c$, in which case it will correspond to a single message). Formally:

$$\begin{aligned} E_{\mathcal{M}} =& \{(s_a^{i,j}, r_b^{i,j}), (s_c^{i,j}, r_b^{i,j}), (s_c^{i,j}, r_a^{i,j}), (s_b^{i,j}, r_a^{i,j}), (s_b^{i,j}, r_c^{i,j}) \mid 1 \le i \le h, 1 \le j \le \ell\} \\ & \cup \{(s_a^{i,j+1}, r_c^{i,j}) \mid 1 \le i \le h, 1 \le j < \ell\}. \end{aligned} \tag{1}$$

[1] We do not explicitly use tree-decompositions, we refer to [4] for their formal definitions.

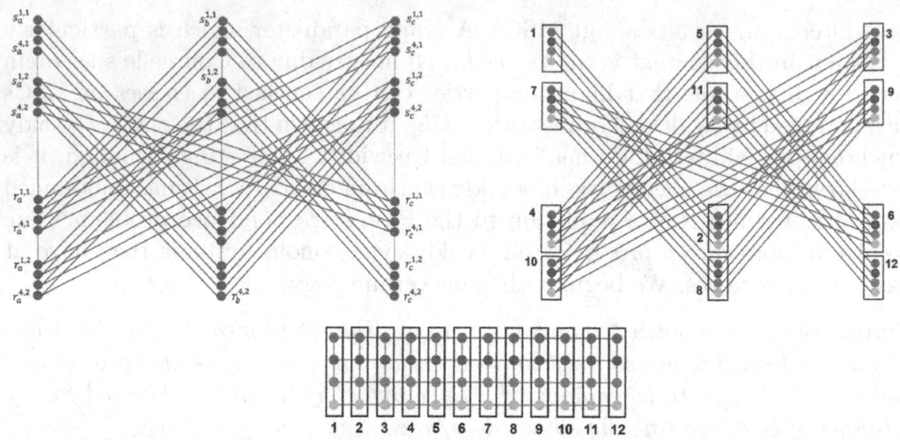

Fig. 2. The undirected graph of $G_{4,2}$ (top left) with a 4×12 grid as a minor (top right and bottom). All arcs go from top to bottom.

Lemma 2. *For any $h, \ell \in \mathbb{N}^+$, $G_{h,\ell}$ is the minor of a graph arising from a weakly synchronous* p2p *MSC $G^*_{h,\ell}$ with 3 processes and a single phase.*

Proof. Fig. 3 exemplifies the transformation below. Note that some vertices of $G_{h,\ell}$ have degree 4 while any MSC is a subcubic graph (i.e., every vertex has degree at most 3). For every $s_x^{i,j}$ with degree 4, let α (resp., β) be the in-neighbor (resp., out-neighbor) of $s_x^{i,j}$ in $\mathcal{P}_x$ and let γ and δ be the other two neighbors of $s_x^{i,j}$. Replace $s_x^{i,j}$ by two vertices $su_x^{i,j}$ and $sd_x^{i,j}$, with the 5 arcs $(\alpha, su_x^{i,j}), (su_x^{i,j}, sd_x^{i,j}), (sd_x^{i,j}, \beta), (su_x^{i,j}, \gamma)$ and $(sd_x^{i,j}, \delta)$. Do a similar transformation for every $r_x^{i,j}$ with degree 4. A similar transformation is done for the four vertices (with degree 3) $s_b^{1,1}, s_c^{1,1}, r_a^{h,\ell}$ and $r_b^{h,\ell}$. Let $G^*_{h,\ell}$ be the obtained digraph. It is clear that $G^*_{h,\ell}$ is an MSC and that $G_{h,\ell}$ is a minor of $G^*_{h,\ell}$.

Note that for any $x \in \{a, b, c\}$, $\mathcal{X} \in \{\mathcal{A}, \mathcal{B}, \mathcal{C}\}$ induces a directed path $\mathcal{P}_x$ with first the vertices $s_x^{i,j}$ (in increasing lexicographical order of (j, i)) and then the vertices $r_x^{i,j}$ (in increasing lexicographical order of (j, i)). The fact that $G^*_{h,\ell}$ is weakly synchronous with one phase directly follows the fact that, for every $x \in \{a, b, c\}$, the vertices s, su and sd (corresponding to sendings) appear before the vertices r, ru and rd (corresponding to receptions) in the directed path $\mathcal{P}_x$.

Moreover, for every $x, y \in \{a, b, c\}$, $x \neq y$, the arcs from $\mathcal{P}_x$ to $\mathcal{P}_y$ are all parallel (i.e., for every arc (u, v) and (u', v') from $\mathcal{P}_x$ to $\mathcal{P}_y$, if u is a predecessor of u' in $\mathcal{P}_x$, then v is a predecessor of v' in $\mathcal{P}_y$). This implies that $G^*_{h,\ell}$ is p2p. □

Note that, for fixed $i \leq h$ and $j < \ell$, $P_{i,j} = (s_a^{i,j}, r_b^{i,j}, s_c^{i,j}, r_a^{i,j}, s_b^{i,j}, r_c^{i,j}, s_a^{i+1,j})$ is a (undirected) path with 6 arcs linking $s_a^{i,j}$ to $s_a^{i,j+1}$. From this, it is not difficult to see that $G_{h,\ell}$ admits a grid of size $h \times 6\ell$ as a minor, which is the content of next lemma (see Fig. 2 for an example).

Let $tw(G_{h,\ell})$ be the treewidth of the underlying undirected graph of $G_{h,\ell}$.

Lemma 3. *For any* $h, \ell \in \mathbb{N}^*$, $tw(G_{h,\ell}) \geq \min\{h, 6\ell\}$.

Proof. The subgraph obtained from $G_{h,\ell}$ by keeping the arcs in item 1 and Eq. 1: $G'_{h,\ell} = (V(G_{h,\ell}), E_{\mathcal{M}} \cup \bigcup_{x\in\{a,b,c\},y\in\{r,s\},1\leq j\leq \ell} Col_{x,y,j})$, is a $h \times 6\ell$ grid. From [4], we know that $tw(G'_{h,\ell}) \geq \min\{h, 6\ell\}$ and, since treewidth is closed under subgraphs [4], $tw(G_{h,\ell}) \geq tw(G'_{h,\ell}) \geq \min\{h, 6\ell\}$. □

We can then easily derive the main result for this section.

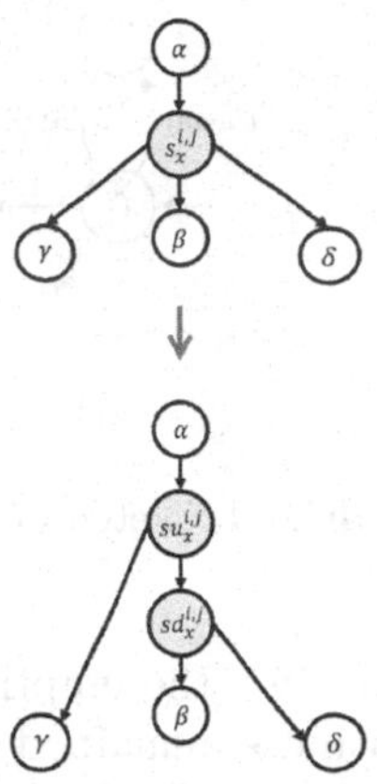

Fig. 3. Transformation of Lemma 2.

Theorem 1. *The class of weakly synchronous* p2p *MSCs with three processes (and a single phase) has unbounded treewidth.*

Proof. From Lemma 2, $G^*_{h,\ell}$ is a weakly synchronous p2p MSC with 3 processes and $G_{h,\ell}$ is a minor of $G^*_{h,\ell}$. Hence, from Lemma 3 and the fact that the treewidth is minor-closed [4], we get that $tw(G^*_{h,\ell}) \geq \min\{h, 6\ell\}$. □

Notice that, a similar technique, this time exploiting four processes instead of three, can be used to show that we can build a weakly synchronous p2p MSC that can be contracted to whatever graph.

Theorem 2. *Let H be any graph. There exists a weakly synchronous* p2p *MSCs with four processes that admits H as minor.*

Proof. Let $V(H) = \{v_1, \cdots, v_h\}$ and $E(H) = \{e_1, \cdots, e_\ell\}$. Take graph $G_{h,\ell}$ defined above. Add a new directed path $(d_1, \cdots, d_\ell)$ (which corresponds to the fourth process). Finally, for every $1 \leq j \leq \ell$, and edge $e_j = \{v_i, v_{i'}\} \in E(H)$, add two arcs $(r_a^{i,j}, d_j)$ and $(r_a^{i',j}, d_j)$. Let G be the obtained graph.

Using similar arguments as in the proof of Lemma 2, G arises from a weakly synchronous p2p MSC with 4 processes. Now, to see that H is a minor of G, first remove all "vertical" arcs from G. Then, for every $1 \leq i \leq h$, contract the path $\bigcup_{1\leq j\leq \ell} P_{i,j}$ into a single vertex (corresponding to v_i), and finally contract the arc $(r_a^{i',j}, d_j)$ for every edge $e_j = \{v_i, v_{i'}\}$. These operations lead to H. □

4 Reachability for Weakly Synchronous p2p Systems with 3 Machines

In [5], it is shown that the control state reachability problem for weakly p2p synchronous systems with at least 4 processes is undecidable. The result is obtained via a reduction of the Post correspondence problem. In the same paper, following from the boundedness of treewidth, it is also shown that reachability is decidable for systems with 2 processes. The arguments easily adapt to show the same

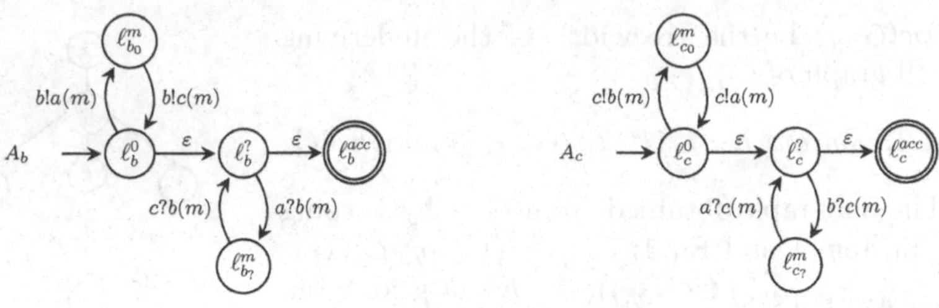

Fig. 4. Sketch of A_b and A_c of $\mathcal{S}_3$ (only a single message m is considered).

results for the emptiness problem instead. The decidability of reachability, or emptiness, remained open for systems with 3 processes. We already showed that the treewidth of weakly synchronous **p2p** MSCs is unbounded for 3 processes. But, this result alone is not enough to prove undecidability, still it gives us a hint on how to conduct the proof. Indeed, inspired by the proof of the unboundedness of the treewidth, we provide a reduction from the emptiness problem for a FIFO automaton $\mathcal{S}_1$ (undecidable, Lemma 1) to the emptiness problem for a weakly synchronous system $\mathcal{S}_3$ with 3 machines. The reduction makes sure that there is an accepting run of $\mathcal{S}_1$ if and only if there is one for $\mathcal{S}_3$.

Let $\mathcal{S}_1 = (A)$, with $A = (Loc, \delta, \ell^0, \ell^{acc})$ be a communicating system with a single process over $\mathbb{M}$. We will consider only automata that, from any state, have at most one non-epsilon outgoing transition, and no self loops (i.e., transitions that start and land in the same state). More precisely, we prove that any system can be encoded into one that satisfies this additional property while accepting the same language.

We provide an encoding of the FIFO automaton $\mathcal{S}_1$ into the system $\mathcal{S}_3 = (A_a, A_b, A_c)$ over $\mathbb{M} \cup \{D\}$, where D is an additional special message called the *dummy message*. We show that $\mathcal{S}_3$ is weakly synchronous, and that $L(\mathcal{S}_1) \neq \emptyset$ if and only if $L(\mathcal{S}_3) \neq \emptyset$. Processes b and c (see Fig. 4) are used as forwarders so that messages circulate as in Fig. 2. Basically, process b (resp., process c) goes through two phases, the first one in which messages are sent to a and c (resp., a and b), and the second in which messages can be received. In Fig. 4, there should be one state $\ell^m_{b_0}$ (resp., $\ell^m_{b_?}$), which is the in and out-neighbor of ℓ^0_b (resp., $\ell^?_b$), per message $m \in \mathbb{M} \cup \{D\}$. Formally, $A_b = (Loc_b, \delta_b, \ell^0_b, \ell^{acc}_b)$ where

$$\begin{aligned} Loc_b &= \{\ell^0_b, \ell^?_b, \ell^{acc}_b\} \cup \{\ell^m_{b_0}, \ell^m_{b_?} \mid m \in \mathbb{M} \cup \{D\}\} \\ \delta_b &= \{(\ell^0_b, \varepsilon, \ell^?_b), (\ell^?_b, \varepsilon, \ell^{acc}_b)\} \;\cup\; \{(\ell^0_b, b!a(m), \ell^m_{b_0}), (\ell^m_{b_0}, b!c(m), \ell^0_b), \\ &\qquad (\ell^?_b, b?a(m), \ell^m_{b_?}), (\ell^m_{b_?}, b?c(m), \ell^?_b) \mid m \in \mathbb{M} \cup \{D\}\} \end{aligned}$$

and symmetrically $A_c = (Loc_c, \delta_c, \ell^0_c, \ell^{acc}_c)$ where

$$\begin{aligned} Loc_c &= \{\ell^0_c, \ell^?_c, \ell^{acc}_c\} \cup \{\ell^m_{c_0}, \ell^m_{c_?} \mid m \in \mathbb{M} \cup \{D\}\} \\ \delta_c &= \{(\ell^0_c, \varepsilon, \ell^?_c), (\ell^?_c, \varepsilon, \ell^{acc}_c)\} \cup \{(\ell^0_c, c!b(m), \ell^m_{c_0}), (\ell^m_{c_0}, c!a(m), \ell^0_c), \\ &\qquad (\ell^?_c, c?b(m), \ell^m_{c_?}), (\ell^m_{c_?}, c?a(m), \ell^?_c) \mid m \in \mathbb{M} \cup \{D\}\}. \end{aligned}$$

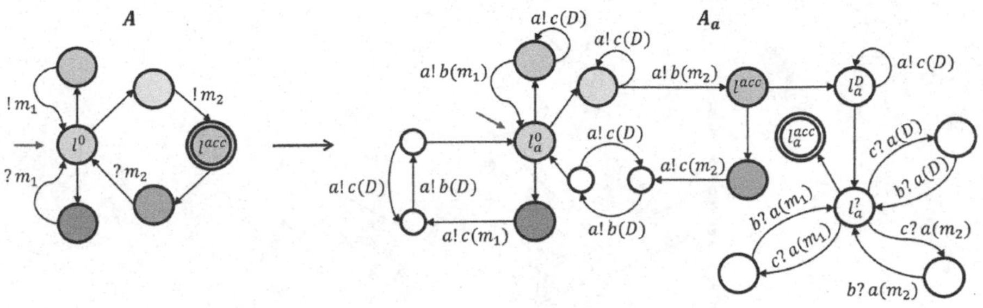

Fig. 5. The automaton A_a for the system $\mathcal{S}_3$, built from the automaton A of $\mathcal{S}_1$. Arcs without actions represent ϵ transitions.

Process a mimics the behavior of A. Figure 5 shows an example of how A_a is built, starting from A. At a high level, A_a is composed of two parts: the first simulates A, and the second (after state ℓ_a^D) receives all messages sent by b and c. In the first part of A_a, each send action of A is replaced by a send action addressed to process b, and each reception of A is replaced by a send action to process c. We then use some dummy messages to ensure that our encoding works properly. Roughly, we force A_a to send a dummy message to b after each message sent to c, and we let A_a send any number of dummy messages to c right before each message sent to b, or right before entering the" receiving phase" of A_a, where messages from b and c are received. Similarly, after A_a sends a dummy message to b, it is also allowed to send two other dummy messages (the first one to c and the second one to b) an unbounded number of times. Formally, $A_a = (Loc_a, \delta_a, \ell^0, \ell_a^{acc})$, where:

$$
\begin{aligned}
Loc_a =& Loc \cup \{\ell_{t_1}, \ell_{t_2} \mid t = (\ell, ?m, \ell') \in \delta\} \cup \{\ell_a^D, \ell_a^?, \ell_a^{acc}\} \cup \{\ell_{a?}^m \mid m \in \mathbb{M} \cup \{D\}\} \\
\delta_a =& \{(\ell, a!b(m), \ell'), (\ell, a!b(D), \ell) \mid (\ell, !m, \ell') \in \delta\} \cup \\
& \{(\ell, a!c(m), \ell_{t_1}), (\ell_{t_1}, a!b(D), \ell_{t_2}), \\
& \quad (\ell_{t_2}, a!c(D), \ell_{t_1}), (\ell_{t_2}, \varepsilon, \ell') \mid t = (\ell, ?m, \ell') \in \delta\} \cup \\
& \{(\ell, \varepsilon, \ell') \mid (\ell, \varepsilon, \ell') \in \delta\} \cup \{(\ell^{acc}, \varepsilon, \ell_a^D), (\ell_a^D, a!c(D), \ell_a^D)\} \cup \\
& \{(\ell_a^D, \varepsilon, \ell_a^?), (\ell_a^?, \varepsilon, \ell_a^{acc})\} \cup \\
& \{(\ell_a^?, a?c(m), \ell_{a?}^m), (\ell_{a?}^m, a?b(m), \ell_a^?) \mid m \in \mathbb{M} \cup \{D\}\}
\end{aligned}
$$

In Fig. 5, colors show the mapping of states from an instance of A to the corresponding automaton A_a. Figure 6 illustrates an accepting run of some system $\mathcal{S}_1$ and one of the corresponding accepting runs of the associated $\mathcal{S}_3$.

Given a sequence of actions $!m$ and $?m$, where m can be any message, we call it a FIFO sequence if (i) all messages are received in the order in which they are sent, and (ii) no message is received before being sent.

We relax this definition to talk about sequences of send actions $a!b(m)$ and $a!c(m)$ taken by a (in the first part of the automaton A_a); in particular, we say that such a sequence γ' is FIFO if, when interpreting each $a!b(m)$ and $a!c(m)$

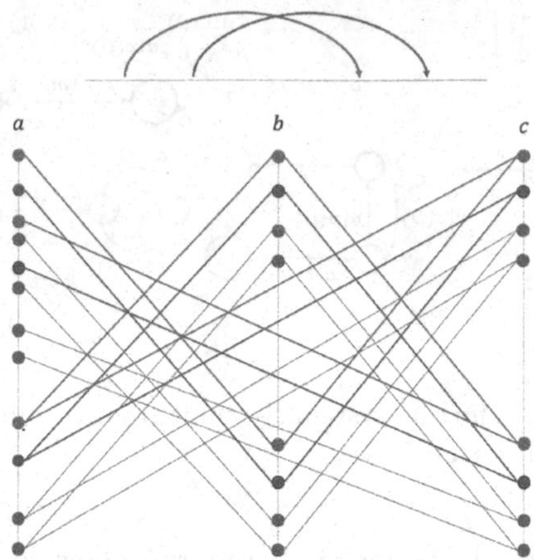

Fig. 6. Above, a run with two messages for some system $\mathcal{S}_1$ with a single process (timeline drawn horizontally). Below, one possible corresponding MSC realized by the associated $\mathcal{S}_3$. Gray lines correspond to dummy messages. (Color figure online)

action as $!m$ and $?m$, respectively, γ' is a FIFO sequence. Dummy messages are used to enforce that the sequence of send actions taken by A_a in an accepting run of $\mathcal{S}_3$ is FIFO.

Theorem 3. *There is an accepting run of* $\mathcal{S}_1$ *if and only if there is an accepting run of* $\mathcal{S}_3$.

Sketch of proof. We only provide a sketch of the proof, which is quite convoluted and requires several intermediate lemmata. The full proof is in [11].

($\Rightarrow$) We design Algorithm 1, which takes an accepting run σ of S_1, and returns an accepting run μ for S_3. At a high level, Algorithm 1 takes the sequence of actions taken by A in σ, rewrites each $!m$ and $?m$ action as $a!b(m)$ and $a!c(m)$, and then adds some actions related to dummy messages. We first show that the sequence of actions γ' returned by Algorithm 1 is a sequence of send actions that takes A_a of $\mathcal{S}_3$ from state ℓ^0 to ℓ^{acc} (note that this is not the final state of A_a, see Fig. 5 for an example). We then show that γ' is a FIFO sequence, and prove that there exists an accepting run of $\mathcal{S}_3$ in which A_a starts by executing exactly the sequence of actions in γ'. Finally, we show that Algorithm 1 always terminates.

($\Leftarrow$) Let μ be an accepting run of $\mathcal{S}_3$, from which we show that it is easy to build a sequence of actions γ taken by A in an accepting run of $\mathcal{S}_1$. Let γ' be the sequence of send actions taken by A_a in the accepting run μ. The first step is to show that γ' is a FIFO sequence. The three automata A_a, A_b, and A_c are built so to ensure that γ' is always a FIFO sequence. This is closely related to the shape of the MSCs associated to accepting runs of $\mathcal{S}_3$; these MSCs exploit the same kind of pattern seen in Sect. 3 to bounce messages back and forth between the three processes. We then prove that, if we ignore actions related to dummy

Algorithm 1. Let σ be an accepting run of $\mathcal{S}_1$, and α^σ be the sequence of n actions taken by A in σ. We use $\alpha^\sigma(i)$ to denote the i-th action of α^σ.

```
γ' ← empty list
Queue ← empty queue
for i from 1 to n do
  action ← α^σ(i)
  if action = !x then
    while first(Queue) = D do
      add a!c(D) to γ'
      dequeue D from Queue
    end while
    add a!b(x) to γ'
    enqueue x in Queue
  else if action = ?x then
    add a!c(x) to γ'
    dequeue x from Queue
    add a!b(D) to γ'
    enqueue D in Queue
    if Queue does not contain only D then
      while first(Queue) = D do
        add a!c(D) to γ'
        dequeue D from Queue
        add a!b(D) to γ'
        enqueue D in Queue
      end while
    end if
  end if
end for
while first(Queue) = D do
  add a!c(D) to γ'
  dequeue D from Queue
end while
return γ';
```

messages in γ' and interpret each $a!b(m)$ and $a!c(m)$ action as $!m$ and $?m$, we get a sequence of actions γ that takes A from its initial state ℓ^0 to its final state ℓ^{acc} in an accepting run of $\mathcal{S}_1$.

The following result immediately follows from Lemma 1 and Theorem 3.

Theorem 4. *The emptiness problem for weakly synchronous communicating systems with three processes is undecidable.*

Notice that our results extend to causally ordered (CO) communication, since an MSC is weakly synchronous if and only if it is weakly synchronous CO.

Corollary 1. *The emptiness problem for causal order communicating systems with three processes is undecidable.*

5 Conclusion

We showed the undecidability of the reachability of a configuration for weakly synchronous systems with three processes or more. The main contribution lies in the technique used to achieve this result. We first show that the treewidth of the class of weakly synchronous MSCs is unbounded, by proving that it is always possible to build such an MSC with an arbitrarily large grid as minor. Then, a similar construction is employed to provide an encoding of a FIFO automaton into a weakly synchronous system with three processes, allowing to show that reachability of a configuration is undecidable.

References

1. Akroun, L., Salaün, G.: Automated verification of automata communicating via FIFO and bag buffers. Formal Methods Syst. Des. **52**(3), 260–276 (2018). https://doi.org/10.1007/s10703-017-0285-8
2. Basu, S., Bultan, T.: Choreography conformance via synchronizability. In: Srinivasan, S., Ramamritham, K., Kumar, A., Ravindra, M.P., Bertino, E., Kumar, R. (eds.) Proceedings of the 20th International Conference on World Wide Web, WWW 2011, Hyderabad, India, March 28–April 1, 2011, pp. 795–804. ACM (2011)
3. Basu, S., Bultan, T.: On deciding synchronizability for asynchronously communicating systems. Theor. Comput. Sci. **656**, 60–75 (2016). https://doi.org/10.1016/j.tcs.2016.09.023
4. Bodlaender, H.L.: A partial k-arboretum of graphs with bounded treewidth. Theor. Comput. Sci. **209**(1–2), 1–45 (1998). https://doi.org/10.1016/S0304-3975(97)00228-4
5. Bollig, B., Di Giusto, C., Finkel, A., Laversa, L., Lozes, É., Suresh, A.: A unifying framework for deciding synchronizability. In: Haddad, S., Varacca, D. (eds.) 32nd International Conference on Concurrency Theory, CONCUR 2021, August 24–27, 2021. LIPIcs, vol. 203, pp. 14:1–14:18. Schloss Dagstuhl - Leibniz-Zentrum für Informatik, Virtual Conference (2021). https://doi.org/10.4230/LIPIcs.CONCUR.2021.14
6. Bollig, B., Di Giusto, C., Finkel, A., Laversa, L., Lozes, É., Suresh, A.: A unifying framework for deciding synchronizability (extended version). Technical report (2021). https://hal.archives-ouvertes.fr/hal-03278370/document
7. Bouajjani, A., Enea, C., Ji, K., Qadeer, S.: On the completeness of verifying message passing programs under bounded asynchrony. In: Chockler, H., Weissenbacher, G. (eds.) CAV 2018. LNCS, vol. 10982, pp. 372–391. Springer, Cham (2018). https://doi.org/10.1007/978-3-319-96142-2_23
8. Brand, D., Zafiropulo, P.: On communicating finite-state machines. J. ACM **30**(2), 323–342 (1983). https://doi.org/10.1145/322374.322380
9. Charron-Bost, B., Mattern, F., Tel, G.: Synchronous, asynchronous, and causally ordered communication. Distributed Comput. **9**(4), 173–191 (1996). https://doi.org/10.1007/s004460050018
10. Di Giusto, C., Ferré, D., Laversa, L., Lozes, É.: A partial order view of message-passing communication models. Proc. ACM Program. Lang. **7**(POPL), 1601–1627 (2023). https://doi.org/10.1145/3571248
11. Di Giusto, C., Ferré, D., Lozes, E., Nisse, N.: Weakly synchronous systems with three machines are Turing powerful. Technical report, Université Côte d'Azur, CNRS, I3S, France; Université Côte d'Azur, Inria, CNRS, I3S, France (2023). https://hal.science/hal-04182953
12. Finkel, A., Lozes, É.: Synchronizability of communicating finite state machines is not decidable. In: Chatzigiannakis, I., Indyk, P., Kuhn, F., Muscholl, A. (eds.) 44th International Colloquium on Automata, Languages, and Programming, ICALP 2017, July 10–14, 2017, Warsaw, Poland. LIPIcs, vol. 80, pp. 122:1–122:14. Schloss Dagstuhl - Leibniz-Zentrum für Informatik (2017)
13. Finkel, A., Sangnier, A.: Reversal-bounded counter machines revisited. In: Ochmański, E., Tyszkiewicz, J. (eds.) MFCS 2008. LNCS, vol. 5162, pp. 323–334. Springer, Heidelberg (2008). https://doi.org/10.1007/978-3-540-85238-4_26
14. Genest, B., Kuske, D., Muscholl, A.: On communicating automata with bounded channels. Fundam. Informaticae **80**(1-3), 147–167 (2007). http://content.iospress.com/articles/fundamenta-informaticae/fi80-1-3-09

15. Heußner, A., Leroux, J., Muscholl, A., Sutre, G.: Reachability analysis of communicating pushdown systems. In: Ong, L. (ed.) FoSSaCS 2010. LNCS, vol. 6014, pp. 267–281. Springer, Heidelberg (2010). https://doi.org/10.1007/978-3-642-12032-9_19
16. Ibarra, O.H.: Reversal-bounded multicounter machines and their decision problems. J. ACM (JACM) **25**(1), 116–133 (1978)
17. ITU-T: Recommendation ITU-T Z.120: Message sequence chart (MSC). Technical report, International Telecommunication Union, Geneva (2011)
18. Lohrey, M., Muscholl, A.: Bounded MSC communication. In: Nielsen, M., Engberg, U. (eds.) FoSSaCS 2002. LNCS, vol. 2303, pp. 295–309. Springer, Heidelberg (2002). https://doi.org/10.1007/3-540-45931-6_21
19. Mayr, E.W.: An algorithm for the general petri net reachability problem. In: Symposium on the Theory of Computing (1981)
20. Torre, S.L., Madhusudan, P., Parlato, G.: Context-bounded analysis of concurrent queue systems. In: Tools and Algorithms for the Construction and Analysis of Systems, 14th International Conference, TACAS 2008, Held as Part of the Joint European Conferences on Theory and Practice of Software, ETAPS 2008, Budapest, Hungary, March 29–April 6, 2008. Proceedings, pp. 299–314 (2008). https://doi.org/10.1007/978-3-540-78800-3_21

On the Identity and Group Problems for Complex Heisenberg Matrices

Paul C. Bell[1], Reino Niskanen[2(✉)], Igor Potapov[3], and Pavel Semukhin[2]

[1] Keele University, Keele, UK
p.c.bell@keele.ac.uk
[2] Liverpool John Moores University, Liverpool, UK
{r.niskanen,p.semukhin}@ljmu.ac.uk
[3] University of Liverpool, Liverpool, UK
potapov@liverpool.ac.uk

Abstract. We study the Identity Problem, the problem of determining if a finitely generated semigroup of matrices contains the identity matrix; see Problem 3 (Chapter 10.3) in "Unsolved Problems in Mathematical Systems and Control Theory" by Blondel and Megretski (2004). This fundamental problem is known to be undecidable for $\mathbb{Z}^{4\times 4}$ and decidable for $\mathbb{Z}^{2\times 2}$. The Identity Problem has been recently shown to be in polynomial time by Dong for the Heisenberg group over complex numbers in any fixed dimension with the use of Lie algebra and the Baker-Campbell-Hausdorff formula. We develop alternative proof techniques for the problem making a step forward towards more general problems such as the Membership Problem. We extend our techniques to show that the fundamental problem of determining if a given set of Heisenberg matrices generates a group, can also be decided in polynomial time.

1 Introduction

Matrices and matrix products can represent dynamics in many systems, from computational applications in linear algebra and engineering to natural science applications in quantum mechanics, population dynamics and statistics, among others [4,5,10,11,15,19,24,28,29]. The analysis of various evolving systems requires solutions of reachability questions in linear systems, which form the essential part of verification procedures, control theory questions, biological systems predictability, security analysis etc.

Reachability problems for matrix products are challenging due to the complexity of this mathematical object and a lack of effective algorithmic techniques. The significant challenge in the analysis of matrix semigroups was initially illustrated by Markov (1947), [27] and later highlighted by Paterson (1970) [30], Blondel and Megretski (2004) [5], and Harju (2009) [21]. The central reachability question is the **Membership Problem:** *Decide whether or not a given matrix M belongs to the matrix semigroup S generated by a set of square matrices G.* By restricting M to be the identity matrix, the problem is known as the *Identity Problem.*

O. Bournez et al. (Eds.): RP 2023, LNCS 14235, pp. 42–55, 2023.
https://doi.org/10.1007/978-3-031-45286-4_4

Problem 1 (Identity Problem). Let S be a matrix semigroup generated by a finite set of $n \times n$ matrices over $\mathbb{K} = \mathbb{Z}, \mathbb{Q}, \mathbb{A}, \mathbb{Q}(\mathrm{i}), \ldots$ Is the identity matrix $\boldsymbol{I}$ in the semigroup, i.e., does $\boldsymbol{I} \in S$ hold?

The Membership Problem is known to be undecidable for integer matrices from dimension three, but the decidability status of the Identity Problem was unknown for a long time for matrix semigroups of any dimension, see Problem 10.3 in "Unsolved Problems in Mathematical Systems and Control Theory" [5]. The Identity Problem was shown to be undecidable for 48 matrices from $\mathbb{Z}^{4 \times 4}$ in [3] and for a generator of eight matrices in [23]. This implies that the *Group Problem* (decide whether a finitely generated semigroup is a group) is also undecidable. The Identity Problem and the Group Problem are open for $\mathbb{Z}^{3 \times 3}$.

The Identity Problem for a semigroup generated by 2×2 matrices was shown to be EXPSPACE decidable in [9] and later improved by showing to be NP-complete in [2]. The only decidability beyond integer 2×2 matrices were shown in [14] for flat rational subsets of $\mathrm{GL}(2, \mathbb{Q})$.

Similarly to [8], the work [23] initiated consideration of matrix decision problems in the Special Linear Group $\mathrm{SL}(3, \mathbb{Z})$, by showing that there is no embedding from pairs of words into matrices from $\mathrm{SL}(3, \mathbb{Z})$. Beyond the 2×2 case, the Identity Problem was shown to be decidable for the discrete Heisenberg group $\mathrm{H}(3, \mathbb{Z})$ which is a subgroup of $\mathrm{SL}(3, \mathbb{Z})$.

The Heisenberg group is widely used in mathematics and physics. This is in some sense the simplest non-commutative group, and has close connections to quantum mechanical systems [6,20,25], harmonic analysis, and number theory [7,13]. It also makes appearances in complexity theory, e.g., the analysis and geometry of the Heiseberg group have been used to disprove the Goemans-Linial conjecture in complexity theory [26]. Matrices in physics and engineering are ordinarily defined with values over $\mathbb{R}$ or $\mathbb{C}$. In this context, we formulate our decision problems and algorithmic solutions over the field of complex numbers with a finite representation, Gaussian rationals $\mathbb{Q}(\mathrm{i})$.

The Identity Problem was recently shown to be decidable in polynomial time for complex Heisenberg matrices in a paper by Dong [18]. They first prove the result for upper-triangular matrices with rational entries and ones on the main diagonal, $\mathrm{UT}(\mathbb{Q})$ and then use a known embedding of the Heisenberg group over algebraic numbers into $\mathrm{UT}(\mathbb{Q})$. Their approach is different from our techniques; the main difference being that [18] uses tools from Lie algebra, and in particular, matrix logarithms and the Baker-Campbell-Hausdorff formula, to reason about matrix products and their properties. In contrast, our approach first characterises matrices which are 'close to' the identity matrix, which we denote Ω-matrices. Such matrices are close to the identity matrix in that they differ only in a single position in the top-right corner. We then argue about the commutator angle of matrices within this set in order to determine whether zero can be reached, in which case the identity matrix is reachable. We believe that these techniques take a step towards proving the decidability of the more general *membership problem*, which we discuss towards the end of the paper. A careful analysis then follows to ensure that all steps require only Polynomial time, and we extend our techniques to show that determining if a given set of matrices forms a group (the *group problem*) is also decidable in P (this result is shown in [16] using different techniques). We thus present polynomial time algorithms for both these problems for Heisenberg matrices over $\mathbb{Q}(\mathrm{i})$ in any dimension n.

These new techniques allow us to extend previous results for the discrete Heisenberg group $\mathrm{H}(n,\mathbb{Z})$ and $\mathrm{H}(n,\mathbb{Q})$ [12,17,23,24] and make a step forward towards proving the decidability of the membership problem for complex Heisenberg matrices.

2 Roadmap

We will give a brief overview of our approach here. Given a Heisenberg matrix $M = \begin{pmatrix} 1 & \boldsymbol{m}_1^T & m_3 \\ \boldsymbol{0} & \boldsymbol{I}_{n-2} & \boldsymbol{m}_2 \\ 0 & \boldsymbol{0}^T & 1 \end{pmatrix} \in \mathrm{H}(n,\mathbb{Q}(\mathrm{i}))$, denote by $\psi(M)$ the triple $(\boldsymbol{m}_1, \boldsymbol{m}_2, m_3) \in \mathbb{Q}(\mathrm{i})^{2n-3}$. We define the set $\Omega \subseteq \mathrm{H}(n,\mathbb{Q}(\mathrm{i}))$ as those matrices where $\boldsymbol{m}_1$ and $\boldsymbol{m}_2$ are zero vectors, i.e., matrices in Ω look like $\boldsymbol{I}_n$ except allowing any element of $\mathbb{Q}(\mathrm{i})$ in the top right element. Such matrices play a crucial role in our analysis.

In particular, given a set of matrices $G = \{G_1, \ldots, G_t\} \subseteq \mathrm{H}(n,\mathbb{Q}(\mathrm{i}))$ generating a semigroup $\langle G \rangle$, we can find a description of $\Omega_{\langle G \rangle} = \langle G \rangle \cap \Omega$. Since $\boldsymbol{I} \in \Omega$, the Identity Problem reduces to determining if $\boldsymbol{I} \in \Omega_{\langle G \rangle}$.

Several problems present themselves, particularly if we wish to solve the problem in Polynomial time (P). The set $\Omega_{\langle G \rangle}$ is described by a linear set $\mathcal{S} \subseteq \mathbb{N}^t$, which is the solution set of a homogeneous system of linear Diophantine equations induced by matrices in G. This is due to the observation that the elements $(\boldsymbol{m}_1, \boldsymbol{m}_2) \in \mathbb{Q}(\mathrm{i})^{2n-4}$ behave in an additive fashion under multiplication of Heisenberg matrices. The main issue is that the size of the basis of $\mathcal{S}$ is exponential in the description size of G. Nevertheless, we can determine *if a solution exists* to such a system in P (Lemma 1), and this proves sufficient.

The second issue is that reasoning about the element $m_3 \in \mathbb{Q}(\mathrm{i})$ (i.e., the top right element) in a product of Heisenberg matrices is much more involved than for elements $(\boldsymbol{m}_1, \boldsymbol{m}_2) \in \mathbb{Q}(\mathrm{i})^{2n-4}$. Techniques to determine if $m_3 = 0$ for an Ω-matrix within $\Omega_{\langle G \rangle}$ take up the bulk of this paper.

The key to our approach is to consider *commutators* of pairs of matrices within G, which in our case can be described by a single complex number. That is, for $M_1, M_2 \in G$, the commutator is $[M_1, M_2] \in \mathbb{Q}(\mathrm{i})$. After removing all *redundant matrices* (those never reaching an Ω-matrix), we have two cases to consider. Either every pair of matrices from G has the same *angle* in the polar form of the commutator or else there are at least two commutators with different angles.

The latter case is used in Lemma 5. It states that the identity matrix can always be constructed using a solution that contains four particular matrices. Let M_1, M_2, M_3 and M_4 be such that $[M_1, M_2] = r\exp(\mathrm{i}\gamma)$ and $[M_3, M_4] = r'\exp(\mathrm{i}\gamma')$, where $\gamma \neq \gamma'$ so that pairs M_1, M_2 and M_3, M_4 have different commutator angles. We may then define four matrix products using the same generators but matrices M_1, M_2, M_3 and M_4 are in a different order. This difference in order and the commutator angles being different, ensures that we can control the top right corner elements in order to construct the identity matrix. Lemma 3 provides details on how to calculate the top right element in these products. We then prove that these top right elements in the four matrices are not contained in an open half-plane and this is sufficient for us to construct the identity matrix.

The above construction does not work when all commutators have the same angle, and indeed in this case the identity may or may not be present. Hence, we need to consider various possible shuffles of matrices in these products. To this end, we extend the result of Lemma 3 to derive a formula for the top right element for any shuffle and prove it as Lemma 4. We observe that there is a *shuffle invariant* part of the product that does not depend on the shuffle, and that shuffles add or subtract commutators. Furthermore, this shuffle invariant component can be calculated from the generators used in the product. As we assume that all commutators have the same angle, γ, different shuffles move the value along the line in the complex plane defined by the common commutator angle which we call the γ-*line*.

It is straightforward to see that if it is not possible to reach the γ-line using the additive semigroup of shuffle invariants, then the identity cannot be generated. Indeed, since different shuffles move the value along the γ-line but the shuffle invariant part never reaches it, then the possible values are never on the γ-line, which includes the origin.

We show that if it *is* possible to reach the γ-line using shuffle invariants and there are at least two non-commuting matrices in the used solution, then the identity matrix is in the semigroup (Lemma 6). Testing this property requires determining the solvability of a polynomially-sized set of non-homogeneous systems of linear Diophantine equations, which can be done in polynomial time by Lemma 1.

If the γ-line can be reached only using commuting matrices, we can construct another system of linear Diophantine equations since the top right element has an explicit formula in terms of generators used (see Lemma 6).

3 Preliminaries

The sets of rational numbers, real numbers and complex numbers are denoted by $\mathbb{Q}$, $\mathbb{R}$ and $\mathbb{C}$. The set of rational complex numbers is denoted by $\mathbb{Q}(\mathrm{i}) = \{a + b\mathrm{i} \mid a, b \in \mathbb{Q}\}$. The set $\mathbb{Q}(\mathrm{i})$ is often called the Gaussian rationals in the literature. A complex number can be written in polar form $a + b\mathrm{i} = r\exp(\mathrm{i}\varphi)$, where $r \in \mathbb{R}$ and $\varphi \in [0, \pi)$. We denote the *angle* of the polar form φ by $\arg(a + b\mathrm{i})$. We also denote $\mathrm{Re}(a + b\mathrm{i}) = a$ and $\mathrm{Im}(a + b\mathrm{i}) = b$. It is worth highlighting that commonly the polar form is defined for a positive real r and an angle between $[0, 2\pi)$. These two definitions are obviously equivalent.

The *identity matrix* is denoted by $\boldsymbol{I}_n$ or, if the dimension n is clear from the context, by $\boldsymbol{I}$. The *Heisenberg group* $\mathrm{H}(n, \mathbb{K})$ is formed by $n \times n$ matrices of the form $M = \begin{pmatrix} 1 & \boldsymbol{m}_1^T & m_3 \\ \boldsymbol{0} & \boldsymbol{I}_{n-2} & \boldsymbol{m_2} \\ 0 & \boldsymbol{0}^T & 1 \end{pmatrix}$, where $\boldsymbol{m}_1, \boldsymbol{m}_2 \in \mathbb{K}^{n-2}$, $m_3 \in \mathbb{K}$ and $\boldsymbol{0} =$

$(0, 0, \ldots, 0)^T \in \mathbb{K}^{n-2}$ is the zero vector. It is easy to see that the Heisenberg group is a non-commutative subgroup of $\mathrm{SL}(n, \mathbb{K}) = \{M \in \mathbb{K}^{n \times n} \mid \det(M) = 1\}$.

We will be interested in subsemigroups of $\mathrm{H}(n, \mathbb{Q}(\mathrm{i}))$ which are finitely generated. Given a set of matrices $G = \{G_1, \ldots, G_t\} \subseteq \mathrm{H}(n, \mathbb{Q}(\mathrm{i}))$, we denote the matrix semigroup generated by G as $\langle G \rangle$.

Let $M = \begin{pmatrix} 1 & \boldsymbol{m}_1^T & m_3 \\ \boldsymbol{0} & \boldsymbol{I}_{n-2} & \boldsymbol{m}_2 \\ 0 & \boldsymbol{0}^T & 1 \end{pmatrix}$, then $(M)_{1,n} = m_3$ is the top right element. To improve readability, by $\psi(M)$ we denote the triple $(\boldsymbol{m}_1, \boldsymbol{m}_2, m_3) \in \mathbb{Q}(\mathrm{i})^{2n-3}$.

The vectors $\boldsymbol{m}_1, \boldsymbol{m}_2$ play a crucial role in our considerations. We define the set $\Omega \subseteq \mathrm{H}(n, \mathbb{Q}(\mathrm{i}))$ as those matrices where $\boldsymbol{m}_1$ and $\boldsymbol{m}_2$ are zero vectors, i.e., matrices in Ω look like $\boldsymbol{I}_n$ except allowing any element of $\mathbb{Q}(\mathrm{i})$ in the top right element. That is, $\Omega = \left\{ \begin{pmatrix} 1 & \boldsymbol{0}^T & m_3 \\ \boldsymbol{0} & \boldsymbol{I}_{n-2} & \boldsymbol{0} \\ 0 & \boldsymbol{0}^T & 1 \end{pmatrix} \mid m_3 \in \mathbb{Q}(\mathrm{i}) \right\}$, where $\boldsymbol{0} = (0, 0, \ldots, 0)^T \in \mathbb{Q}(\mathrm{i})^{n-2}$ is the zero vector.

Let us define a shuffling of a product of matrices. Let $M_1, \ldots, M_k \in G$. The set of permutations of a product of these matrices is denoted by $\mathsf{shuffle}(M_1, \ldots, M_k) = \{M_{\sigma(1)} \cdots M_{\sigma(k)} \mid \sigma \in \mathcal{S}_k\}$, where $\mathcal{S}_k$ is the set of permutations on k elements. If some matrix appears multiple times in the list, say M_1 appears x times, we write $\mathsf{shuffle}(M_1^x, M_2, \ldots, M_k)$ instead of $\mathsf{shuffle}(\underbrace{M_1, \ldots, M_1}_{x \text{ times}}, M_2, \ldots, M_k)$.

Let $M_1 = \begin{pmatrix} 1 & \boldsymbol{a}_1^T & c_1 \\ \boldsymbol{0} & \boldsymbol{I}_{n-2} & \boldsymbol{b}_1 \\ 0 & \boldsymbol{0}^T & 1 \end{pmatrix}$ and $M_2 = \begin{pmatrix} 1 & \boldsymbol{a}_2^T & c_2 \\ \boldsymbol{0} & \boldsymbol{I}_{n-2} & \boldsymbol{b}_2 \\ 0 & \boldsymbol{0}^T & 1 \end{pmatrix}$. By an abuse of notation, we define the *commutator* $[M_1, M_2]$ of M_1 and M_2 by $[M_1, M_2] = \boldsymbol{a}_1^T \boldsymbol{b}_2 - \boldsymbol{a}_2^T \boldsymbol{b}_1 \in \mathbb{Q}(\mathrm{i})$. Note that the commutator of two arbitrary matrices A, B is ordinarily defined as $[A, B] = AB - BA$, i.e., a matrix. However, for matrices $M_1, M_2 \in \mathrm{H}(n, \mathbb{Q}(\mathrm{i}))$, it is clear that $M_1 M_2 - M_2 M_1 = \begin{pmatrix} 0 & \boldsymbol{0}^T & \boldsymbol{a}_1^T \boldsymbol{b}_2 - \boldsymbol{a}_2^T \boldsymbol{b}_1 \\ \boldsymbol{0} & O & \boldsymbol{0} \\ 0 & \boldsymbol{0}^T & 0 \end{pmatrix}$, where O is the $(n-2) \times (n-2)$ zero matrix, thus justifying our notation which will be used extensively. Observe that the matrices M_1, M_2 commute if and only if $[M_1, M_2] = 0$.

Note that the commutator is antisymmetric, i.e., $[M_1, M_2] = -[M_2, M_1]$. We further say that γ is the *angle of the commutator* if $[M_1, M_2] = r \exp(\mathrm{i}\gamma)$ for some $r \in \mathbb{R}$ and $\gamma \in [0, \pi)$. If two commutators $[M_1, M_2]$, $[M_3, M_4]$ have the same angles, that is, $[M_1, M_2] = r \exp(\mathrm{i}\gamma)$ and $[M_3, M_4] = r' \exp(\mathrm{i}\gamma)$ for some $r, r' \in \mathbb{R}$, then we denote this property by $[M_1, M_2] \stackrel{\gamma}{=} [M_3, M_4]$. If they have different angles, then we write $[M_1, M_2] \stackrel{\gamma}{\neq} [M_3, M_4]$. By convention, if $[M_1, M_2] = 0$, then $[M_1, M_2] \stackrel{\gamma}{=} [M_3, M_4]$ for every $M_3, M_4 \in \mathrm{H}(n, \mathbb{Q}(\mathrm{i}))$.

To show that our algorithms run in polynomial time, we will need the following lemma.

Lemma 1. *(i) Let $A \in \mathbb{Q}^{n\times m}$ be a rational matrix, and $\boldsymbol{b} \in \mathbb{Q}^n$ be an n-dimensional rational vector with non-negative coefficients. Then we can decide in polynomial time whether the system of inequalities $A\boldsymbol{x} \geq \boldsymbol{b}$ has an integer solution $\boldsymbol{x} \in \mathbb{Z}^m$.*

(ii) Let $A_1 \in \mathbb{Q}^{n_1\times m}$ and $A_2 \in \mathbb{Q}^{n_2\times m}$ be a rational matrices. Then we can decide in polynomial time whether the system of inequalities $A_1\boldsymbol{x} \geq \boldsymbol{0}^{n_1}$ and $A_2\boldsymbol{x} > \boldsymbol{0}^{n_2}$ has an integer solution $\boldsymbol{x} \in \mathbb{Z}^m$.

4 Properties of Ω-Matrices

To solve the Identity Problem for subsemigroups of $\mathrm{H}(n, \mathbb{Q}(\mathrm{i}))$ (Problem 1), we will be analysing matrices in Ω (matrices with all zero elements, except possibly the top-right corner value). Let us first discuss how to construct Ω-matrices from a given set of generators $G \subseteq \mathrm{H}(n, \mathbb{Q}(\mathrm{i}))$.

As observed earlier, when multiplying Heisenberg matrices of the form $\begin{pmatrix} 1 & \boldsymbol{m}_1^T & m_3 \\ \boldsymbol{0} & \boldsymbol{I}_{n-2} & \boldsymbol{m}_2 \\ 0 & \boldsymbol{0}^T & 1 \end{pmatrix}$, elements $\boldsymbol{m}_1$ and $\boldsymbol{m}_2$ are *additive*. We can thus construct a homogeneous system of linear Diophantine equations (SLDEs) induced by matrices in G. Each Ω-matrix then corresponds to a solution to this system.

Let $G = \{G_1, \ldots, G_t\}$, where $\psi(G_i) = (\boldsymbol{a}_i, \boldsymbol{b}_i, c_i)$. For a vector $\boldsymbol{a} \in \mathbb{Q}(\mathrm{i})^{n-2}$, define $\mathrm{Re}(\boldsymbol{a}) = (\mathrm{Re}(\boldsymbol{a}(1)), \ldots, \mathrm{Re}(\boldsymbol{a}(n-2)))$ (similarly for $\mathrm{Im}(\boldsymbol{a})$). We consider system $A\boldsymbol{x} = \boldsymbol{0}$, where

$$A = \begin{pmatrix} \mathrm{Re}(\boldsymbol{a}_1) & \mathrm{Re}(\boldsymbol{a}_2) & \cdots & \mathrm{Re}(\boldsymbol{a}_t) \\ \mathrm{Im}(\boldsymbol{a}_1) & \mathrm{Im}(\boldsymbol{a}_2) & \cdots & \mathrm{Im}(\boldsymbol{a}_t) \\ \mathrm{Re}(\boldsymbol{b}_1) & \mathrm{Re}(\boldsymbol{b}_2) & \cdots & \mathrm{Re}(\boldsymbol{b}_t) \\ \mathrm{Im}(\boldsymbol{b}_1) & \mathrm{Im}(\boldsymbol{b}_2) & \cdots & \mathrm{Im}(\boldsymbol{b}_t) \end{pmatrix}, \tag{1}$$

$\boldsymbol{x} \in \mathbb{N}^t$ and $\boldsymbol{0}$ is the $4(n-2)$-dimensional zero vector; noting that $A \in \mathbb{Q}^{4(n-2)\times t}$. Let $\mathcal{S} = \{\boldsymbol{s}_1, \ldots, \boldsymbol{s}_p\}$ be the set of minimal solutions to the system. Recall that elements of $\mathcal{S}$ are irreducible. That is, a minimal solution cannot be written as a sum of two nonzero solutions. The set $\mathcal{S}$ is always finite and constructable [31].

A matrix $M_i \in G$ is *redundant* if the ith component is 0 in every minimal solution $\boldsymbol{s} \in \mathcal{S}$. Non-redundant matrices can be recognized by checking whether a non-homogeneous SLDE has a solution. More precisely, to check whether M_i is non-redundant, we consider the system $A\boldsymbol{x} = \boldsymbol{0}$ together with the constraint that $\boldsymbol{x}(i) \geq 1$, where $\boldsymbol{x}(i)$ is the ith component of $\boldsymbol{x}$. Using Lemma 1, we can determine in polynomial time whether such a system has an integer solution.

For the remainder of the paper, we assume that G is the set of non-redundant matrices. This implicitly assumes that for this G, the set $\mathcal{S} \neq \emptyset$. Indeed, if there are no solutions to the corresponding SLDEs, then all matrices are redundant. Hence $G = \emptyset$ and $\boldsymbol{I} \notin \langle G \rangle$ holds trivially.

Let $M_1, \ldots, M_k \in G$ be such that $X = M_1 \cdots M_k \in \Omega$. The Parikh vector of occurrences of each matrix from G in product X may be written as $\boldsymbol{x} = (m_1, \ldots, m_t) \in \mathbb{N}^t$. This Parikh vector $\boldsymbol{x}$ is a linear combination of elements of

$\mathcal{S}$, i.e., $\boldsymbol{x} = \sum_{j=1}^{p} y_j \boldsymbol{s}_j$, with $y_j \in \mathbb{N}$, because $\boldsymbol{x}$ is a solution to the SLDEs. Each element of $\mathsf{shuffle}(M_1, \ldots, M_k)$ has the same Parikh vector, but their product is not necessarily the same matrix; potentially differing in the top right element.

Let us state some properties of Ω-matrices.

Lemma 2. *The Ω-matrices are closed under matrix product; the top right element is additive under the product of two matrices; and Ω-matrices commute with Heisenberg matrices. In other words, let $A, B \in \Omega$ and $M \in \mathrm{H}(n, \mathbb{Q}(\mathrm{i}))$, then*

$$(i)\quad AB \in \Omega; \qquad (ii)\quad (AB)_{1,n} = A_{1,n} + B_{1,n}; \qquad (iii)\quad AM = MA.$$

Furthermore, if $N = M_1 M_2 \cdots M_{k-1} M_k \in \Omega$ for some $M_1, \ldots, M_k \in \mathrm{H}(n, \mathbb{Q}(\mathrm{i}))$, then every cyclic permutation of matrices results in the same matrix, N. That is, $N = M_2 M_3 \cdots M_k M_1 = \cdots = M_k M_1 \cdots M_{k-2} M_{k-1}$.

We require the following technical lemma that allows us to calculate the value in top right corner for particular products. The claim is proven by a direct computation.

Lemma 3. *Let $M_1, M_2, \ldots, M_k \in \mathrm{H}(n, \mathbb{Q}(\mathrm{i}))$ such that $M_1 M_2 \cdots M_k \in \Omega$ and let $\ell \geq 1$. Then,*

$$(M_1^\ell M_2^\ell \cdots M_k^\ell)_{1,n} = \ell \sum_{i=1}^{k} \left(c_i - \frac{1}{2} \boldsymbol{a}_i^T \boldsymbol{b}_i \right) + \frac{\ell^2}{2} \sum_{1 \leq i < j \leq k-1} [M_i, M_j],$$

where $\psi(M_i) = (\boldsymbol{a}_i, \boldsymbol{b}_i, c_i)$ for each $i = 1, \ldots, k$.

If we further assume that the matrices from the previous lemma commute, then for every $M \in \mathsf{shuffle}(M_1^\ell, M_2^\ell, \ldots, M_k^\ell)$:

$$M_{1,n} = \ell \sum_{i=1}^{k} \left(c_i - \frac{1}{2} \boldsymbol{a}_i^T \boldsymbol{b}_i \right) + \frac{\ell^2}{2} \sum_{1 \leq i < j \leq k-1} [M_i, M_j] = \ell \sum_{i=1}^{k} \left(c_i - \frac{1}{2} \boldsymbol{a}_i^T \boldsymbol{b}_i \right), \quad (2)$$

noting that $[M_i, M_j] = 0$ when matrices M_i and M_j commute.

In Lemma 3, the matrix product has an ordering which yielded a simple presentation of the value in the top right corner. In the next lemma, we consider an arbitrary shuffle of the product and show that the commutators are important when expressing the top right corner element.

Lemma 4. *Let $M_1, M_2, \ldots, M_k \in \mathrm{H}(n, \mathbb{Q}(\mathrm{i}))$ such that $M_1 M_2 \cdots M_k \in \Omega$ and let $\ell \geq 1$. Let M be a shuffle of the product $M_1^\ell M_2^\ell \cdots M_k^\ell$ by a permutation σ that acts on $k\ell$ elements. Then*

$$(M)_{1,n} = \ell \sum_{i=1}^{k} \left(c_i - \frac{1}{2} \boldsymbol{a}_i^T \boldsymbol{b}_i \right) + \frac{\ell^2}{2} \sum_{1 \leq i < j \leq k-1} [M_i, M_j] - \sum_{1 \leq i < j \leq k} z_{ji} [M_i, M_j],$$

where $\psi(M_i) = (\boldsymbol{a}_i, \boldsymbol{b}_i, c_i)$ for $i = 1, \ldots, k$, and z_{ji} is the number of times M_j appears before M_i in the product; so z_{ji} is the number of inversions of i, j in σ.

The crucial observation is that regardless of the shuffle, the top right corner element has a common term, namely $\sum_{i=1}^{k}(c_i - \frac{1}{2}\boldsymbol{a}_i^T\boldsymbol{b}_i)$, plus some linear combination of commutators. We call the common term the *shuffle invariant.* Note that the previous lemmas apply to any Heisenberg matrices, even those in $\mathrm{H}(n, \mathbb{C})$. For the remainder of the section, we restrict considerations to matrices in G.

Definition 1 (Shuffle Invariant). *Let $M_1, \ldots, M_k \in G$ be such that $X = M_1 \cdots M_k \in \Omega$. The Parikh vector of occurrences of each matrix from G in product X may be written as $\boldsymbol{x} = (m_1, \ldots, m_t) \in \mathbb{N}^t$ where $t = |G|$ as before. Define $\Lambda_{\boldsymbol{x}} = \sum_{i=1}^{t} m_i(c_i - \frac{1}{2}\boldsymbol{a}_i^T\boldsymbol{b}_i)$ as the shuffle invariant of Parikh vector $\boldsymbol{x}$.*

Note that the shuffle invariant is dependant only on the generators used in the product and the Parikh vector $\boldsymbol{x}$.

Let $\mathcal{S} = \{\boldsymbol{s}_1, \ldots, \boldsymbol{s}_p\} \subseteq \mathbb{N}^k$ be the set of minimal solutions to the system of linear Diophantine equations for G giving an Ω-matrix, as described in the beginning of the section. Each $\boldsymbol{s}_j$ thus induces a shuffle invariant that we denote $\Lambda_{\boldsymbol{s}_j} \in \mathbb{Q}(\mathrm{i})$ as shown in Definition 1. The Parikh vector of any $X = M_1M_2 \cdots M_k$ with $X \in \Omega$, denoted $\boldsymbol{x}$, is a linear combination of elements of $\mathcal{S}$, i.e., $\boldsymbol{x} = \sum_{j=1}^{p} y_j\boldsymbol{s}_j$. We then note that the shuffle invariant $\Lambda_{\boldsymbol{x}}$ of $\boldsymbol{x}$ is $\Lambda_{\boldsymbol{x}} = \sum_{j=1}^{p} y_j\Lambda_{\boldsymbol{s}_j}$, i.e., a linear combination of shuffle invariants of $\mathcal{S}$.

Finally, it follows from Lemma 4 that for any $X \in \mathsf{shuffle}(M_1, M_2, \ldots, M_k)$, where as before $M_1M_2 \cdots M_k \in \Omega$ and whose Parikh vector is $\boldsymbol{x} = \sum_{j=1}^{p} y_j\boldsymbol{s}_j$, the top right entry of X is equal to

$$X_{1,n} = \Lambda_{\boldsymbol{x}} + \sum_{1\leq i<j\leq k} \alpha_{ij}[M_i, M_j] = \sum_{j=1}^{p} y_j\Lambda_{\boldsymbol{s}_j} + \sum_{1\leq i<j\leq k} \alpha_{ij}[M_i, M_j], \quad (3)$$

where each $\alpha_{ij} \in \mathbb{Q}$ depends on the shuffle.

Furthermore, if a product of Heisenberg matrices is an Ω-matrix and all matrix pairs share a common angle γ for their commutators, then shuffling the matrix product only modifies the top right element of the matrix by a real multiple of $\exp(\mathrm{i}\gamma)$. This drastically simplifies our later analysis.

5 The Identity Problem for Subsemigroups of $\mathrm{H}(n, \mathbb{Q}(\mathrm{i}))$

In this section, we prove our main result.

Theorem 1. *Let $G \subseteq \mathrm{H}(n, \mathbb{Q}(\mathrm{i}))$ be a finite set of matrices. Then it is decidable in polynomial time if $\boldsymbol{I} \in \langle G \rangle$.*

The proof relies on analysing generators used in a product that results in an Ω-matrix. There are two distinct cases to consider: either there is a pair of commutators with distinct angles, or else all commutators have the same angle. The former case is considered in Lemma 5 and the latter in Lemma 6. More precisely, we will prove that in the former case, the identity matrix is always in the generated semigroup and that the latter case reduces to deciding whether shuffle invariants reach the line defined by the angle of the commutator.

The two cases are illustrated in Fig. 1. On the left, is a depiction of the case where there are at least two commutators with different angles, γ_1 and γ_2. We will construct a sequence of products where the top right element tends to $r_1 \exp(\mathrm{i}\gamma_1)$ with positive r_1 and another product that tends to $r_2 \exp(\mathrm{i}\gamma_1)$ with negative r_2. This is achieved by changing the order of matrices whose commutator has angle γ_1. Similarly, we construct two sequences of products where the top right elements tend to $r_3 \exp(\mathrm{i}\gamma_2)$ and $r_4 \exp(\mathrm{i}\gamma_2)$, where r_3 and r_4 have the opposite signs. Together these sequences ensure, that eventually, the top right elements do not lie in the same open half-planes. On the right, is a depiction of the other case, where all commutators lie on γ-line. In this case, the shuffle invariants of products need to be used to reach the line.

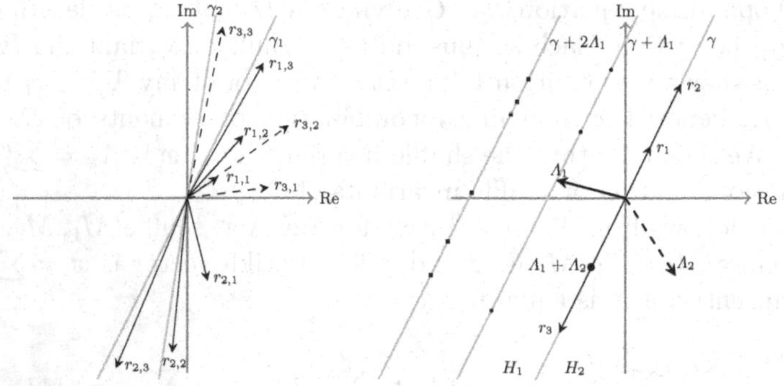

Fig. 1. Illustrations of Lemma 5 and Lemma 6. Left shows two lines defined by two different commutators and how the values $r_{1,\ell}$ and $r_{2,\ell}$ tend to γ_1-line in opposite directions, while $r_{3,\ell}$ tends to γ_2-line ($r_{4,\ell}$ is omitted for clarity). Eventually, they are not all within the same closed half-plane. Right shows that if there is only one shuffle invariant, say Λ_1, then all reachable values are on lines parallel to the γ-line, namely, $\gamma + k\Lambda_1$ for $k > 0$. But if there exists Λ_2 in the opposite half-plane, then the γ-line itself is reachable.

Lemma 5. *Let* $G = \{G_1, \ldots, G_t\} \subseteq \mathrm{H}(n, \mathbb{Q}(\mathrm{i}))$*, where each* G_i *is non-redundant. Suppose there exist* $M_1, M_2, M_3, M_4 \in G$ *such that* $[M_1, M_2] \not\eqsim [M_3, M_4]$*. Then* $\boldsymbol{I} \in \langle G \rangle$*.*

It remains to consider the case when the angles of commutators coincide for each pair of non-redundant matrices. Our aim is to prove that, under this condition, it is decidable whether the identity matrix is in the generated semigroup.

Lemma 6. *Let* $G = \{G_1, \ldots, G_t\} \subseteq \mathrm{H}(n, \mathbb{Q}(\mathrm{i}))$ *be a set of non-redundant matrices such that the angle of commutator* $[G_i, G_{i'}]$ *is* γ *for all* $1 \leq i, i' \leq t$*, then we can determine in polynomial time if* $\boldsymbol{I} \in \langle G \rangle$*.*

Proof. Let $\{\boldsymbol{s}_1, \ldots, \boldsymbol{s}_p\} \subseteq \mathbb{N}^t$ be the set of minimal solutions to the SLDEs for G giving zeros in $\boldsymbol{a}$ and $\boldsymbol{b}$ elements. Each $\boldsymbol{s}_j$ induces a shuffle invariant $\Lambda_{\boldsymbol{s}_j} \in \mathbb{Q}(\mathrm{i})$ as explained in Definition 1.

Consider a product $X = M_1 \cdots M_k \in \Omega$, where each $M_i \in G$. Let $\boldsymbol{x} = (m_1, m_2, \ldots, m_t) \in \mathbb{N}^t$ be the Parikh vector of the number of occurrences of each matrix from G in product X. Since $X \in \Omega$, we have $\boldsymbol{x} = \sum_{j=1}^{p} y_j \boldsymbol{s}_j$, where each $y_j \in \mathbb{N}$. Notice that $X \in \mathsf{shuffle}(G_1^{m_1}, \ldots, G_t^{m_t})$. Hence, by Eq. (3), we have

$$X_{1,n} = \Lambda_{\boldsymbol{x}} + \sum_{1 \le i < j \le k} \alpha_{ij}[M_i, M_j] = \sum_{j=1}^{p} y_j \Lambda_{\boldsymbol{s}_j} + r\exp(\mathrm{i}\gamma), \tag{4}$$

where $\alpha_{ij} \in \mathbb{Q}$ and $r \in \mathbb{R}$ depend on the shuffle. In other words, any shuffle of the product X will change the top right entry $X_{1,n}$ by a real multiple of $\exp(\mathrm{i}\gamma)$.

Let H_1, H_2 be the two *open* half-planes of the complex plane induced by $\exp(\mathrm{i}\gamma)$, that is, the union $H_1 \cup H_2$ is the complement of the γ-line; thus $0 \notin H_1 \cup H_2$. We now prove that if $\{\Lambda_{\boldsymbol{s}_1}, \ldots, \Lambda_{\boldsymbol{s}_p}\} \subseteq H_1$ or $\{\Lambda_{\boldsymbol{s}_1}, \ldots, \Lambda_{\boldsymbol{s}_p}\} \subseteq H_2$ then we cannot reach the identity matrix.

Assume that $\{\Lambda_{\boldsymbol{s}_1}, \ldots, \Lambda_{\boldsymbol{s}_p}\} \subseteq H_1$, renaming H_1, H_2 if necessary. Assume that there exists some product $X = X_1 X_2 \cdots X_k$ equal to the identity matrix, where $k > 0$ and $X_j \in G$. Then since $X \in \Omega$, we see from Eq. (4) that $X_{1,n} = \sum_{j=1}^{p} y_j \Lambda_{\boldsymbol{s}_j} + r\exp(\mathrm{i}\gamma)$, where $r \in \mathbb{R}$.

Clearly, $\sum_{j=1}^{p} y_j \Lambda_{\boldsymbol{s}_j} \in H_1$, and since $y_j \neq 0$ for at least one i, we have $\sum_{j=1}^{p} y_j \Lambda_{\boldsymbol{s}_j} \neq 0$. Now, since $r\exp(\mathrm{i}\gamma)$ is on the γ-line, which is the boundary of H_1, the value $X_{1,n}$ belongs to H_1 and cannot equal zero. This contradicts the assumption that X is the identity matrix.

If $\{\Lambda_{\boldsymbol{s}_1}, \ldots, \Lambda_{\boldsymbol{s}_p}\}$ is not fully contained in either H_1 or H_2, then there are two possibilities. Either there exists some $\Lambda_{\boldsymbol{s}_j} \in \mathbb{Q}(\mathrm{i})$ such that the angle of $\Lambda_{\boldsymbol{s}_j}$ is equal to γ (in which case such a $\Lambda_{\boldsymbol{s}_j}$ lies on the line defined by $\exp(\mathrm{i}\gamma)$), or else there exist $\Lambda_{\boldsymbol{s}_i}, \Lambda_{\boldsymbol{s}_j}$ such that $1 \le i < j \le p$ and $\Lambda_{\boldsymbol{s}_i}$ and $\Lambda_{\boldsymbol{s}_j}$ lie in different open half-planes, say $\Lambda_{\boldsymbol{s}_i} \in H_1$ and $\Lambda_{\boldsymbol{s}_j} \in H_2$.

In the latter case, note that there exists $x, y \in \mathbb{N}$ such that $x\Lambda_{\boldsymbol{s}_i} + y\Lambda_{\boldsymbol{s}_j} = r\exp(\mathrm{i}\gamma)$ for some $r \in \mathbb{R}$ since $\Lambda_{\boldsymbol{s}_i}, \Lambda_{\boldsymbol{s}_j}$ and the commutators that define the γ-line have rational components. It means that in both cases there exist $z_1, \ldots, z_p \in \mathbb{N}$ such that $\sum_{j=1}^{p} z_j \Lambda_{\boldsymbol{s}_j} = r\exp(\mathrm{i}\gamma)$ for some $r \in \mathbb{R}$.

Consider a product $T = T_1 \cdots T_k \in \Omega$, where each $T_j \in G$ and whose Parikh vector is equal to $\sum_{j=1}^{p} z_j \boldsymbol{s}_j$, where $z_1, \ldots, z_p \in \mathbb{N}$ are as above. It follows from Eq. (4) that $T_{1,n} = \sum_{j=1}^{p} z_j \Lambda_{\boldsymbol{s}_j} + r'\exp(\mathrm{i}\gamma) = r\exp(\mathrm{i}\gamma) + r'\exp(\mathrm{i}\gamma)$, where $r, r' \in \mathbb{R}$ and shuffles of such a product change only r'.

We have two possibilities. Either $T = T_1 \cdots T_k$ is a product only consisting of commuting matrices from G, or else two of the matrices in the product of T do not commute. In the latter case, let us write $T' = N_1 N_2 X' \in \mathsf{shuffle}(T_1, \ldots, T_k)$, where $N_1 \in G$ and $N_2 \in G$ do not commute and X' is the product of the remaining matrices in any order. We observe that Lemma 3 implies

$$(N_1^{\ell_1} N_2^{\ell_1} X'^{\ell_1})_{1,n} = \ell_1 r \exp(\mathrm{i}\gamma) + \frac{\ell_1^2}{2}[N_1, N_2] = \ell_1 r \exp(\mathrm{i}\gamma) + \frac{\ell_1^2}{2} r' \exp(\mathrm{i}\gamma) \quad \text{and}$$

$$(N_2^{\ell_2} N_1^{\ell_2} X'^{\ell_2})_{1,n} = \ell_2 r \exp(\mathrm{i}\gamma) + \frac{\ell_2^2}{2}[N_2, N_1] = \ell_2 r \exp(\mathrm{i}\gamma) - \frac{\ell_2^2}{2} r' \exp(\mathrm{i}\gamma),$$

for some $0 \neq r' \in \mathbb{R}$. We then notice that $\left((N_1^{\ell_1} N_2^{\ell_1} X'^{\ell_1})^{d_1} (N_2^{\ell_2} N_1^{\ell_2} X'^{\ell_2})^{d_2}\right)_{1,n} = d_1 \left(\ell_1 r \exp(\mathrm{i}\gamma) + \frac{\ell_1^2}{2} r' \exp(\mathrm{i}\gamma)\right) + d_2 \left(\ell_2 r \exp(\mathrm{i}\gamma) - \frac{\ell_2^2}{2} r' \exp(\mathrm{i}\gamma)\right)$. Now,

$$\begin{aligned} d_1 \left(\ell_1 r \exp(\mathrm{i}\gamma) + \frac{\ell_1^2}{2} r' \exp(\mathrm{i}\gamma)\right) + d_2 \left(\ell_2 r \exp(\mathrm{i}\gamma) - \frac{\ell_2^2}{2} r' \exp(\mathrm{i}\gamma)\right) &= 0 \\ \iff d_1(2\ell_1 r + \ell_1^2 r') + d_2(2\ell_2 r - \ell_2^2 r') &= 0 \\ \iff d_1(2\frac{r}{r'}\ell_1 + \ell_1^2) + d_2(2\frac{r}{r'}\ell_2 - \ell_2^2) &= 0. \end{aligned}$$

By our assumption, the vectors $r \exp(\mathrm{i}\gamma)$ and $r' \exp(\mathrm{i}\gamma)$ have rational coordinates and the same angle γ. It follows that $\frac{r}{r'} \in \mathbb{Q}$. Hence we may choose sufficiently large $\ell_1, \ell_2 > 1$ such that $2\frac{r}{r'}\ell_1 + \ell_1^2$ and $2\frac{r}{r'}\ell_2 - \ell_2^2$ have different signs, and then integers $d_1, d_2 > 1$ can be chosen that satisfy the above equation. This choice of ℓ_1, ℓ_2, d_1, d_2 is then such that $(N_1^{\ell_1} N_2^{\ell_1} X'^{\ell_1})^{d_1} (N_2^{\ell_2} N_1^{\ell_2} X'^{\ell_2})^{d_2} = \boldsymbol{I}$ as required. Thus if such non-commuting matrices are present, we can reach the identity.

Otherwise, our final case is that only commuting matrices can be used to reach the γ-line. In this case we can compute in polynomial time a subset $C \subseteq G$ of these matrices. Then we can check if the identity matrix is in $\langle C \rangle$ in polynomial time as follows.

Since C consists only of commuting matrices, by Eq. (2), the top corner value $M_{1,n}$ of any $M \in \langle C \rangle \cap \Omega$ can be expressed as a linear combination of $c_i - \frac{1}{2}\boldsymbol{a}_i^T \boldsymbol{b}_i$, where $G_i \in C$. We now construct a new homogeneous system of linear Diophantine equations. Let $C = \{G_1, \ldots, G_{t'}\}$, and let $A \in \mathbb{Q}^{4(n-2)\times t'}$ be defined as in Eq. (1) using only matrices present in C. Also, let $A_2 = (c_1 - \frac{1}{2}\boldsymbol{a}_1^T \boldsymbol{b}_1, \ldots, c_{t'} - \frac{1}{2}\boldsymbol{a}_{t'}^T \boldsymbol{b}_{t'})$. Now construct a system $\begin{pmatrix} A \\ A_2 \end{pmatrix} \boldsymbol{x} = \boldsymbol{0}$, where $\boldsymbol{x} \in \mathbb{N}^{t'}$ and $\boldsymbol{0}$ is the $(4(n-2)+1)$-dimensional zero vector. Note that if this system has a solution $\boldsymbol{x}$, then $G_1^{x(1)} G_2^{x(2)} \cdots G_{t'}^{x(t')} = \boldsymbol{I}$. By Lemma 1 (see also [22]), we can decide if such a system has a non-zero solution in polynomial time.[1]

The proof is concluded by showing that the whole procedure is in P. Namely, we first decide if there is a pair G_i, G_j of non-commuting matrices such that the γ-line can be reached using G_i and G_j, in which case $\boldsymbol{I} \in \langle G \rangle$ by the above argument. This requires constructing a polynomially sized set of non-homogeneous systems of linear Diophantine equations and deciding whether they have solutions. This can be done in polynomial time.

[1] Note by a result of [1], the Membership Problem is decidable in polynomial time for commuting matrices. However, the authors prefer to have a self-contained proof.

If the γ-line can be reached only using commuting matrices, then we can compute the set $C \subseteq G$ of these matrices and check whether $\boldsymbol{I} \in \langle \mathcal{C} \rangle$ in polynomial time. □

Lemmata 5 and 6 allow us to prove the main result, Theorem 1.

The decidability of the Identity Problem implies that the Subgroup Problem is also decidable. That is, whether the semigroup generated by the generators G contains a non-trivial subgroup. However, the decidability of the Group Problem, i.e., whether $\langle G \rangle$ is a group, does not immediately follow. Our result can be extended to show decidability of the Group Problem.

Corollary 1. *It is decidable in polynomial time whether a finite set of matrices* $G \subseteq \mathrm{H}(n, \mathbb{Q}(\mathrm{i}))$ *forms a group.*

Proof. We give a brief overview of the proof. For $\langle G \rangle$ to be a group, each element of G must have a multiplicative inverse in $\langle G \rangle$. If $\boldsymbol{I} \in \langle G \rangle$, then each element used in a factorization of $\boldsymbol{I}$ has such an inverse. E.g., if $M_1 \cdots M_k = \boldsymbol{I}$, then $M_1^{-1} = M_2 \cdots M_k$ etc. The difficulty is that perhaps $\boldsymbol{I} \in \langle G \rangle$, but this does not imply that every matrix in G has a multiplicative inverse (since not every matrix may be used within a product equal to $\boldsymbol{I}$).

We therefore proceed by first ensuring there are no redundant matrices (carried out in P) since a redundant matrix cannot even be used to reach an Ω-matrix. Assuming all matrices are non-redundant, we then adapt the proofs of Lemmata 5 and 6 to ensure that not only can we reach the identity matrix, but we can do so with a product that uses every matrix from G. Both lemmata use Ω-matrices as part of their proofs, and we know there is a product containing all matrices giving an Ω-matrix since all matrices are non-redundant. Lemma 5 can then be adapted to say that if two pairs have different commutator angles, then we can reach the identity matrix using all matrices within the product. If all commutator angles of pairs of matrices in G are identical, then we can adapt the non-homogeneous system of linear Diophantine equations from the proof of Lemma 6 to enforce that all matrices are used at least once. This gives us a polynomial time algorithm for deciding whether $\langle G \rangle$ is a group. □

6 Future Research

We believe that the techniques, and the general approach, presented in the previous chapters can act as stepping stones for related problems. In particular, consider the Membership Problem, i.e., where the target matrix can be any matrix rather than the identity matrix. Let $\mathrm{M} = \begin{pmatrix} 1 & \boldsymbol{m}_1^T & m_3 \\ \mathbf{0} & \boldsymbol{I}_{n-2} & \boldsymbol{m}_2 \\ 0 & \mathbf{0}^T & 1 \end{pmatrix}$ be the target matrix and let $G = \{G_1, \ldots, G_t\}$, where $\psi(G_i) = (\boldsymbol{a}_i, \boldsymbol{b}_i, c_i)$. Following the idea of Sect. 4, we can consider system $A\boldsymbol{x} = (\boldsymbol{m}_1, \boldsymbol{m}_2)$, where $\boldsymbol{x} \in \mathbb{N}^t$. This system is a non-homogeneous system of linear Diophantine equations that can be solved in NP. The solution set is a union of two finite solution sets, S_0 and S_1. The set S_0 being the solutions to the corresponding homogeneous system that can be repeated any number of times as they add up to $\mathbf{0}$ on the right-hand side. The other set, S_1, corresponds to reaching the vector $(\boldsymbol{m}_1, \boldsymbol{m}_2)$. The matrices corresponding to the solutions in S_1 have to be used exactly this number of times.

The techniques developed in Sect. 4 allow us to manipulate matrices corresponding to solutions in S_0 in order to obtain the desired value in the top right corner. However, this is not enough as the main technique relies on repeated use of Ω-matrices. These can be interspersed with matrices corresponding to a solution in S_1 affecting the top right corner in uncontrollable ways.

References

1. Babai, L., Beals, R., Cai, J., Ivanyos, G., Luks, E.M.: Multiplicative equations over commuting matrices. In: Proceedings of SODA 1996, pp. 498–507. SIAM (1996). http://dl.acm.org/citation.cfm?id=313852.314109
2. Bell, P.C., Hirvensalo, M., Potapov, I.: The identity problem for matrix semigroups in SL(2, $\mathbb{Z}$) is NP-complete. In: Proceedings of SODA 2017, pp. 187–206. SIAM (2017). https://doi.org/10.1137/1.9781611974782.13
3. Bell, P.C., Potapov, I.: On the undecidability of the identity correspondence problem and its applications for word and matrix semigroups. Int. J. Found. Comput. Sci. **21**(6), 963–978 (2010). https://doi.org/10.1142/S0129054110007660
4. Blaney, K.R., Nikolaev, A.: A PTIME solution to the restricted conjugacy problem in generalized Heisenberg groups. Groups Complex. Cryptol. **8**(1), 69–74 (2016). https://doi.org/10.1515/gcc-2016-0003
5. Blondel, V.D., Megretski, A. (eds.): Unsolved Problems in Mathematical Systems and Control Theory. Princeton University Press, Princeton (2004)
6. Brylinski, J.L.: Loop spaces, characteristic classes, and geometric quantization. Birkhäuser (1993)
7. Bump, D., Diaconis, P., Hicks, A., Miclo, L., Widom, H.: An exercise(?) in Fourier analysis on the Heisenberg group. Ann. Fac. Sci. Toulouse Math. (6) **26**(2), 263–288 (2017). https://doi.org/10.5802/afst.1533
8. Cassaigne, J., Harju, T., Karhumäki, J.: On the undecidability of freeness of matrix semigroups. Int. J. Algebra Comput. **9**(03n04), 295–305 (1999). https://doi.org/10.1142/S0218196799000199
9. Choffrut, C., Karhumäki, J.: Some decision problems on integer matrices. RAIRO - Theor. Inf. Appl. **39**(1), 125–131 (2005). https://doi.org/10.1051/ita:2005007
10. Chonev, V., Ouaknine, J., Worrell, J.: The orbit problem in higher dimensions. In: Proceedings of STOC 2013, pp. 941–950. ACM (2013). https://doi.org/10.1145/2488608.2488728
11. Chonev, V., Ouaknine, J., Worrell, J.: On the complexity of the orbit problem. J. ACM **63**(3), 23:1–23:18 (2016). https://doi.org/10.1145/2857050
12. Colcombet, T., Ouaknine, J., Semukhin, P., Worrell, J.: On reachability problems for low-dimensional matrix semigroups. In: Proceedings of ICALP 2019. LIPIcs, vol. 132, pp. 44:1–44:15. Schloss Dagstuhl - Leibniz-Zentrum für Informatik (2019). https://doi.org/10.4230/LIPIcs.ICALP.2019.44
13. Diaconis, P., Malliaris, M.: Complexity and randomness in the Heisenberg groups (and beyond) (2021). https://doi.org/10.48550/ARXIV.2107.02923

14. Diekert, V., Potapov, I., Semukhin, P.: Decidability of membership problems for flat rational subsets of GL(2, $\mathbb{Q}$) and singular matrices. In: Emiris, I.Z., Zhi, L. (eds.) ISSAC 2020: International Symposium on Symbolic and Algebraic Computation, Kalamata, Greece, 20-23 July 2020, pp. 122–129. ACM (2020). https://doi.org/10.1145/3373207.3404038
15. Ding, J., Miasnikov, A., Ushakov, A.: A linear attack on a key exchange protocol using extensions of matrix semigroups. IACR Cryptology ePrint Archive **2015**, 18 (2015)
16. Dong, R.: On the identity problem and the group problem for subsemigroups of unipotent matrix groups. CoRR abs/2208.02164 (2022). https://doi.org/10.48550/arXiv.2208.02164
17. Dong, R.: On the identity problem for unitriangular matrices of dimension four. In: Proceedings of MFCS 2022. LIPIcs, vol. 241, pp. 43:1–43:14 (2022). https://doi.org/10.4230/LIPIcs.MFCS.2022.43
18. Dong, R.: Semigroup intersection problems in the Heisenberg groups. In: Proceedings of STACS 2023. LIPIcs, vol. 254, pp. 25:1–25:18 (2023). https://doi.org/10.4230/LIPIcs.STACS.2023.25
19. Galby, E., Ouaknine, J., Worrell, J.: On matrix powering in low dimensions. In: Proceedings of STACS 2015. LIPIcs, vol. 30, pp. 329–340 (2015). https://doi.org/10.4230/LIPIcs.STACS.2015.329
20. Gelca, R., Uribe, A.: From classical theta functions to topological quantum field theory. In: The Influence of Solomon Lefschetz in Geometry and Topology, Contemprorary Mathematics, vol. 621, pp. 35–68. American Mathematical Society (2014). https://doi.org/10.1090/conm/621
21. Harju, T.: Post correspondence problem and small dimensional matrices. In: Diekert, V., Nowotka, D. (eds.) DLT 2009. LNCS, vol. 5583, pp. 39–46. Springer, Heidelberg (2009). https://doi.org/10.1007/978-3-642-02737-6_3
22. Khachiyan, L.G.: Polynomial algorithms in linear programming. USSR Comput. Math. Math. Phys. **20**(1), 53–72 (1980)
23. Ko, S.K., Niskanen, R., Potapov, I.: On the identity problem for the special linear group and the Heisenberg group. In: Proceedings of ICALP 2018. LIPIcs, vol. 107, pp. 132:1–132:15 (2018). https://doi.org/10.4230/lipics.icalp.2018.132
24. König, D., Lohrey, M., Zetzsche, G.: Knapsack and subset sum problems in nilpotent, polycyclic, and co-context-free groups. Algebra Comput. Sci. **677**, 138–153 (2016). https://doi.org/10.1090/conm/677/13625
25. Kostant, B.: Quantization and unitary representations. In: Taam, C.T. (ed.) Lectures in Modern Analysis and Applications III. LNM, vol. 170, pp. 87–208. Springer, Heidelberg (1970). https://doi.org/10.1007/BFb0079068
26. Lee, J.R., Naor, A.: LP metrics on the Heisenberg group and the Goemans-Linial conjecture. In: 2006 47th Annual IEEE Symposium on Foundations of Computer Science (FOCS 2006), pp. 99–108 (2006). https://doi.org/10.1109/FOCS.2006.47
27. Markov, A.A.: On certain insoluble problems concerning matrices. Dokl. Akad. Nauk SSSR **57**(6), 539–542 (1947)
28. Mishchenko, A., Treier, A.: Knapsack problem for nilpotent groups. Groups Complex. Cryptol. **9**(1), 87–98 (2017). https://doi.org/10.1515/gcc-2017-0006
29. Ouaknine, J., Pinto, J.A.S., Worrell, J.: On termination of integer linear loops. In: Proceedings of SODA 2015, pp. 957–969. SIAM (2015). https://doi.org/10.1137/1.9781611973730.65
30. Paterson, M.S.: Unsolvability in 3 × 3 matrices. Stud. Appl. Math. **49**(1), 105 (1970). https://doi.org/10.1002/sapm1970491105
31. Schrijver, A.: Theory of Linear and Integer Programming. Wiley, Hoboken (1998)

Reachability Analysis of a Class of Hybrid Gene Regulatory Networks

Honglu Sun[1(✉)], Maxime Folschette[2], and Morgan Magnin[1]

[1] Nantes Université, École Centrale Nantes, CNRS, LS2N, UMR 6004, 44000 Nantes, France
honglu.sun@ls2n.fr

[2] Univ. Lille, CNRS, Centrale Lille, UMR 9189 CRIStAL, 59000 Lille, France

Abstract. In this work, we study the reachability analysis method of a class of hybrid systems called HGRN which is a special case of hybrid automata. The reachability problem concerned in this work is, given a singular state and a region (a set of states), to determine whether the trajectory from this singular state can reach this region. This problem is undecidable for general hybrid automata, and is decidable only for a restricted class of hybrid automata, but this restricted class does not include HGRNs. A priori, reachability in HGRNs is not decidable; however, we show in this paper that it is decidable in certain cases, more precisely if there is no chaos. Based on this fact, the main idea of this work is that if the decidable cases can be determined automatically, then the reachability problem can be solved partially. The two major contributions are the following: firstly, we classify trajectories into different classes and provide theoretical results about decidability; then based on these theoretical results, we propose a reachability analysis algorithm which always stops in finite time and answers the reachability problem partially (meaning that it can stop with the inconclusive result, for example with the presence of chaos).

Keywords: Reachability · Hybrid system · Decidability · Gene regulatory networks · Limit cycle

1 Introduction

Reachability problem of dynamical system has been investigated on different formalisms, majorly on discrete systems [6,14,23] and hybrid systems [2,3,8,15,19,25]. In this work, we study a reachability analysis method on a class of hybrid system called hybrid gene regulatory network (HGRN) [4,7], which is an extension of Thomas' discrete modeling framework [27,28]. This hybrid system is proposed to model gene regulatory networks, which are networks of genes describing the regulation relations between genes.

Supported by China Scholarship Council.

O. Bournez et al. (Eds.): RP 2023, LNCS 14235, pp. 56–69, 2023.
https://doi.org/10.1007/978-3-031-45286-4_5

HGRNs are similar to piecewise-constant derivative systems (PCD systems) [2] which is a special case of hybrid automata [1]. The major difference between HGRNs and PCD systems of the works [2,3,25] is the existence of sliding mode, which means that when a trajectory reaches a black wall (a boundary of the discrete state which can be reached but cannot be crossed by trajectories), it is forced to move along the black wall. There exist other methods to define behaviors of trajectories on a black wall [17,24] which are different from the sliding mode in HGRNs.

The reachability problem concerned in this work is to determine whether the trajectory from certain state can reach a certain region (a set of states). We mainly focus on the decidability problem, that is, whether we can find an algorithm to determine the reachability problem such that this algorithm always stops in finite time and gives a correct answer.

The decidability problem among hybrid systems that are close to HGRNs is already studied in the literature. It has been proved that, for PCD systems, it is decidable in 2 dimensions [21] but it is undecidable in 3 dimensions [2]. For general hybrid automata, there exists a restricted class called initialized rectangular automata which is decidable in any dimension [19], but this class does not include HGRNs.

Up to now, there is no theoretical results of the decidability of this problem on HGRNs. A priori, we can expect that it is not decidable because of the existence of chaos. However, if we can show that it is decidable in certain cases, for example, when the trajectory considered in a reachability problem converges asymptotically to a n-dimensional limit cycle, and if these cases can be identified automatically, then the reachability problem can be answered partially, which is the main idea of this work. In order to prove the existence of chaos in HGRNs, we exhibit a HGRN with a chaotic attractor based on a different pre-existing hybrid system [18]. This work has the following contributions:

- We classify trajectories of HGRNs into three classes: trajectories halting in finite time, trajectories attracted by regularly oscillating cycles and chaotic trajectories. For the first two classes, we prove that the reachability problem is decidable and we provide methods to determine automatically their classes. For the third class, a priori, it is undecidable, and we provide a necessary condition for that a trajectory is chaotic.
- Based on the above theoretical results, we propose a reachability analysis algorithm for HGRNs which always stops in finite time and once it stops, it returns whether the set of target states is reached, not reached or if the result is unknown. The unknown result is related to the existence of chaos. To our knowledge, this is the first reachability analysis algorithm for HGRNs and it can be applied to HGRNs in any dimension.

This paper is organized as follows. In Sect. 2, we introduce basic notions of HGRNs. In Sect. 3, we present our reachability analysis method, including theoretical results and the reachability analysis algorithm. And finally in Sect. 4, we make a conclusion by discussing the merits and limits of this method and our future work.

2 Preliminary Definitions

In this section, we present HGRNs and its basic notions. Consider a gene regulatory network with N genes; the set of genes is denoted $G = \{G_1, G_2, ..., G_N\}$. A *discrete state* is an integer vector of length N, noted by d_s, which assigns the discrete level d_s^i to gene G_i, where $i \in \{1, 2, 3, ..., N\}$ and d_s^i is the ith component of d_s. The set of all discrete states is denoted by E_d.

A *hybrid gene regulatory network* (HGRN) is noted $\mathcal{H} = (E_d, c)$. c is a function from E_d to $\mathbb{R}^N$. For each $d_s \in E_d$, $c(s)$, also noted c_s, is called the *celerity* of discrete state d_s and describes the temporal derivative of the system in d_s. A 2-dimensional HGRN is shown in Fig. 1. In this system, each of these two genes $(A = G_1, B = G_2)$ has two discrete levels: 0 and 1, so there are 4 discrete states: $00, 01, 10, 11$. Black arrows represent the celerities (temporal derivatives) of each discrete state.

In HGRNs, a *state* is also called a *hybrid state*, which is a couple $h = (\pi, d_s)$ containing a *fractional part* π, which is a real vector $[0, 1]^N$, and a discrete state d_s. The set of all hybrid states is denoted by E_h.

A *(hybrid) trajectory* τ of HGRN is a function from a time interval $[0, t_0]$ to $E_\tau = E_h \cup E_{sh}$, where $t_0 \in \mathbb{R}^+ \cup \{\infty\}$, and E_{sh} is the set of all finite or infinite sequences of states: $E_{sh} = \{(h_0, h_1, ..., h_m) \in (E_h)^{m+1} \mid m \in \mathbb{N} \cup \{\infty\}\}$. A trajectory τ is called a *closed trajectory* if it is defined on $[0, \infty[$ and $\exists T > 0, \forall t \in [0, \infty[, \tau(t) = \tau(t+T)$. In Fig. 1, red arrows represent a possible trajectory of this system, which happens, in this particular case, to be a closed trajectory.

A *boundary* in a discrete state d_s is a set of states defined by $e(G_i, \pi_0, d_s) = \{(\pi, d_s) \in E_h \mid \pi^i = \pi_0, \}$, where $i \in \{1, 2, ..., N\}, d_s \in E_d$ and $\pi_0 \in \{0, 1\}$. In the rest of this paper, we simply use e to represent a boundary.

In Fig. 1, the state $h_M = ((\pi_M^1, 1), (1, 1))$ of point M belongs to $e_1 = (B, 1, (1, 1))$, that is, the upper boundary in the second dimension (the dimension of gene B) of the discrete state 11. Since there is no other discrete state on the

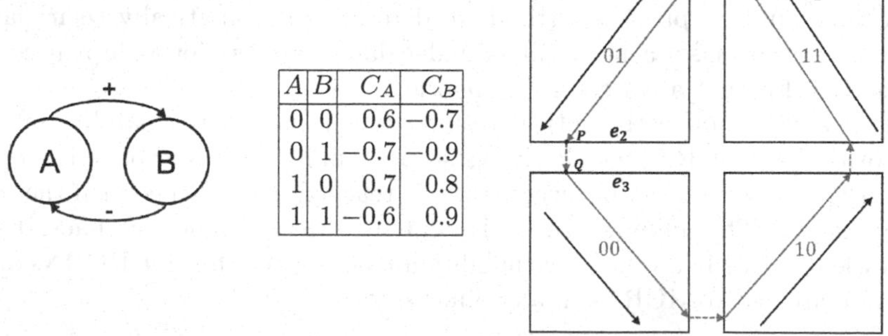

A	B	C_A	C_B
0	0	0.6	−0.7
0	1	−0.7	−0.9
1	0	0.7	0.8
1	1	−0.6	0.9

Fig. 1. Example of a HGRN in 2 dimensions. Left: Influence graph (negative feedback loop with 2 genes). Middle: Example of corresponding parameters (celerities). Right: Corresponding example of dynamics; abscissa represents gene A and ordinate represents gene B.

other side of e_1, the trajectory from h_M cannot cross e_1 and has to slide along e_1 (e_1 can be called a black wall). The existence of such sliding mode is a speciality of HGRNs. Boundaries like e_1, which can be reached by trajectories but cannot be crossed, are defined as *attractive boundaries.* The state $h_P = ((\pi_P^1, 0), (0, 1))$ of point P belongs to $e_2 = (B, 0, (0, 1))$, the lower boundary in the second dimension of the discrete state 01. The trajectory from h_P reaches instantly $h_Q = ((\pi_Q^1, 1), (0, 0))$, which belongs to $e_3 = (B, 1, (0, 0))$, the upper boundary in the second dimension of discrete state 00, because the celerities on both sides allow this (instant) discrete transition. e_2 is called an *output boundary* of 01 and e_3 is called an *input boundary* of 00.

When a trajectory reaches several output boundaries at the same time (Fig. 2 left), it can cross any of them but can only cross one boundary at a time, which causes non-deterministic behaviors. The simulation of HGRNs is presented more formally in the Appendix.

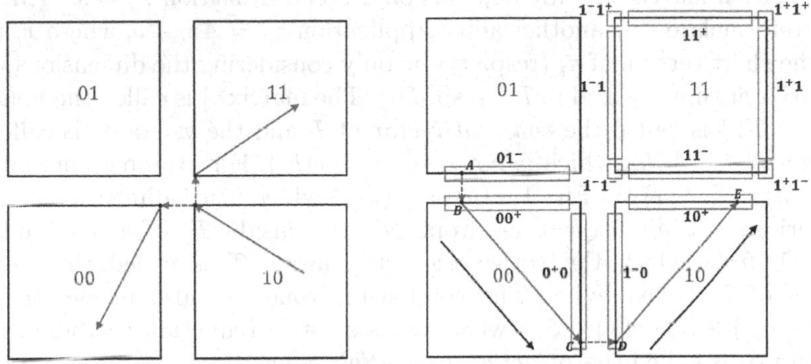

Fig. 2. Left: Illustration of a non-deterministic behavior. Right: Illustration of all discrete domains of state 11, and a sequence of discrete domains in the other states.

In order to analyze dynamical properties of HGRNs, the concepts of *discrete domain, transition matrix* and *compatible zone* are introduced in [26]. A *discrete domain* $\mathcal{D}(d_s, S_-, S_+)$ is a set of states inside one discrete state d_s, defined by:

$$\mathcal{D}(d_s, S_-, S_+) = \{(\pi, d_s) \mid \forall i \in \{1, 2, ..., N\}, \pi^i \in \begin{cases} \{1\} & \text{if } i \in S_+ \\ \{0\} & \text{if } i \in S_- \\ ]0, 1[& \text{if } i \notin S_- \cup S_+ \end{cases} \}$$

where S_+ and S_- are power sets of $\{1, 2, ..., N\}$ such that $S_+ \cap S_- = \emptyset$ and $S_+ \cup S_- \neq \emptyset$. In fact, S_+ (S_-) represents the dimensions in which the upper (lower) boundaries are reached by any state $h \in \mathcal{D}(d_s, S_-, S_+)$. In the rest of this paper, we simply use $\mathcal{D}$ to represent a discrete domain when there is no ambiguity.

Some discrete domains are illustrated in Fig. 2 right. For example, 11^+ denotes the discrete domain inside discrete state 11 where the upper boundary is reached for the second dimension and no boundary is reached for the first dimension, that is: $\mathcal{D}((1, 1), \emptyset, \{2\}) = \{(\pi, (1, 1)) \mid \pi^1 \in]0, 1[\wedge \pi^2 = 1\}$. The state D

in this figure belongs to the discrete domain 1^-0. The discrete state 11 contains 8 discrete domains: 1^-1^-, 11^-, 1^+1^-, 1^-1^+, 11^+, 1^+1^+, 1^-1 and 1^+1, which are depicted in Fig. 2 right. Note that, for instance, 1^+1^+ is represented by a small red rectangle for readability, but in fact it only contains one singular hybrid state $((1,1),(1,1))$.

To order to introduce the concepts of *transition matrix* and *compatible zone*, consider a sequence of discrete domains $\mathcal{T} = (\mathcal{D}_i, \mathcal{D}_{i+1}, \mathcal{D}_{i+2}, ..., \mathcal{D}_j)$ in the rest of this section and assume that there is a trajectory τ which starts from $h_i = (\pi_i, d_{s_i}) \in \mathcal{D}_i$, reaches all discrete domains of $\mathcal{T}$ in order without reaching any other discrete domain, and finally reaches $h_j = (\pi_j, d_{s_j}) \in \mathcal{D}_j$. In this case, we say that τ is inside $\mathcal{T}$. For example, in Fig. 2 right, the red trajectory is inside the sequence of discrete domains $(01^-, 00^+, 0^+0, 1^-0, 10^+)$.

The relation between π_i and π_j can be described by a *transition matrix* M: $\pi_j = s^{-1}(Ms(\pi_i))$, where s is a function that adds an extra dimension and the value in the extra dimension is always 1: $s((a_1, a_2, ..., a_N)) = (a_1, a_2, ..., a_N, 1)$. The transition matrix M only depends on $\mathcal{T}$. The transition $\pi_j = s^{-1}(Ms(\pi_i))$ can be reformulated by another affine application $x_j = Ax_i + b$, where x_i (resp. x_j) is the short version of π_i (resp. π_j) by only considering the dimensions where boundaries are not reached in $\mathcal{D}_i$ (resp. $\mathcal{D}_j$). The matrix A is called the *reduction matrix* of $\mathcal{T}$, b is called the *constant vector* of $\mathcal{T}$ and the vector x_i is called the *reduction vector* of h_i, which is noted by $x_i = r(h_i)$. For example, for the state $h_M = ((\pi_M^1, 1), (1,1))$ in Fig. 1, $r(h_M) = (\pi_M^1)$ which is a 1-dimensional vector.

A priori, not all trajectories from $\mathcal{D}_i$ stay inside $\mathcal{T}$. The maximal subset of $\mathcal{D}_i$ from which the trajectories stay inside $\mathcal{T}$ is called the *compatible zone* of $\mathcal{T}$, noted by $\mathcal{S}$. The compatible zone can also be described by $\mathcal{S} = \{(\pi, d_{s_i}) \in \mathcal{D}_i \mid r(\pi) \in \mathcal{S}_r\}$ where $\mathcal{S}_r$ is a set of reduction vectors of states in $\mathcal{D}_i$ and $\mathcal{S}_r$ is called the *reduction compatible zone.*

3 Reachability Analysis Method

In this section, we firstly define the reachability problem concerned in this work.

Problem 1 (Reachability). Consider a hybrid state $h_1 = (\pi_1, d_{s_1})$ and a region $R_2 = \{(\pi, d_{s_2}) \mid \pi^i \in [a_i, b_i], i \in \{1, 2, ..., N\}\}$, where $a_i, b_i \in \mathbb{R}$ and $0 \leq a_i \leq b_i \leq 1, \forall i \in \{1, 2, ..., N\}$. Does the trajectory τ from h_1 enter the region R_2? In other words, does there exist t_0 such that $\tau(t_0) \in R_2$?

Problem 1 is illustrated in the examples of Fig. 3, where the initial state of the trajectory (red arrows) is h_1 and the blue rectangle represents R_2.

The following assumptions are made in this work.

Assumption 1. For any sequence of discrete domains $\mathcal{T}$ of which the compatible zone is not empty, we assume that all eigenvalues of the reduction matrix of $\mathcal{T}$ are real.

For now, we have not found such reduction matrix with complex eigenvalues.

Assumption 2. The trajectory from h_1 has no non-deterministic behavior.

Generally, trajectories with non-deterministic behaviors exist, but among state-of-the-art HGRNs of gene regulatory networks, the probability of a randomly chosen initial state that leads to non-deterministic behaviors is almost 0. Therefore, we ignore this kind of trajectory in this work. In fact, the method of this work could also be adapted for non-deterministic trajectories (each time when a non-deterministic state is reached, the current trajectory splits into two or several trajectories, and same method is applied on each of these new trajectories).

Assumption 3. Any non-instant transition on a limit cycle does not reach more than one new boundary at the same time.

In real-life systems, it is indeed very unlikely for parameters to be that constrained due to the existence of noise.

3.1 Different Classes of Hybrid Trajectories

In this section, we classify trajectories of HGRNs into three classes: trajectories halting in finite time, trajectories attracted by cycles of discrete domains and chaotic trajectories. And we provide some theoretical results regarding this reachability problem.

Trajectories Halting in Finite Time. A trajectory τ is a *trajectory halting in finite time* if $\exists t_0$ such that the derivative of $\tau(t_0)$ is 0 in any dimension, in other words $\tau(t_0)$ is a fixed point. The trajectory in Fig. 3 left is a trajectory halting in finite time. We can easily see that Problem 1 is decidable if the trajectory from h_1 is a trajectory halting in finite time, because, in this case, the trajectory is a composition of a finite number of n-dimensional "straight lines"; to verify if this trajectory reaches R_2, we only need to verify if any of these "straight lines" cross R_2, which can be verified in finite time.

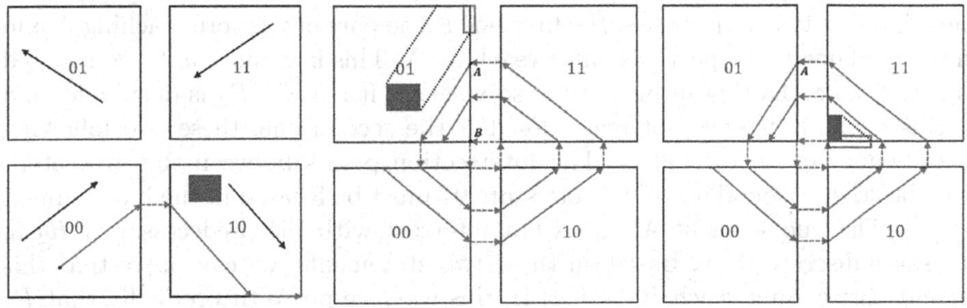

Fig. 3. Left: Illustration of Problem 1 and trajectory halting in finite time. Blue rectangle represents R_2 of Problem 1. Middle and right: Illustration of trajectories attracted by cycles of discrete domains and predecessor in the same discrete state. Blues rectangles represent R_2 of Problem 1 and blue boxes represent their predecessors in the same discrete state.

Trajectories Attracted by Cycles of Discrete Domains. A trajectory τ is a *trajectory attracted by a cycle of discrete domains* if $\exists t_0$ such that after t_0, τ always stays inside a cycle of discrete domains $\mathcal{C}_T = (\mathcal{D}_0, \mathcal{D}_1, \mathcal{D}_2, ..., \mathcal{D}_p, \mathcal{D}_0)$, meaning that τ crosses this cycle an infinite number of times without leaving it. Intuitively, if a trajectory τ is attracted by a cycle of discrete domains, then τ converges to or reaches a limit cycle. In Fig. 3 middle and right, both trajectories are attracted by a cycle of discrete domains: indeed, these trajectories converge to the limit cycle in the center of the figure (which only has instant transitions).

To prove the decidability of trajectories attracted by cycles of discrete domains, we introduce the notion of *predecessor in the same discrete state*: for any set of hybrid states in the same discrete state defined by $R = \{(\pi, d_s) \mid \pi \in E\}$ where $E \subseteq [0,1]^N$ is a closed set, the predecessor of R in the same discrete state, noted by $Pre_{d_s}(R)$, is the union of sets of hybrid states: $Pre_{d_s}(R) = \bigcup_{i \in \{1,2,...,q\}} Z_i$, such that: 1) each Z_i belongs to a different discrete domain on an input boundary of d_s, 2) any trajectory from $Pre_{d_s}(R)$ reaches R directly ("reach R directly" means that reach R before reaching a new discrete state), 3) any trajectory from an input boundary of d_S but not from $Pre_{d_s}(R)$ does not reach directly R. For Problem 1, we can see that if the trajectory τ from h_1 has already crossed at least one discrete state (we say τ has already crossed a discrete state at t_0 if there exists $t < t_0$ such that $\tau(t_0)$ and $\tau(t)$ do not belong to the same discrete state) without reaching the region R_2, then Problem 1 is equivalent to "Does τ reach $Pre_{d_{s_2}}(R_2)$?".

Examples of predecessors in the same discrete state are illustrated in Fig. 3 middle and right where blues rectangles represent R_2 and blue boxes present their predecessors in the same discrete state.

Theorem 1. *Problem 1 is decidable if the trajectory from h_1 is a trajectory attracted by a cycle of discrete domains.*

The proof of Theorem 1 is given in the Appendix. The idea of this proof can be explained intuitively by 2-dimensional examples in Fig. 3 middle and right. In Fig. 3 middle, the trajectory which reaches state A, noted by τ, can be considered as two trajectories: the first one is the part of τ before reaching A and the second one is the part of τ after reaching A. This first one can be considered as a trajectory halting in finite time so whether it reaches R_2 is decidable, and in this example it does not reach R_2. For the second one, these two following statements can be verified: 1. The intersection points between this trajectory and the "right" boundary of discrete state 01 must be located in the line segment AB. 2. The line segment AB does not intersect with the predecessor of R_2 in the same discrete state. Based on these two statements, we can prove that this second part cannot reach R_2 either. In this way, we prove theoretically that R_2 is not reached by τ, and since this process can be done automatically in finite time, the problem is decidable. Note that in the general case, this "line segment AB" is a $(n{-}1)$-dimensional region such that the trajectory always returns to this region and this region does not intersect the predecessor of R_2 in the same discrete state. In Fig. 3 right, it can be verified automatically in finite time that the limit cycle with only instant transitions (at the center) reaches R_2, and that τ converges to this limit cycle, so we can prove that τ finally reaches R_2, and this case is thus decidable too.

We also develop the following theorem to determine if a trajectory is attracted by a cycle of discrete domains. In order to simplify this theorem, for the cycle of discrete domains $\mathcal{C}_{\mathcal{T}} = (\mathcal{D}_0, \mathcal{D}_1, \mathcal{D}_2, ..., \mathcal{D}_p, \mathcal{D}_0)$ and the hybrid state $h_0 \in \mathcal{D}_0$ considered in this theorem, we note that:

- The reduction matrix and the constant vector of $\mathcal{C}_{\mathcal{T}}$ are A and b respectively.
- The reduction compatible zone of $\mathcal{C}_{\mathcal{T}}$ is described by linear constraints $\{x \mid Wx > c\}$ where c is a vector and W is a matrix. W is of size $n_0 \times n_1$, where n_1 is the number of dimensions of $r(h_0)$. W_i is the ith line of matrix W (W_i is of size $1 \times n_1$) and c_i is the ith component of vector c.
- $r_\infty = \lim_{n\to\infty} f^n(r(h_0))$ where $f(x) = Ax + b$.
- The eigenvalues and eigenvectors of A are $\{\lambda_i \mid i \in \{1, 2, ..., n_1\}\}$ and $\{v_i \mid i \in \{1, 2, ..., n_1\}\}$ respectively. λ_1 is chosen as the eigenvalue with the maximum absolute value among the eigenvalues that differ from 1.
- The decomposition of $r(h_0) - r_\infty$ in the directions of eigenvectors of the reduction matrix A is noted as $r(h_0) - r_\infty = \sum_{i=1}^{n_1} \alpha_i v_i$.

Theorem 2. *A trajectory τ is attracted by a cycle of discrete domains if and only if τ reaches h_0 which belongs to the compatible zone of a cycle of discrete domains $\mathcal{C}_{\mathcal{T}} = (\mathcal{D}_0, \mathcal{D}_1, \mathcal{D}_2, ..., \mathcal{D}_p, \mathcal{D}_0)$ such that $\mathcal{D}_0$ has no free dimension (meaning that, in $\mathcal{D}_0$, boundaries are reached in all dimensions) or the following conditions are satisfied.*

- *$\mathcal{D}_0$ has at least one free dimension.*
- *$\forall i \in \{1, 2, ..., n_1\}, |\lambda_i| \leq 1 \wedge \lambda_i \neq -1$.*
- *$\forall i \in \{1, 2, ..., n_0\}$, we have either $W_i r_\infty = c_i$ or $W_i r_\infty > c_i$. We use I_e to represent the maximum set of integers such that $\forall i \in I_e, W_i r_\infty = c_i$ and we use I_n to represent the maximum set of integers such that $\forall i \in I_n, W_i r_\infty > c_i$.*
- *If $\lambda_1 \neq 0$ (we assume that λ_1 is unique if $\lambda_1 \neq 0$) and I_e is not empty, then λ_1 is positive.*
- *If $\lambda_1 \neq 0$, then $\forall i \in I_e, \forall j \in \{2, ..., n_1\}, |W_i v_1 \alpha_1| > n_1 |W_i v_j \alpha_j|$ (we ignore the case that $\exists i \in I_e, W_i v_1 = 0$).*
- *If $\lambda_1 \neq 0$, then $\forall i \in I_n, \max_{\beta \in \{-1,1\}^{n_1}} \|\sum_{j=1}^{n_1} \beta_j \alpha_j v_j\|_2 < \frac{W_i r_\infty - c_i}{\|W_i\|_2}$.*

The proof of Theorem 2 is given in the Appendix. The main idea of Theorem 2 is illustrated in Fig. 4 where the huge rectangle represents a discrete domain $\mathcal{D}$ which has two free dimensions and the zone surrounded by dashed lines represents the compatible zone $\mathcal{S}$ (which is a open set) of a certain cycle of discrete domains $\mathcal{C}_{\mathcal{T}}$. Each dashed line l_{ci} represents a linear constraint of the form $w^T x > c$ where w, x are vectors and c is a real number. The fact that a trajectory τ is attracted by $\mathcal{C}_{\mathcal{T}}$ is equivalent to the fact that the intersection points between τ and $\mathcal{D}$, noted by the sequence $(h_1, h_2, ...)$, always stay inside $\mathcal{S}$ and converge to ($\lambda_1 \neq 0$) or reach ($\lambda_1 = 0$) h_∞, which belongs to the closure of $\mathcal{S}$. Need to mention that this idea of using the intersection points between a trajectory and a hyperplan to study the properties of this trajectory is based on the idea of Poincaré map. Similar ideas have been widely used in the literature to study limit cycles of other hybrid systems [5,10–13,16,20,22,29] and also have been applied to analyze the stability of limit cycles of HGRNs in [26].

Whether h_∞ belongs to the closure of $\mathcal{S}$ or not can be easily verified by using these linear constraints. A necessary condition for this sequence to always satisfy these linear constraints is that the absolute values of all eigenvalues of the reduction matrix of $\mathcal{C}_\mathcal{T}$ are less than or equal to 1. In case that these eigenvalues satisfy this necessary condition, to verify if this sequence always satisfies these linear constraints, we separate these constraints on two classes: the first class contains all constraints which are not reached by h_∞: l_{c2}, l_{c3}, l_{c4}, the second class contains all constraints which are reached by h_∞: l_{c1}, l_{c5}. To verify if l_{c2}, l_{c3}, l_{c4} are always satisfied, we can verify if this sequence enters and stays in a circle centered by h_∞ which only contains states satisfying constraints l_{c2}, l_{c3}, l_{c4} (this is related to the condition: if $\lambda_1 \neq 0$, then $\forall i \in I_n, \max_{\beta\in\{-1,1\}^{n_1}} \|\sum_{j=1}^{n_1} \beta_j\alpha_j v_j\|_2 < \frac{W_i r_\infty - c_i}{\|W_i\|_2}$), such circle can always be found if it is sufficiently small, for example, the circle in Fig. 4. To verify if l_{c1}, l_{c5} are always satisfied, we can verify if this sequence is sufficiently "close" to v_1 which is the eigenvector related to the eigenvalue with the maximum absolute value among the eigenvalues that differ from 1 and which also "points into" $\mathcal{S}$ (this is related to the condition: if $\lambda_1 \neq 0$, then $\forall i \in I_e, \forall j \in \{2, ..., n_1\}, |W_i v_1 \alpha_1| > n_1 |W_i v_j \alpha_j|$). Here, sufficiently "close" to v_1 means intuitively that the angle between $\overrightarrow{h_\infty h_i}$ and v_1 is sufficiently small.

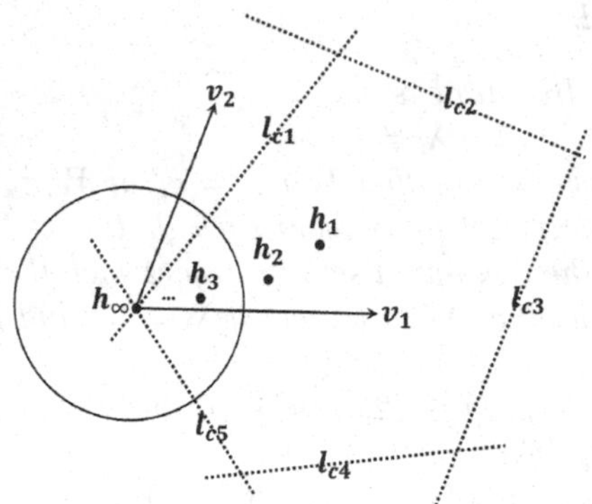

Fig. 4. Illustration of the idea of Theorem 2.

Chaotic Trajectories. In this work, a trajectory of HGRN is called a chaotic trajectory if it does not reach a fixed point and it is not attracted by a cycle of discrete domains. So all trajectories which are not included in the previous two classes are chaotic trajectories. Need to mention that the dynamics of chaotic trajectories, a priori, can be different from the chaotic dynamics of classic nonlinear dynamical systems. The reason why we still use the terminology "chaotic" is that similar concept of chaos has been used in some pre-existing works of other hybrid systems [9,18].

To prove such chaotic trajectories exist, we have constructed a HGRN with chaotic trajectories based on a pre-existing model of circuit with a chaotic attractor [18]. Parameters of this HGRN are given in the code of our work.

In our work, we have not yet found a method to check reachability for chaotic trajectories, which, a priori, can be undecidable. So, in this subsection, we only introduce a method to predict whether a trajectory is chaotic, based on a necessary condition.

For a chaotic trajectory τ, there exist t_0 and a finite set of discrete domains L_D, such that after t_0, τ cannot reach any discrete domain which does not belong to L_D, and for any discrete domain $\mathcal{D}_0 \in L_D$, $\mathcal{D}_0$ is reached by τ an infinite number of times. This is a result of the fact that the number of discrete domains is finite and the trajectory does not stay in a particular discrete domain.

For any $\mathcal{D}_0 \in L_D$, we can find $t_1 > t_0$ such that, from t_1, τ returns to $\mathcal{D}_0$ an infinite number of times, and each time it stays inside a sequence of discrete domains of the form $(\mathcal{D}_0, ..., \mathcal{D}_0)$. The set of all such sequences of discrete domains is noted by L_T. We assume that L_T is a finite set, which is based on the fact that the number of discrete domains is limited and the dynamics in the discrete states is simple (a constant vector). Based on this, if t_1 is sufficiently big, then we can derive that from t_1, $\forall \mathcal{T} \in L_T$ is crossed by τ an infinite number of times.

Now the sequence of discrete domains crossed by τ from t_1 can be described by the infinite sequence $(\mathcal{T}_1, \mathcal{T}_2, \mathcal{T}_3, ...)$, where $\forall i \in \mathbb{N}, \mathcal{T}_i \in L_T$. And we can get the following property of chaotic trajectories, which is used in the following section to predict whether a trajectory is chaotic.

Property 1. $\exists i \in \mathbb{N}, \exists k \in \mathbb{N}, k \neq 1$, such that $\mathcal{T}_i \neq \mathcal{T}_{i+1}$ and $\mathcal{T}_i = \mathcal{T}_{i+k}$.

Proof. This can be derived from the two facts: 1) $\exists i \in \mathbb{N}$, such that $\mathcal{T}_i \neq \mathcal{T}_{i+1}$; 2) $\forall i \in \mathbb{N}, \exists k \in \mathbb{N}$, such that $\mathcal{T}_i = \mathcal{T}_{i+k}$. The first one is a direct result of the fact that all elements of L_T must appear in the sequence $(\mathcal{T}_1, \mathcal{T}_2, \mathcal{T}_3, ...)$ and L_T has at least two elements. If the second one is not true, then $\mathcal{T}_i$ is crossed by τ for finite times, which contradicts with the result that $\forall \mathcal{T} \in L_T$ is crossed by τ an infinite number of times. □

3.2 Reachability Analysis Algorithm

In this section, we present our reachability analysis algorithm, see Algorithm 1, where we call a *transition from h to h′*, noted $h \to h'$, a minimal trajectory from state h that reaches a new boundary in state h'. In other words, $h \to h'$ can be considered an atomic step of simulation, either instant (change of discrete state) or not (with continuous time elapsed).

Algorithm 1. Reachability analysis algorithm

Input 1: A hybrid state $h_1 = (\pi_1, d_{s_1})$
Input 2: A region $R_2 = \{(\pi, d_{s_2}) \mid \pi^i \in [a_i, b_i], i \in \{1, 2, ..., N\}\}$
Output: "R_2 is reached", "R_2 is not reached" or "unknown result"

1: Current state $h := h_1$
2: **while** h is not a fixed point **do**
3: h' := next state so that $h \rightarrow h'$ is a transition
4: **if** Transition $h \rightarrow h'$ reaches R_2 **then**
5: **return** "R_2 is reached"
6: **else**
7: $h := h'$
8: **if** Current simulation is attracted by a cycle of discrete domains **then**
9: **if** *Stop_condition*($Cycle_h, Cycle_\mathcal{D}, R_2$) returns Yes **then**
10: **return** "R_2 is not reached"
11: **else if** *Stop_condition*($Cycle_h, Cycle_\mathcal{D}, R_2$) returns Reached **then**
12: **return** "R_2 is reached"
13: **end if**
14: **else if** Current simulation is probably a chaotic trajectory **then**
15: **return** "unknown result"
16: **end if**
17: **end if**
18: **end while**
19: **return** "R_2 is not reached"

To determine if the current simulation is attracted by a cycle of discrete domains (line 8) or if the current simulation is probably a chaotic trajectory (line 14), we use Theorem 2 or Property 1 respectively.

The objective of the function *Stop_condition* is, knowing that this trajectory is attracted by a cycle of discrete domains, to determine if the trajectory can reach R_2 after an infinite number of transitions (see Fig. 3 right). If it is the case, the function returns "Reached". Otherwise, if from the current state, there is no more chance to reach R_2 (see Fig. 3 middle), then the function returns "Yes". For both cases, this function can give the right answer in finite time, and the result stops the algorithm. However, if both cases do not apply, the function returns "No" and the algorithm continues. The idea of the function *Stop_condition* is similar to the proof of Theorem 1. Details about the function *Stop_condition* are given in the Appendix.

It can be proved that Algorithm 1 always stops in finite time. Firstly, if the trajectory from h_1 is a trajectory halting in finite time, then the algorithm stops after a finite number of transitions. Secondly, if the trajectory is a chaotic trajectory, then Property 1 will be satisfied after a finite number of transitions, and once it is satisfied, the algorithm stops. Thirdly, if the trajectory is attracted by a cycle of discrete domains, then there are three cases: 1. The trajectory reaches R_2 in finite time; 2. The trajectory reaches R_2 after an infinite number of transitions; 3 The trajectory does not reach R_2. We assume here that Property 1 is not satisfied before the trajectory reaching the attractive cycle of discrete domains (the cycle of discrete domains which attracts the trajectory from h_1). For case 1, the algorithm must stop in finite time, as the trajectory will eventually reach R_2. For case 2, the function *Stop_condition* returns "Reached" in finite time. For case 3, the function *Stop_condition* returns "Yes" in finite time. Need to mention that, since Property 1 is a necessary condition for that a trajectory is chaotic, the algorithm might return inconclusive results ("unknown result") even in the cases that are decidable (trajectories are non-chaotic). In fact, among HGRNs of gene regulatory networks, the cases that satisfy this necessary condition are likely to be very rare: there is no identified HGRN of a gene regulatory network with either chaos or non-chaotic trajectory that satisfies this condition. So, for now, this algorithm is sufficient for checking reachability in practice.

4 Conclusion

In this work, we propose a reachability analysis method for HGRNs. In the first part of this work, we classify trajectories of HGRNs into different classes: trajectories halting in finite time, trajectories attracted by cycles of discrete domains and chaotic trajectories, and provide some theoretical results about these trajectories regarding the reachability problem. Then, based on these theoretical results, we provide the first reachability analysis algorithm for HGRNs.

This algorithm always stops, and it returns the correct answer to the reachability problem if it does not stop with the inconclusive result ("unknown result"). In the presence of chaos, the algorithm always stops with this inconclusive result. However, so far, no model with such chaotic behavior has been identified in the model repositories we use from real-life case studies. But the fact that a HGRN with a chaotic trajectory has been identified is a motivation to investigate more.

In our future work, we will try to find other applications of this reachability analysis method and mainly focus on the development of control strategies of gene regulatory networks. Moreover, we are interested in improving the current method to analyze reachability problems in chaotic trajectories.

Additional Information. Link to the code: https://github.com/Honglu42/Reachability_HGRN/. Link to the Appendix: https://hal.science/hal-04182253.

References

1. Alur, R., Courcoubetis, C., Henzinger, T.A., Ho, P.H.: Hybrid automata: an algorithmic approach to the specification and verification of hybrid systems. Technical report, Cornell University (1993)
2. Asarin, E., Maler, O., Pnueli, A.: Reachability analysis of dynamical systems having piecewise-constant derivatives. Theoret. Comput. Sci. **138**(1), 35–65 (1995)
3. Asarin, E., Mysore, V.P., Pnueli, A., Schneider, G.: Low dimensional hybrid systems-decidable, undecidable, don't know. Inf. Comput. **211**, 138–159 (2012)
4. Behaegel, J., Comet, J.P., Bernot, G., Cornillon, E., Delaunay, F.: A hybrid model of cell cycle in mammals. J. Bioinform. Comput. Biol. **14**(01), 1640001 (2016)

5. Belgacem, I., Gouzé, J.L., Edwards, R.: Control of negative feedback loops in genetic networks. In: 2020 59th IEEE Conference on Decision and Control (CDC), pp. 5098–5105. IEEE (2020)
6. Chai, X., Ribeiro, T., Magnin, M., Roux, O., Inoue, K.: Static analysis and stochastic search for reachability problem. Electron. Notes Theor. Comput. Sci. **350**, 139–158 (2020)
7. Cornillon, E., Comet, J.P., Bernot, G., Enée, G.: Hybrid gene networks: a new framework and a software environment. Adv. Syst. Synthetic Biol., 57–84 (2016)
8. Dang, T., Testylier, R.: Reachability analysis for polynomial dynamical systems using the Bernstein expansion. Reliab. Comput. **17**(2), 128–152 (2012)
9. Edwards, R., Glass, L.: A calculus for relating the dynamics and structure of complex biological networks. Adv. Chem. Phys. **132**, 151–178 (2006)
10. Edwards, R.: Analysis of continuous-time switching networks. Phys. D **146**(1–4), 165–199 (2000)
11. Edwards, R., Glass, L.: A calculus for relating the dynamics and structure of complex biological networks. Adv. Chem. Phys.: Spec. Vol. Adv. Chem. Phys. **132**, 151–178 (2005)
12. Firippi, E., Chaves, M.: Topology-induced dynamics in a network of synthetic oscillators with piecewise affine approximation. Chaos Interdisc. J. Nonlinear Sci. **30**(11), 113128 (2020)
13. Flieller, D., Riedinger, P., Louis, J.P.: Computation and stability of limit cycles in hybrid systems. Nonlinear Anal.: Theory Methods Appl. **64**(2), 352–367 (2006)
14. Folschette, M., Paulevé, L., Magnin, M., Roux, O.: Sufficient conditions for reachability in automata networks with priorities. Theoret. Comput. Sci. **608**, 66–83 (2015)
15. Frehse, G., et al.: SpaceEx: scalable verification of hybrid systems. In: Gopalakrishnan, G., Qadeer, S. (eds.) CAV 2011. LNCS, vol. 6806, pp. 379–395. Springer, Heidelberg (2011). https://doi.org/10.1007/978-3-642-22110-1_30
16. Girard, A.: Computation and stability analysis of limit cycles in piecewise linear hybrid systems. IFAC Proc. Vol. **36**(6), 181–186 (2003)
17. Gouzé, J.L., Sari, T.: A class of piecewise linear differential equations arising in biological models. Dyn. Syst. **17**(4), 299–316 (2002)
18. Hamatani, S., Tsubone, T.: Analysis of a 3-dimensional piecewise-constant chaos generator without constraint. IEICE Proc. Ser. **48**(A2L-B-3), 11–14 (2016)
19. Henzinger, T.A., Kopke, P.W., Puri, A., Varaiya, P.: What's decidable about hybrid automata? In: Proceedings of the Twenty-Seventh Annual ACM Symposium on Theory of Computing, pp. 373–382 (1995)
20. Hiskens, I.A.: Stability of hybrid system limit cycles: application to the compass gait biped robot. In: Proceedings of the 40th IEEE Conference on Decision and Control (Cat. No. 01CH37228), vol. 1, pp. 774–779. IEEE (2001)
21. Maler, O., Pnueli, A.: Reachability analysis of planar multi-linear systems. In: Courcoubetis, C. (ed.) CAV 1993. LNCS, vol. 697, pp. 194–209. Springer, Heidelberg (1993). https://doi.org/10.1007/3-540-56922-7_17
22. Mestl, T., Lemay, C., Glass, L.: Chaos in high-dimensional neural and gene networks. Phys. D **98**(1), 33–52 (1996)
23. Paulevé, L.: Reduction of qualitative models of biological networks for transient dynamics analysis. IEEE/ACM Trans. Comput. Biol. Bioinf. **15**(4), 1167–1179 (2017)
24. Plahte, E., Kjøglum, S.: Analysis and generic properties of gene regulatory networks with graded response functions. Phys. D **201**(1–2), 150–176 (2005)

25. Sandler, A., Tveretina, O.: Deciding reachability for piecewise constant derivative systems on orientable manifolds. In: Filiot, E., Jungers, R., Potapov, I. (eds.) RP 2019. LNCS, vol. 11674, pp. 178–192. Springer, Cham (2019). https://doi.org/10.1007/978-3-030-30806-3_14
26. Sun, H., Folschette, M., Magnin, M.: Limit cycle analysis of a class of hybrid gene regulatory networks. In: Petre, I., Păun, A. (eds.) CMSB 2022. LNCS, vol. 13447, pp. 217–236. Springer, Cham (2022). https://doi.org/10.1007/978-3-031-15034-0_11
27. Thomas, R.: Boolean formalization of genetic control circuits. J. Theor. Biol. **42**(3), 563–585 (1973)
28. Thomas, R.: Regulatory networks seen as asynchronous automata: a logical description. J. Theor. Biol. **153**(1), 1–23 (1991)
29. Znegui, W., Gritli, H., Belghith, S.: Design of an explicit expression of the poincaré map for the passive dynamic walking of the compass-gait biped model. Chaos Solitons Fractals **130**, 109436 (2020)

Quantitative Reachability Stackelberg-Pareto Synthesis Is NEXPTIME-Complete

Thomas Brihaye[1], Véronique Bruyère[1(✉)], and Gaspard Reghem[2]

[1] University of Mons, Mons, Belgium
{Thomas.Brihaye,Veronique.Bruyere}@umons.ac.be
[2] ENS Paris-Saclay, Université Paris-Saclay, Gif-sur-Yvette, France
gaspard.reghem@ens-paris-saclay.fr

Abstract. In this paper, we deepen the study of two-player Stackelberg games played on graphs in which Player 0 announces a strategy and Player 1, having several objectives, responds rationally by following plays providing him Pareto-optimal payoffs given the strategy of Player 0. The Stackelberg-Pareto synthesis problem, asking whether Player 0 can announce a strategy which satisfies his objective, whatever the rational response of Player 1, has been recently investigated for ω-regular objectives. We solve this problem for weighted graph games and quantitative reachability objectives such that Player 0 wants to reach his target set with a total cost less than some given upper bound. We show that it is NEXPTIME-complete, as for Boolean reachability objectives.

Keywords: Two-player Stackelberg games played on graphs · Strategy synthesis · Quantitative reachability objectives · Pareto-optimal costs

1 Introduction

Formal verification, and more specifically *model-checking*, is a branch of computer science which offers techniques to check automatically whether a system is correct [3,18]. This is essential for systems responsible for critical tasks like air traffic management or control of nuclear power plants. Much progress has been made in model-checking both theoretically and in tool development, and the technique is now widely used in industry.

Nowadays, it is common to face more complex systems, called *multi-agent systems*, that are composed of heterogeneous components, ranging from traditional pieces of reactive code, to wholly autonomous robots or human users. Modelling and verifying such systems is a challenging problem that is far from being solved. One possible approach is to rely on *game theory*, a branch of mathematics that

Thomas Brihaye – Partly supported by the F.R.S.- FNRS under grant n°T.0027.21.
Véronique Bruyère – Partly supported by the F.R.S.- FNRS under grant n°T.0023.22.

O. Bournez et al. (Eds.): RP 2023, LNCS 14235, pp. 70–84, 2023.
https://doi.org/10.1007/978-3-031-45286-4_6

studies mathematical models of interaction between agents and the understanding of their decisions assuming that they are *rational* [32,33]. Typically, each agent (i.e. player) composing the system has his own objectives or preferences, and the way he manages to achieve them is influenced by the behavior of the other agents.

Rationality can be formalized in several ways. A famous model of agents' rational behavior is the concept of *Nash equilibrium* (NE) [31] in a multiplayer non-zero sum game graph that represents the possible interactions between the players [38]. Another model is the one of *Stackelberg games* [35], in which one designated player – the leader, announces a strategy to achieve his goal, and the other players – the followers, respond rationally with an optimal response depending on their goals (e.g. with an NE). This framework is well-suited for the verification of correctness of a controller intending to enforce a given property, while interacting with an environment composed of several agents each having his own objective. In practical applications, a strategy for interacting with the environment is committed before the interaction actually happens.

Our contribution. In this paper, we investigate the recent concept of two-player Stackelberg games, where the environment is composed of *one player* aiming at satisfying *several objectives*, and its related *Stackelberg-Pareto synthesis* (SPS) problem [13,14]. In this framework, for Boolean objectives, given the strategy announced by the leader, the follower responses rationally with a strategy that ensures him a vector of Boolean payoffs that is *Pareto-optimal*, that is, with a maximal number of satisfied objectives. This setting encompasses scenarios where, for instance, several components of the environment can collaborate and agree on trade-offs. The SPS problem is to decide whether the leader can announce a strategy that guarantees him to satisfy his own objective, whatever the rational response of the follower.

The SPS problem is solved in [14] for ω*-regular* objectives. We here solve this problem for weighted game graphs and *quantitative reachability* objectives for both players. Given a target of vertices, the goal is to reach this target with a cost as small as possible. In this quantitative context, the follower responds to the strategy of the leader with a strategy that ensures him a Pareto-optimal cost vector given his series of targets. The aim of the leader is to announce a strategy in a way to reach his target with a total cost less than some given upper bound, whatever the rational response of the follower. We show that the SPS problem is NEXPTIME-complete (Theorem 1), as for Boolean reachability objectives. The proofs of our results are available in the long version of this paper [11].

It is well-known that moving from Boolean objectives to quantitative ones allows to model *richer properties*. This paper is a first step in this direction for the SPS problem for two-player Stackelberg games with multiple objectives for the follower. Our proof follows the same pattern as for Boolean reachability [14]: if there is a solution to the SPS problem, then there is one that is finite-memory whose memory size is at most exponential. The non-deterministic algorithm thus guesses such a strategy and checks whether it is a solution. However, a crucial intermediate step is to prove that if there exists a solution, then there exists one whose Pareto-optimal costs for the follower are *exponential* in the size of the instance (Theorem 2). The proof of this non trivial step (which is meaningless in the Boolean case) is the main contribution of the paper. Given a solution, we first present some hypotheses and techniques that allow to locally modify it into a solution with smaller Pareto-optimal cost vectors. We then conduct a proof by induction on the number of follower's targets, to globally decrease the cost vectors and to get an exponential number of Pareto-optimal cost vectors. The NEXPTIME-hardness of the SPS problem is trivially obtained by reduction from this problem for Boolean reachability. Indeed, the Boolean version is equivalent to the quantitative one with all weights put to zero and with the given upper bound equal to zero. Notice that the two versions differ: we exhibit an example of game that has a solution to the SPS problem for quantitative reachability, but none for Boolean reachability.

Related Work. During the last decade, multiplayer non-zero sum games and their applications to reactive synthesis have raised a growing attention, see for instance the surveys [4,12,24]. When several players (like the followers) play with the aim to satisfy their objectives, several *solution concepts* exist such as NE, subgame perfect equilibrium (SPE) [34], secure equilibria [16,17], or admissibility [2,5]. Several results have been obtained, for Boolean and quantitative objectives, about the constrained existence problem which consists in deciding whether there exists a solution concept such that the payoff obtained by each player is larger than some given threshold. Let us mention [19,38,39] for results on the constrained existence for NEs and [7,8,10,37] for SPEs. Some of them rely on a recent elegant characterization of SPE outcomes [6,22].

Stackelberg games with *several followers* have been recently studied in the context of *rational synthesis*: in [21] in a setting where the followers are cooperative with the leader, and later in [29] where they are adversarial. Rational responses of the followers are, for instance, to play an NE or an SPE. The rational synthesis problem and the SPS problem are incomparable, as illustrated in [36, Section 4.3.2]: in rational synthesis, each component of the environment acts selfishly, whereas in SPS, the components cooperate in a way to obtain a Pareto-optimal cost. In [30], the authors solve the rational synthesis problem that consists in deciding whether the leader can announce a strategy satisfying his objective, when the objectives of the players are specified by LTL formulas. Complexity classes for various ω-regular objectives are established in [19] for both cooperative and adversarial settings. Extension to quantitative payoffs, like mean-payoff or discounted sum, is studied in [25,26] in the cooperative setting and in [1,20] in the adversarial setting.

The concept of *rational verification* has been introduced in [27], where instead of deciding the existence of a strategy for the leader, one verifies that some given leader's strategy satisfies his objective, whatever the NE responses of the followers. An algorithm and its implementation in the EVE system are presented in [27] for objectives specified by LTL formulas. This verification problem is studied in [28] for mean-payoff objectives for the followers and an omega-regular objective for the leader, and it is solved in [9] for both NE and SPE responses of the followers and for a variety of objectives including quantitative objectives.

The Stackelberg-Pareto verification problem is solved in [15] for some ω-regular or LTL objectives.

2 Preliminaries and Studied Problem

We introduce the concept of Stackelberg-Pareto games with quantitative reachability costs. We present the related Stackelberg-Pareto synthesis problem and state our main result.

2.1 Graph Games

Game Arenas. A *game arena* is a tuple $A = (V, V_0, V_1, E, v_0, w)$ where: *(1)* (V, E) is a finite directed graph with V as set of vertices and E as set of edges (it is supposed that every vertex has a successor), *(2)* V is partitioned as $V_0 \cup V_1$ such that V_0 (resp. V_1) represents the vertices controlled by Player 0 (resp. Player 1), *(3)* $v_0 \in V$ is the initial vertex, and *(4)* $w\colon E \to \mathbb{N}$ is a weight function that assigns a non-negative integer[1] to each edge, such that $W = \max_{e \in E} w(e)$ denotes the maximum weight. An arena is *binary* if $w(e) \in \{0,1\}$ for all $e \in E$.

Plays and Histories. A *play* in an arena A is an infinite sequence of vertices $\rho = \rho_0\rho_1 \ldots \in V^\omega$ such that $\rho_0 = v_0$ and $(\rho_k, \rho_{k+1}) \in E$ for all $k \in \mathbb{N}$. *Histories* are finite sequences $h = h_0 \ldots h_k \in V^+$ defined similarly. We denote $\mathsf{last}(h)$ the last vertex h_k of the history h and by $|h|$ its length (equal to k). Let Play_A denote the set of all plays in A, Hist_A the set of all histories in A, and Hist^i_A the set of all histories in A ending on a vertex in V_i, $i = 0, 1$. The mention of the arena will be omitted when it is clear from the context. If a history h is prefix of a play ρ, we denote it by $h\rho$. Given a play $\rho = \rho_0\rho_1 \ldots$, we denote by $\rho_{\leq k}$ the prefix $\rho_0 \ldots \rho_k$ of ρ, and by $\rho_{\geq k}$ its suffix $\rho_k\rho_{k+1} \ldots$. We also write $\rho_{[k,\ell]}$ for $\rho_k \ldots \rho_\ell$. The *weight* of $\rho_{[k,\ell]}$ is equal to $w(\rho_{[k,\ell]}) = \Sigma_{j=k}^{\ell-1} w(\rho_j, \rho_{j+1})$.

Strategies. Let $i \in \{0,1\}$, a *strategy* for Player i is a function $\sigma_i\colon \mathsf{Hist}^i \to V$ assigning to each history $h \in \mathsf{Hist}^i$ a vertex $v = \sigma_i(h)$ such that $(\mathsf{last}(h), v) \in E$. We denote by Σ_i the set of all strategies for Player i. We say that a strategy σ_i is *memoryless* if for all $h, h' \in \mathsf{Hist}^i$, if $\mathsf{last}(h) = \mathsf{last}(h')$, then $\sigma_i(h) = \sigma_i(h')$. A strategy is considered *finite-memory* if it can be encoded by a Mealy machine and its *memory size* is the number of states of the machine [23].

A play ρ is *consistent* with a strategy σ_i if for all $k \in \mathbb{N}$, $\rho_k \in V_i$ implies that $\rho_{k+1} = \sigma_i(\rho_{\leq k})$. Consistency is extended to histories as expected. We denote Play_{σ_i} (resp. Hist_{σ_i}) the set of all plays (resp. histories) consistent with σ_i. Given a couple of strategies (σ_0, σ_1) for Players 0 and 1, there exists a single play that is consistent with both of them, that we denote by $\mathsf{out}(\sigma_0, \sigma_1)$ and call the *outcome* of (σ_0, σ_1).

Reachability Costs. Given an arena A, let us consider a subset $T \subseteq V$ of vertices called *target*. We say that a play $\rho = \rho_0\rho_1 \ldots$ *visits* the target T, if $\rho_k \in T$

[1] Notice that null weights are allowed.

for some k. We define a *cost function* $\mathsf{cost}_T \colon \mathsf{Play} \to \overline{\mathbb{N}}$, where $\overline{\mathbb{N}} = \mathbb{N} \cup \{\infty\}$, that assigns to every play ρ the quantity $\mathsf{cost}_T(\rho) = \min\{w(\rho_{\leq k}) \mid \rho_k \in T\}$, that is, the weight to the first visit of T if ρ visits T, and ∞ otherwise. The cost function is extended to histories in the expected way.

2.2 Stackelberg-Pareto Synthesis Problem

Stackelberg-Pareto Games. Let $t \in \mathbb{N} \setminus \{0\}$, a *Stackelberg-Pareto reachability game* (SP game) is a tuple $G = (A, T_0, T_1, \ldots, T_t)$ where A is a game arena and T_i are targets for all $i \in \{0, \ldots, t\}$, such that T_0 is Player 0's target and $T_1, \ldots, T_t$ are the t targets of Player 1. When A is binary, we say that G is *binary*. The *dimension* t of G is the number of Player 1's targets, and we denote by Games_t (resp. $\mathsf{BinGames}_t$) the set of all (resp. binary) SP games with dimension t. The notations Play_G and Hist_G may be used instead of Play_A and Hist_A.

To distinguish the two players with respect to their targets, we introduce the following terminology. The *cost* of a play ρ is the tuple $\mathsf{cost}(\rho) \in \overline{\mathbb{N}}^t$ such that $\mathsf{cost}(\rho) = (\mathsf{cost}_{T_1}(\rho), \ldots, \mathsf{cost}_{T_t}(\rho))$. The *value* of a play ρ is a non-negative integer or ∞ defined by $\mathsf{val}(\rho) = \mathsf{cost}_{T_0}(\rho)$. The value can be viewed as the score of Player 0 and the cost as the score of Player 1. Both functions are extended to histories in the expected way. In the sequel, given a cost $c \in \overline{\mathbb{N}}^t$, we denote by c_i the i-th component of c and by c_{min} the component of c that is minimum, i.e. $c_{min} = \min\{c_i \mid i \in \{1, \ldots, t\}\}$.

In an SP game, Player 0 wishes to minimize the value of a play with respect to the usual order $<$ on $\mathbb{N}$ extended to $\overline{\mathbb{N}}$ such that $n < \infty$ for all $n \in \mathbb{N}$. To compare the costs of Player 1, the following component-wise order is introduced. Let $c, c' \in \overline{\mathbb{N}}^t$ be two costs, we say that $c \leq c'$ if $c_i \leq c'_i$ for all $i \in \{1, \ldots, t\}$. Moreover, we write $c < c'$ if $c \leq c'$ and $c \neq c'$. Notice that the order defined on costs is not total. Given two plays with respective costs c and c', if $c < c'$, then Player 1 prefers the play with lower cost c.

Stackelberg-Pareto Synthesis Problem. Given an SP game and a strategy σ_0 for Player 0, we consider the set C_{σ_0} of costs of plays consistent with σ_0 that are *Pareto-optimal* for Player 1, i.e., minimal with respect to the order $\leq$ on costs. Hence, $C_{\sigma_0} = \min\{\mathsf{cost}(\rho) \mid \rho \in \mathsf{Play}_{\sigma_0}\}$. Notice that C_{σ_0} is an antichain. A cost c is said to be *σ_0-fixed Pareto-optimal* if $c \in C_{\sigma_0}$. Similarly, a play is said to be σ_0-fixed Pareto-optimal if its cost is σ_0-fixed Pareto-optimal. We will omit the mention of σ_0 when it is clear from context.

The problem we study is the following one: given an SP game G and a bound $B \in \mathbb{N}$, is there a strategy σ_0 for Player 0 such that, for all strategies σ_1 for Player 1, if the outcome $\mathsf{out}(\sigma_0, \sigma_1)$ is Pareto-optimal, then the value of the outcome is below B. It is equivalent to say that for all $\rho \in \mathsf{Play}_{\sigma_0}$, if $\mathsf{cost}(\rho)$ is σ_0-fixed Pareto-optimal, then $\mathsf{val}(\rho)$ is below B.

Problem 1. The *Stackelberg-Pareto Synthesis problem* (SPS problem) is to decide, given an SP game G and a bound B, whether

$$\exists \sigma_0 \in \Sigma_0, \forall \sigma_1 \in \Sigma_1, \mathsf{cost}(\mathsf{out}(\sigma_0, \sigma_1)) \in C_{\sigma_0} \Rightarrow \mathsf{val}(\mathsf{out}(\sigma_0, \sigma_1)) \leq B. \quad (1)$$

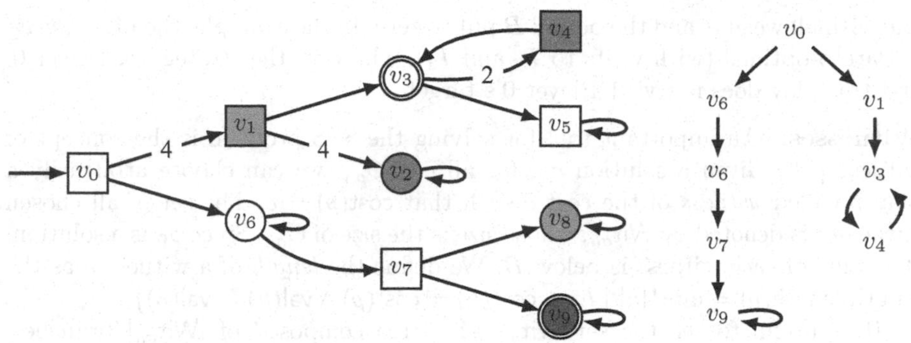

Fig. 1. Arena A (on the left) – Witness tree (on the right)

Any strategy σ_0 satisfying (1) is called a *solution* and we denote it by $\sigma_0 \in \text{SPS}(G, B)$. Our main result is the following theorem.

Theorem 1. *The SPS problem is* NEXPTIME*-complete.*

The non-deterministic algorithm is exponential in the number of targets t and in the size of the binary encoding of the maximum weight W and the bound B. The general approach to obtain NEXPTIME-membership is to show that when there is a solution $\sigma_0 \in \text{SPS}(G, B)$, then there exists one that is finite-memory and whose memory size is exponential. An important part of this paper is devoted to this proof. Then we show that such a strategy can be guessed and checked to be a solution in exponential time.

Example. To provide a better understanding of the SPS problem, let us solve it on a specific example. The arena A is displayed on Fig. 1 where the vertices controlled by Player 0 (resp. Player 1) are represented as circles (resp. squares). The weights are indicated only if they are different from 1 (e.g., the edge (v_0, v_6) has a weight of 1). The initial vertex is v_0. The target of Player 0 is $T_0 = \{v_3, v_9\}$ and is represented by doubled vertices. Player 1 has three targets: $T_1 = \{v_1, v_8\}$, $T_2 = \{v_9\}$ and $T_3 = \{v_2, v_4\}$, that are represented using colors (green for T_1, red for T_2, blue for T_3). Let us exhibit a solution σ_0 in $\text{SPS}(G, 5)$.

We define σ_0 as the strategy that always moves from v_3 to v_4, and that loops once on v_6 and then moves to v_7. The plays consistent with σ_0 are $v_0v_1v_2^\omega$, $v_0v_1(v_3v_4)^\omega$, $v_0v_6v_6v_7v_8^\omega$, and $v_0v_6v_6v_7v_9^\omega$. The Pareto-optimal plays are the plays $v_0v_1(v_3v_4)^\omega$ and $v_0v_6v_6v_7v_9^\omega$ with respective costs $(4, \infty, 7)$ and $(\infty, 4, \infty)$, and they both yield a value less than or equal to 5. Notice that σ_0 has to loop once on v_6, i.e., it is not memoryless[2], otherwise the consistent play $v_0v_6v_7v_8^\omega$ has a Pareto-optimal cost of $(3, \infty, \infty)$ and an infinite value.

Interestingly, the Boolean version of this game does not admit any solution. In this case, given a target, the player's goal is simply to visit it (and not to minimize the cost to reach it). That is, the Boolean version is equivalent to the quantitative

[2] One can prove that there exists no memoryless solution.

one with all weights and the bound B put to zero. In the example, the play $v_0 v_1 v_2^\omega$ is Pareto-optimal (with visits to T_1 and T_3), whatever the strategy of Player 0, and this play does not visit Player 0's target.

Witnesses. An important tool for solving the SPS problem is the concept of witness [14]. Given a solution σ_0, for all $c \in C_{\sigma_0}$, we can choose arbitrarily a play ρ called *witness* of the cost c such that $\mathsf{cost}(\rho) = c$. The set of all chosen witnesses is denoted by Wit_{σ_0}, whose size is the size of C_{σ_0}. Since σ_0 is a solution, the value of each witness is below B. We define the *length* of a witness ρ as the length $\mathsf{length}(\rho) = \min\{|h| \mid h\rho \wedge \mathsf{cost}(h) = \mathsf{cost}(\rho) \wedge \mathsf{val}(h) = \mathsf{val}(\rho)\}$.

It is useful to see the set Wit_{σ_0} as a tree composed of $|\mathsf{Wit}_{\sigma_0}|$ branches. Moreover, given $h \in \mathsf{Hist}_{\sigma_0}$, we write $\mathsf{Wit}_{\sigma_0}(h)$ the set of witnesses for which h is a prefix, i.e., $\mathsf{Wit}_{\sigma_0}(h) = \{\rho \in \mathsf{Wit}_{\sigma_0} \mid h\rho\}$. Notice that $\mathsf{Wit}_{\sigma_0}(h) = \mathsf{Wit}_{\sigma_0}$ when $h = v_0$, and that the size of $\mathsf{Wit}_{\sigma_0}(h)$ decreases as the size of h increases, until it contains a single play or becomes empty.

The following notions about the tree Wit_{σ_0} will be useful. We say that a history h is a *branching point* if there are two witnesses whose greatest common prefix is h, that is, there exists $v \in V$ such that $0 < |\mathsf{Wit}_{\sigma_0}(hv)| < |\mathsf{Wit}_{\sigma_0}(h)|$. We define the following equivalence relations $\sim$ on histories that are prefixes of a witness: $h \sim h'$ if and only if $(\mathsf{val}(h), \mathsf{cost}(h), \mathsf{Wit}_{\sigma_0}(h)) = (\mathsf{val}(h'), \mathsf{cost}(h'), \mathsf{Wit}_{\sigma_0}(h'))$. Notice that if $h \sim h'$, then either hh' or $h'h$ and no new target is visited and no branching point is crossed from the shortest history to the longest one. We call *region* of h its equivalence class. This leads to a *region decomposition* of each witness, such that the first region is the region of the initial state v_0 and the last region is the region of $h\rho$ such that $|h| = \mathsf{length}(\rho)$. A *deviation* is a history hv with $h \in \mathsf{Hist}^1$, $v \in V$, such that h is prefix of some witness, but hv is prefix of no witness.

We illustrate these notions on the previous example and its solution σ_0. A set of witnesses is $\mathsf{Wit}_{\sigma_0} = \{v_0 v_1 (v_3 v_4)^\omega, v_0 v_6 v_6 v_7 v_9^\omega\}$ depicted on Fig. 1. We have that $\mathsf{length}(v_0 v_6 v_6 v_7 v_9^\omega) = |v_0 v_6 v_6 v_7 v_9| = 4$, v_0 is a branching point, $v_0 v_1 v_2$ is a deviation, and the region decomposition of the witness $v_0 v_6 v_6 v_7 v_9^\omega$ is $\{v_0\}$, $\{v_0 v_6, v_0 v_6 v_6, v_0 v_6 v_6 v_7\}$, $\{v_0 v_6 v_6 v_7 v_9^k \mid k \geq 1\}$.

Reduction to Binary Arenas. Working with general arenas requires to deal with the parameter W in most of the proofs. To simplify the arguments, we reduce the SPS problem to *binary* arenas, by replacing each edge with a weight $w \geq 2$ by a path of w edges of weight 1. This (standard) reduction is exponential, but only in the size of the binary encoding of W.

Lemma 1. *Let $G = (A, T_0, \ldots, T_t)$ be an SP game and $B \in \mathbb{N}$. Then one can construct in exponential time an SP game $G' = (A', T_0, \ldots, T_t)$ with a binary arena A' such that*

- *the set of vertices V' of A' contains V and has size $|V'| \leq |V| \cdot W$,*
- *there exists a solution in SPS(G, B) if and only if there exists a solution in SPS(G', B).*

The transformation of the arena A into a binary arena A' has consequences on the size of the SPS problem instance. Since the weights are encoded in binary, the size $|V'|$ could be exponential in the size $|V|$ of the original instance. However, this will have *no impact* on our main result because $|V|$ never appears in the exponent in our calculations.

3 Bounding Pareto-Optimal Payoffs

In this section, we show that if there exists a solution to the SPS problem, then there exists one whose Pareto-optimal costs are exponential in the size of the instance (see Theorem 2 below). It is a *crucial step* to prove that the SPS problem is in NEXPTIME. This is the main contribution of this paper.

3.1 Improving a Solution

We begin by presenting some techniques that allow to modify a solution to the SPS problem into a solution with smaller Pareto-optimal costs.

Order on Strategies and Subgames. Given two strategies σ_0, σ'_0 for Player 0, we say that $\sigma'_0 \leq \sigma_0$ if for all $c \in C_{\sigma_0}$, there exists $c' \in C_{\sigma'_0}$ such that $c' \leq c$. This relation $\leq$ on strategies is a preorder (it is reflexive and transitive). We define $\sigma'_0 < \sigma_0$ when $\sigma'_0 \leq \sigma_0$ and $C_{\sigma'_0} \neq C_{\sigma_0}$, and we say that σ'_0 is *better* than σ_0 whenever $\sigma'_0 \leq \sigma_0$. In the sequel, we modify solutions σ_0 to the SPS problem to get better solutions $\sigma'_0 \leq \sigma_0$, and we say that σ'_0 *improves* the given solution σ_0.

A *subgame* of an SP game G is a couple (G, h), denoted $G_{|h}$, where $h \in \mathsf{Hist}$. In the same way that G can be seen as the set of its plays, $G_{|h}$ is seen as the restriction of G to plays with prefix h. In particular, we have $G_{|v_0} = G$ where v_0 is the initial vertex of G. The value and cost of a play ρ in $G_{|h}$ are the same as those of ρ as a play in G. The *dimension* of $G_{|h}$ is the dimension of G minus the number of targets visited[3] by h' such that $h'\mathsf{last}(h) = h$.

A strategy for Player 0 on $G_{|h}$ is a strategy τ_0 that is only defined for the histories $h' \in \mathsf{Hist}$ such that hh'. We denote $\Sigma_{0|h}$ the set of those strategies. Given a strategy σ_0 for Player 0 in G and $h \in \mathsf{Hist}_{\sigma_0}$, we denote the restriction of σ_0 to $G_{|h}$ by the strategy $\sigma_{0|h}$. Moreover, given $\tau_0 \in \Sigma_{0|h}$, we can define a new strategy $\sigma_0[h \to \tau_0]$ from σ_0 as the strategy on G which consists in playing the strategy σ_0 everywhere, except in the subgame $G_{|h}$ where τ_0 is played. That is, $\sigma_0[h \to \tau_0](h') = \sigma_0(h')$ if $h \not h'$, and $\sigma_0[h \to \tau_0](h') = \tau_0(h')$ otherwise.

As done with $\mathrm{SPS}(G, B)$, we denote by $\mathrm{SPS}(G_{|h}, B)$ the set of all solutions $\tau_0 \in \Sigma_{0|h}$ to the SPS problem for the subgame $G_{|h}$ and the bound B.

Improving a Solution. A natural way to improve a strategy is to improve it on a subgame. Moreover, if it is a solution to the SPS problem, it is also the case for the improved strategy.

---

[3] Notice that we do not include $\mathsf{last}(h)$ in h', as it can be seen as the initial vertex of $G_{|h}$.

Lemma 2. *Let G be a binary SP game, $B \in \mathbb{N}$, and $\sigma_0 \in SPS(G,B)$ be a solution. Consider a history $h \in \mathsf{Hist}_{\sigma_0}$ and a strategy $\tau_0 \in \Sigma_{0|h}$ in the subgame $G_{|h}$ such that $\tau_0 < \sigma_{0|h}$ and $\tau_0 \in SPS(G_{|h}, B)$. Then the strategy $\sigma'_0 = \sigma_0[h \to \tau_0]$ is a solution in $SPS(G,B)$ and $\sigma'_0 < \sigma_0$.*

Another way to improve solutions to the SPS problem is to delete some particular cycles occurring in witnesses as explained in the next lemma.

Lemma 3. *Let G be a binary SP game, $B \in \mathbb{N}$, and $\sigma_0 \in SPS(G,B)$ be a solution. Suppose that in a witness $\rho = \rho_0\rho_1 \ldots \in \mathsf{Wit}_{\sigma_0}$, there exist $m, n \in \mathbb{N}$ such that*

- *$m < n < \mathsf{length}(\rho)$ and $\rho_m = \rho_n$,*
- *$\rho_{\leq m}$ and $\rho_{\leq n}$ belong to the same region, and*
- *if $\mathsf{val}(\rho_{\leq m}) = \infty$, then the weight $w(\rho_{[m,n]})$ is null.*

Then the strategy $\sigma'_0 = \sigma_0[\rho_{\leq m} \to \sigma_{0|\rho_{\leq n}}]$ is a solution in $SPS(G,B)$ such that $\sigma'_0 \leq \sigma_0$.

The first condition means that $\rho_{[m,n]}$ is a cycle and that it appears before the last visit of a target by ρ. The second one says that $\rho_{\leq m} \sim \rho_{\leq n}$, i.e., no new target is visited and no branching point is crossed from history $\rho_{\leq m}$ to history $\rho_{\leq n}$. The third one says that if $\rho_{\leq m}$ does not visit Player 0's target, then the cycle $\rho_{[m,n]}$ must have a null weight. The new strategy σ'_0 is obtained from σ_0 by playing after $\rho_{\leq m}$ as playing after $\rho_{\leq n}$ (thus deleting the cycle $\rho_{[m,n]}$).

From now on, we say that we can *eliminate cycles* according to this lemma[4] without explicitly building the new strategy. We also say that a solution σ_0 is *without cycles* if it does not satisfy the hypotheses of Lemma 3, i.e., if it is impossible to eliminate cycles to get a better solution.

3.2 Crucial Step

We can now state the theorem announced at the beginning of Sect. 3 and provide some ideas about its proof.

Theorem 2. *Let $G \in \mathsf{BinGames}_t$ be a binary SP game with dimension t, $B \in \mathbb{N}$, and $\sigma_0 \in SPS(G,B)$ be a solution. Then there exists a solution $\sigma'_0 \in SPS(G,B)$ without cycles such that $\sigma'_0 \leq \sigma_0$, and*

$$\forall c' \in C_{\sigma'_0}, \forall i \in \{1,\ldots,t\}: \quad c'_i \leq 2^{\Theta(t^2)} \cdot |V|^{\Theta(t)} \cdot (B+3) \quad \vee \quad c'_i = \infty \qquad (2)$$

In case of any general SP game $G \in \mathsf{Games}_t$, the same result holds with $|V|$ replaced by $|V| \cdot W$ in the inequality.

[4] These are the cycles satisfying the lemma, and not just any cycle.

In view of this result, a solution to the SPS problem is said to be *bounded* when its Pareto-optimal costs are bounded as stated in the theorem.

Theorem 2 is proved by induction on the dimension t, with the calculation of a *function* $f(B,t)$ *depending on both* B *and* t, that bounds the components $c'_i \neq \infty$. This function is defined by induction on t through the proofs, and afterwards made explicit and upper bounded by the bound given in Theorem 2. Notice that the function f can be considered as *increasing in* t.[5] The proof of Theorem 2 is sketched in the next lemmas for binary SP games; it is then easily adapted to any SP games by Lemma 1.

The base case is dimension $t = 1$. In this case, the order on costs is total and we have the next lemma (notice that C_{σ_0} is a singleton).

Lemma 4. *Let* $G \in \mathit{BinGames}_1$ *be a binary SP game with dimension* 1, $B \in \mathbb{N}$, *and* $\sigma_0 \in \mathit{SPS}(G,B)$ *be a solution. Then there exists a solution* $\sigma'_0 \in \mathit{SPS}(G,B)$ *without cycles such that* $\sigma'_0 \leq \sigma_0$ *and*

$$\forall c' \in C_{\sigma'_0}: \quad c' \leq f(B,1) = B + |V| \quad \vee \quad c' = \infty. \tag{3}$$

Notice that $f(B,1)$ respects the bound given in Theorem 2 when $t = 1$. Lemma 4 is proved by showing that if the unique Pareto-optimal cost of σ_0 is finite but greater than $B + |V|$, then we can eliminate a cycle according to Lemma 3 and get a better solution.

The next lemma considers the case of dimension $t+1$, with $t \geq 1$. It is proved by using the *induction hypothesis*. Recall that c_{min} is the minimum component of the cost c.

Lemma 5. *Let* $G \in \mathit{BinGames}_{t+1}$ *be a binary SP game with dimension* $t+1$, $B \in \mathbb{N}$, *and* $\sigma_0 \in \mathit{SPS}(G,B)$ *be a solution. Then there exists a solution* $\sigma'_0 \in \mathit{SPS}(G,B)$ *without cycles such that* $\sigma'_0 \leq \sigma_0$, *and*

$$\forall c' \in C_{\sigma'_0}, \forall i \in \{1,\dots,t+1\}: \; c'_i \leq \max\{c'_{min}, B\} + 1 + f(0,t) \; \vee \; c'_i = \infty. \tag{4}$$

The idea of the proof is as follows. If there exists $c \in C_{\sigma_0}$ such that for some $i \in \{1,\dots,t+1\}$, c_i does not satisfy (4), then we consider a witness ρ with $\mathsf{cost}(\rho) = c$ and the history h of minimal length such that $h\rho$ and $w(h) = \max\{c_{min}, B\} + 1$. It follows that the subgame $G_{|h}$ has dimension $k \leq t$ and we can thus apply the induction hypothesis in $G_{|h}$ with $B = 0$ as h has already visited Player 0's target. Hence, by Theorem 2, we get a better solution in the subgame $G_{|h}$, and then a better solution in the whole game G by Lemma 2.

To prove Theorem 2, in view of Lemma 5, our last step is to provide a bound on c_{min}, the minimum component of each Pareto-optimal cost $c \in C_{\sigma_0}$. Notice that if $c_{min} = \infty$, then all the components of c are equal to ∞. In this case, $C_{\sigma_0} = \{(\infty,\dots,\infty)\}$, i.e., there is no play in Play_{σ_0} visiting Player 1's targets. The bound on c_{min} is provided in the next lemma, when $C_{\sigma_0} \neq \{(\infty,\dots,\infty)\}$. It depends on $|C_{\sigma_0}|$, a bound of which is also given in this lemma.

[5] We could artificially duplicate some targets in a way to increase the dimension.

Lemma 6. *Let $G \in BinGames_{t+1}$ be a binary SP game with dimension $t+1$, $B \in \mathbb{N}$, and $\sigma_0 \in SPS(G, B)$ be a solution satisfying (4). Suppose that $C_{\sigma_0} \neq \{(\infty, \ldots, \infty)\}$. Then*

$$|C_{\sigma_0}| \leq \left(f(0,t) + B + 3\right)^{t+1} \tag{5}$$

$$\forall c \in C_{\sigma_0} : \; c_{min} \leq B + 2^{t+1}\left(|V| \cdot \log_2(|C_{\sigma_0}|) + 1 + f(0,t)\right) \tag{6}$$

While (5) is a corollary of Lemma 5, the proof of (6) is rather technical.

Finally, thanks to Lemmas 4–6, calculations can be done in a way to have an explicit formula for $f(B,t)$ and a bound on its value. This completes the proof of Theorem 2. Moreover, with Lemmas 1, 6 and Theorem 2, we easily get a bound for $|C_{\sigma_0}|$ depending on G and B, as stated in the next proposition.

Proposition 1. *For all games $G \in Games_t$ and for all bounded*[6] *solutions $\sigma_0 \in SPS(G, B)$, the size $|C_{\sigma_0}|$ is either equal to 1 or bounded exponentially by $2^{\Theta(t^3)} \cdot (|V| \cdot W)^{\Theta(t^2)} \cdot (B+3)^{\Theta(t)}$.*

4 Complexity of the SPS Problem

In this section, we sketch the proof that the SPS problem is NEXPTIME-complete (Theorem 1). It follows the same pattern as for Boolean reachability [14], however it requires the results of Sect. 3 (which are meaningless in the Boolean case) and some modifications to handle quantitative reachability.

Finite-Memory Solutions. We first show that if there exists a solution to the SPS problem, then there is one that is finite-memory and whose memory size is bounded exponentially.

Proposition 2. *Let G be an SP game, $B \in \mathbb{N}$, and $\sigma_0 \in SPS(G, B)$ be a solution. Then there exists a bounded solution $\sigma_0' \in SPS(G, B)$ such that σ_0' is finite-memory and its memory size is bounded exponentially.*

When $C_{\sigma_0} \neq \{(\infty, \ldots, \infty)\}$, the proof of this proposition is based on the following ideas (the case $C_{\sigma_0} = \{(\infty, \ldots, \infty)\}$ is easier to handle and not detailed).

- We first transform the arena of G into a binary arena and adapt the given solution $\sigma_0 \in \text{SPS}(G, B)$ to the new game. We keep the same notations G and σ_0. We can suppose that σ_0 is bounded by Theorem 2. We consider a set of witnesses Wit_{σ_0} for which we thus know that the costs c_i, $i \in \{1, \ldots, t\}$, are either infinite or exponentially bounded by $f(B,t)$. We also know that the size of C_{σ_0} is exponentially bounded by Proposition 1.
- We show that at any deviation[7], Player 0 can switch to a *punishing strategy* that imposes that the consistent plays π either satisfy $\mathsf{val}(\pi) \leq B$ or $\mathsf{cost}(\pi)$ is

[6] The notion of bounded solution has been defined below Theorem 2.

[7] We recall that a deviation is a history hv with $h \in \mathsf{Hist}^1$, $v \in V$, such that h is prefix of some witness, but hv is prefix of no witness.

not Pareto-optimal. One can prove that this punishing strategy is a winning strategy in a two-player zero-sum game H with an exponential arena and an ω-regular objective. The arena of H is the initial arena extended with information that keeps track of the weight, value and cost of the current history (truncated to $f(B,t)$) and the objective of H is the disjunction of a reachability objective and a safety objective. It follows that this punishing strategy is finite-memory with an exponential memory.

- We then show how to transform the witnesses into lassos. Recall that as the solution σ_0 is bounded, it is impossible to eliminate cycles in any witness ρ. In view of Lemma 3 and by considering the region decomposition of ρ, this means that after the visit by ρ of Player 0's target, each of its regions contains no cycle, except in the last one. In the last region, as soon as a vertex is repeated, we replace the suffix of ρ by the infinite repetition of this cycle. As each resulting lasso traverses an exponential number of regions thanks to Proposition 1, all of them can be produced by a finite-memory strategy with exponential memory. We also show that we need at most exponentially many different punishing strategies.

NEXPTIME-Completeness. We now briefly explain the proof of Theorem 1. For the NEXPTIME-membership, let G be an SP game and $B \in \mathbb{N}$. Proposition 2 states the existence of a solution σ_0 that uses a finite memory bounded exponentially. We can guess such a strategy σ_0 as a Mealy machine $\mathcal{M}$. To verify in exponential time that the guessed strategy σ_0 is a solution to the SPS problem, we proceed as follows. We construct H as the cartesian product $G \times \mathcal{M}$ extended with information that keeps track of the current weight, value, and cost. It can be proved that this information can be truncated to $\max\{B, |V| \cdot |M| \cdot t \cdot W\}$ with $|M|$ the memory size of $\mathcal{M}$. We then compute the set C_{σ_0} of Pareto-optimal costs by testing for the existence of plays in H with a given cost c, beginning with the smallest possible cost $c = (0, \ldots, 0)$, and finishing with the largest possible one $c = (\infty, \ldots, \infty)$. As there is at most an exponential number of costs c to consider, the set C_{σ_0} can be computed in exponential time. Finally, we check whether σ_0 is *not* a solution, i.e., there exists a play ρ in H with a cost $c \in C_{\sigma_0}$ such that $\mathsf{val}(\rho) > B$. This can be done in deterministic exponential time.

Let us now comment on the NEXPTIME-hardness. In [14], the Boolean variant of the SPS problem is proved to be NEXPTIME-complete. It can be reduced to its quantitative variant by labeling each edge with a weight equal to 0 and by considering a bound B equal to 0. Hence the value and cost components are either equal to 0 or ∞. It follows that the (quantitative) SPS problem is NEXPTIME-hard.

5 Conclusion and Future Work

In [14], the SPS problem is proved to be NEXPTIME-complete for Boolean reachability. In this paper, we proved that the same result holds for quantitative reachability (with non-negative weights). The difficult part was to show that when there exists a solution to the SPS problem, there is one whose Pareto-optimal costs are exponentially bounded.

Considering negative weights is a non-trivial extension that is deferred to future work. It will require to study how cycles with a negative cost are useful to improve a solution. Considering multiple objectives for Player 0 (instead of one) is also a non-trivial problem. The order on the tuples of values becomes partial and we could consider several weight functions.

It is well-known that quantitative objectives make it possible to model richer properties than with Boolean objectives. This paper studied quantitative reachability. It would be very interesting to investigate the SPS problem for other quantitative payoffs, like mean-payoff or discounted sum.

References

1. Balachander, M., Guha, S., Raskin, J.: Fragility and robustness in mean-payoff adversarial stackelberg games. In: Haddad, S., Varacca, D. (eds.) 32nd International Conference on Concurrency Theory, CONCUR 2021, 24–27 August 2021, Virtual Conference. LIPIcs, vol. 203, pp. 9:1–9:17. Schloss Dagstuhl - Leibniz-Zentrum für Informatik (2021). https://doi.org/10.4230/LIPIcs.CONCUR.2021.9
2. Berwanger, D.: Admissibility in infinite games. In: Thomas, W., Weil, P. (eds.) STACS 2007. LNCS, vol. 4393, pp. 188–199. Springer, Heidelberg (2007). https://doi.org/10.1007/978-3-540-70918-3_17
3. Bloem, R., Chatterjee, K., Jobstmann, B.: Graph games and reactive synthesis. In: Handbook of Model Checking, pp. 921–962. Springer, Cham (2018). https://doi.org/10.1007/978-3-319-10575-8_27
4. Brenguier, R., et al.: Non-zero sum games for reactive synthesis. In: Dediu, A.-H., Janoušek, J., Martín-Vide, C., Truthe, B. (eds.) LATA 2016. LNCS, vol. 9618, pp. 3–23. Springer, Cham (2016). https://doi.org/10.1007/978-3-319-30000-9_1
5. Brenguier, R., Raskin, J., Sankur, O.: Assume-admissible synthesis. In: Aceto, L., de Frutos-Escrig, D. (eds.) 26th International Conference on Concurrency Theory, CONCUR 2015, Madrid, Spain, 14 September 2015. LIPIcs, vol. 42, pp. 100–113. Schloss Dagstuhl - Leibniz-Zentrum für Informatik (2015). https://doi.org/10.4230/LIPIcs.CONCUR.2015.100
6. Brice, L., Raskin, J., van den Bogaard, M.: Subgame-perfect equilibria in mean-payoff games. In: Haddad, S., Varacca, D. (eds.) 32nd International Conference on Concurrency Theory, CONCUR 2021, 24–27 August 2021, Virtual Conference. LIPIcs, vol. 203, pp. 8:1–8:17. Schloss Dagstuhl - Leibniz-Zentrum für Informatik (2021). https://doi.org/10.4230/LIPIcs.CONCUR.2021.8
7. Brice, L., Raskin, J., van den Bogaard, M.: The complexity of SPEs in mean-payoff games. In: Bojanczyk, M., Merelli, E., Woodruff, D.P. (eds.) 49th International Colloquium on Automata, Languages, and Programming, ICALP 2022, Paris, France, 4–8 July 2022. LIPIcs, vol. 229, pp. 116:1–116:20. Schloss Dagstuhl - Leibniz-Zentrum für Informatik (2022). https://doi.org/10.4230/LIPIcs.ICALP.2022.116
8. Brice, L., Raskin, J., van den Bogaard, M.: On the complexity of SPEs in parity games. In: Manea, F., Simpson, A. (eds.) 30th EACSL Annual Conference on Computer Science Logic, CSL 2022, Göttingen, Germany, 14–19 February 2022, (Virtual Conference). LIPIcs, vol. 216, pp. 10:1–10:17. Schloss Dagstuhl - Leibniz-Zentrum für Informatik (2022). https://doi.org/10.4230/LIPIcs.CSL.2022.10

9. Brice, L., Raskin, J., van den Bogaard, M.: Rational verification and checking for Nash and subgame-perfect equilibria in graph games. CoRR abs/2301.12913 (2023). https://doi.org/10.48550/arXiv.2301.12913
10. Brihaye, T., Bruyère, V., Goeminne, A., Raskin, J., van den Bogaard, M.: The complexity of subgame perfect equilibria in quantitative reachability games. Log. Methods Comput. Sci. **16**(4) (2020). https://lmcs.episciences.org/6883
11. Brihaye, T., Bruyère, V., Reghem, G.: Quantitative reachability stackelberg-pareto synthesis is NEXPTIME-complete. CoRR abs/2308.09443 (2023). https://doi.org/10.48550/arXiv.2308.09443
12. Bruyère, V.: Synthesis of equilibria in infinite-duration games on graphs. ACM SIGLOG News **8**(2), 4–29 (2021). https://doi.org/10.1145/3467001.3467003
13. Bruyère, V., Fievet, B., Raskin, J., Tamines, C.: Stackelberg-pareto synthesis (extended version). CoRR **abs/2203.01285** (2022). https://doi.org/10.48550/arXiv.2203.01285
14. Bruyère, V., Raskin, J., Tamines, C.: Stackelberg-pareto synthesis. In: Haddad, S., Varacca, D. (eds.) 32nd International Conference on Concurrency Theory, CONCUR 2021, 24–27 August 2021, Virtual Conference. LIPIcs, vol. 203, pp. 27:1–27:17. Schloss Dagstuhl - Leibniz-Zentrum für Informatik (2021). https://doi.org/10.4230/LIPIcs.CONCUR.2021.27
15. Bruyère, V., Raskin, J., Tamines, C.: Pareto-rational verification. In: Klin, B., Lasota, S., Muscholl, A. (eds.) 33rd International Conference on Concurrency Theory, CONCUR 2022, Warsaw, Poland, 12–16 September 2022. LIPIcs, vol. 243, pp. 33:1–33:20. Schloss Dagstuhl - Leibniz-Zentrum für Informatik (2022). https://doi.org/10.4230/LIPIcs.CONCUR.2022.33
16. Chatterjee, K., Henzinger, T.A.: Assume-guarantee synthesis. In: Grumberg, O., Huth, M. (eds.) TACAS 2007. LNCS, vol. 4424, pp. 261–275. Springer, Heidelberg (2007). https://doi.org/10.1007/978-3-540-71209-1_21
17. Chatterjee, K., Henzinger, T.A., Jurdzinski, M.: Games with secure equilibria. Theor. Comput. Sci. **365**(1-2), 67–82 (2006). https://doi.org/10.1016/j.tcs.2006.07.032
18. Clarke, E.M., Grumberg, O., Kroening, D., Peled, D.A., Veith, H.: Model Checking, 2nd Edn. MIT Press (2018). https://mitpress.mit.edu/books/model-checking-second-edition
19. Condurache, R., Filiot, E., Gentilini, R., Raskin, J.: The complexity of rational synthesis. In: Chatzigiannakis, I., Mitzenmacher, M., Rabani, Y., Sangiorgi, D. (eds.) 43rd International Colloquium on Automata, Languages, and Programming, ICALP 2016, Rome, Italy, 11–15 July 2016. LIPIcs, vol. 55, pp. 121:1–121:15. Schloss Dagstuhl - Leibniz-Zentrum für Informatik (2016). https://doi.org/10.4230/LIPIcs.ICALP.2016.121
20. Filiot, E., Gentilini, R., Raskin, J.: The adversarial Stackelberg value in quantitative games. In: Czumaj, A., Dawar, A., Merelli, E. (eds.) 47th International Colloquium on Automata, Languages, and Programming, ICALP 2020, Saarbrücken, Germany, 8–11 July 2020, (Virtual Conference). LIPIcs, vol. 168, pp. 127:1–127:18. Schloss Dagstuhl - Leibniz-Zentrum für Informatik (2020). https://doi.org/10.4230/LIPIcs.ICALP.2020.127
21. Fisman, D., Kupferman, O., Lustig, Y.: Rational synthesis. In: Esparza, J., Majumdar, R. (eds.) TACAS 2010. LNCS, vol. 6015, pp. 190–204. Springer, Heidelberg (2010). https://doi.org/10.1007/978-3-642-12002-2_16
22. Flesch, J., Predtetchinski, A.: A characterization of subgame-perfect equilibrium plays in Borel games of perfect information. Math. Oper. Res. **42**(4), 1162–1179 (2017). https://doi.org/10.1287/moor.2016.0843

23. Grädel, E., Thomas, W., Wilke, T. (eds.): Automata Logics, and Infinite Games. LNCS, vol. 2500. Springer, Heidelberg (2002). https://doi.org/10.1007/3-540-36387-4
24. Grädel, E., Ummels, M.: Solution Concepts and Algorithms for Infinite Multiplayer Games, pp. 151–178. Amsterdam University Press (2008). http://www.jstor.org/stable/j.ctt46mwfz.11
25. Gupta, A., Schewe, S.: Quantitative verification in rational environments. In: Cesta, A., Combi, C., Laroussinie, F. (eds.) 21st International Symposium on Temporal Representation and Reasoning, TIME 2014, Verona, Italy, 8–10 September 2014, pp. 123–131. IEEE Computer Society (2014). https://doi.org/10.1109/TIME.2014.9
26. Gupta, A., Schewe, S., Wojtczak, D.: Making the best of limited memory in multi-player discounted sum games. In: Esparza, J., Tronci, E. (eds.) Proceedings Sixth International Symposium on Games, Automata, Logics and Formal Verification, GandALF 2015, Genoa, Italy, 21–22 September 2015. EPTCS, vol. 193, pp. 16–30 (2015). https://doi.org/10.4204/EPTCS.193.2
27. Gutierrez, J., Najib, M., Perelli, G., Wooldridge, M.J.: Automated temporal equilibrium analysis: verification and synthesis of multi-player games. Artif. Intell. **287**, 103353 (2020). https://doi.org/10.1016/j.artint.2020.103353
28. Gutierrez, J., Steeples, T., Wooldridge, M.J.: Mean-payoff games with ω-regular specifications. Games **13**(1), 19 (2022). https://doi.org/10.3390/g13010019
29. Kupferman, O., Perelli, G., Vardi, M.Y.: Synthesis with rational environments. Ann. Math. Artif. Intell. **78**(1), 3–20 (2016). https://doi.org/10.1007/s10472-016-9508-8
30. Kupferman, O., Shenwald, N.: The complexity of LTL rational synthesis. In: TACAS 2022. LNCS, vol. 13243, pp. 25–45. Springer, Cham (2022). https://doi.org/10.1007/978-3-030-99524-9_2
31. Nash, J.F.: Equilibrium points in n-person games. In: PNAS, vol. 36, pp. 48–49. National Academy of Sciences (1950)
32. von Neumann, J., Morgenstern, O.: Theory of Games and Economic Behavior. Princeton University Press (1944)
33. Osborne, M.J., Rubinstein, A.: A Course in Game Theory. MIT Press, Cambridge, MA (1994)
34. Selten, R.: Spieltheoretische Behandlung eines Oligopolmodells mit Nachfrageträgheit. Zeitschrift für die gesamte Staatswissenschaft **121**, 301–324, 667–689 (1965)
35. von Stackelberg, H.F.: Marktform und Gleichgewicht (1937)
36. Tamines, C.: On pareto-optimality for verification and synthesis in games played on graphs. Ph.D. thesis, University of Mons (2022)
37. Ummels, M.: Rational behaviour and strategy construction in infinite multiplayer games. In: Arun-Kumar, S., Garg, N. (eds.) FSTTCS 2006. LNCS, vol. 4337, pp. 212–223. Springer, Heidelberg (2006). https://doi.org/10.1007/11944836_21
38. Ummels, M.: The complexity of Nash equilibria in infinite multiplayer games. In: Amadio, R. (ed.) FoSSaCS 2008. LNCS, vol. 4962, pp. 20–34. Springer, Heidelberg (2008). https://doi.org/10.1007/978-3-540-78499-9_3
39. Ummels, M., Wojtczak, D.: The complexity of Nash equilibria in limit-average games. In: Katoen, J.-P., König, B. (eds.) CONCUR 2011. LNCS, vol. 6901, pp. 482–496. Springer, Heidelberg (2011). https://doi.org/10.1007/978-3-642-23217-6_32

Multi-weighted Reachability Games

Thomas Brihaye[1] and Aline Goeminne[2(✉)]

[1] UMONS - Université de Mons, Mons, Belgium
thomas.brihaye@umons.ac.be
[2] F.R.S.-FNRS and UMONS - Université de Mons, Mons, Belgium
aline.goeminne@umons.ac.be

Abstract. We study two-player multi-weighted reachability games played on a finite directed graph, where an agent, called $\mathcal{P}_1$, has several quantitative reachability objectives that he wants to optimize against an antagonistic environment, called $\mathcal{P}_2$. In this setting, we ask what cost profiles $\mathcal{P}_1$ can ensure regardless of the opponent's behavior. Cost profiles are compared thanks to: *(i)* a lexicographic order that ensures the unicity of an upper value and *(ii)* a componentwise order for which we consider the Pareto frontier. We synthesize *(i)* lexico-optimal strategies and *(ii)* Pareto-optimal strategies. The strategies are obtained thanks to a fixpoint algorithm which also computes the upper value in polynomial time and the Pareto frontier in exponential time. Finally, the constrained existence problem is proved in PTIME for the lexicographic order and PSPACE-complete for the componentwise order.

Keywords: two-player games on graphs · multi-weighted reachability games · Pareto-optimal strategies · lexico-optimal strategies

1 Introduction

Two-player zero-sum games played on graphs are commonly used in the endeavor to *synthesize* systems that are *correct by construction*. In the two-player zero-sum setting the *system* wants to achieve a given objective whatever the behavior of the *environment*. This situation is modeled by a two-player game in which $\mathcal{P}_1$ (resp. $\mathcal{P}_2$) represents the system (resp. the environment). Each vertex of the graph is owned by one player and they take turn by moving a token from vertex to vertex by following the graph edges. This behavior leads to an infinite sequence of vertices called a *play*. The choice of a player's next move is dictated by its *strategy*. In a quantitative setting, edges are equipped with a *weight function* and a *cost function* assigns a cost to each play. This cost depends on the weights of the edges along the play. With this quantitative perspective, $\mathcal{P}_1$ wants to *minimize* the cost function. We say that $\mathcal{P}_1$ can ensure a cost of x if there exists a strategy of $\mathcal{P}_1$ such that, whatever the strategy followed by $\mathcal{P}_2$, the corresponding cost is less than or equal to x. An interesting question is thus to determine what are

T. Brihaye—Partly supported by the F.R.S.- FNRS under grant n°T.0027.21.
A. Goeminne—F.R.S.-FNRS postdoctoral researcher.

O. Bournez et al. (Eds.): RP 2023, LNCS 14235, pp. 85–97, 2023.
https://doi.org/10.1007/978-3-031-45286-4_7

the costs that can be ensured by $\mathcal{P}_1$. In this document, these costs are called the *ensured values*. Other frequently studied questions are: Given a threshold x, does there exist a strategy of $\mathcal{P}_1$ that ensures a cost less than or equal to x? Is it possible to synthesize such a strategy, or even better, if it exists, a strategy that ensures the best ensured value, *i.e.*, an *optimal strategy*?

A well-known studied quantitative objective is the one of *quantitative reachability objective*. A player who wants to achieve such an objective has a subset of vertices, called *target set*, that he wants to reach as quickly as possible. In terms of edge weights, that means that he wants to minimize the cumulative weights until a vertex of the target set is reached. In this setting it is proved that the best ensured value is computed in polynomial time and that optimal strategies exist and do not require memory [9].

Considering systems with only one cost to minimize may seem too restrictive. Indeed, $\mathcal{P}_1$ may want to optimize different quantities while reaching his objective. Moreover, optimizing these different quantities may lead to antagonistic behaviors, for instance when a vehicle wants to reach his destination while minimizing both the delay and the energy consumption. This is the reason why in this paper, we study two-player multi-weighted reachability games, where $\mathcal{P}_1$ aims at reaching a target while minimizing several costs. In this setting each edge of the graph is labeled by a d-tuple of d natural numbers, one per quantity to minimize. Given a sequence of vertices in the game graph, the *cost profile* of $\mathcal{P}_1$ corresponds to the sum of the weights of the edges, component by component, until a given target set is reached. We consider the multi-dimensional counterpart of the previous studied problems: we wonder what cost profiles are ensured by $\mathcal{P}_1$. Thus $\mathcal{P}_1$ needs to arbitrate the trade-off induced by the multi-dimensional setting. In order to do so, we consider two alternatives: the cost profiles can be compared either via *(i)* a lexicographic order that ranks the objectives *a priori* and leads to a unique minimal ensured value; or via *(ii)* a componentwise order. In this second situation, $\mathcal{P}_1$ takes his decision *a posteriori* by choosing an element of the Pareto frontier (the set of minimal ensured values, which is not necessarily a singleton).

Contributions. Our contributions are threefold. First, in Sect. 3.1, given a two-player multi-weighted reachability game, independently of the order considered, we provide a fixpoint algorithm, which computes the minimal cost profiles that can be ensured by $\mathcal{P}_1$. In Sect. 3.2, we study the time complexity of this algorithm, depending on the order considered. When considering the lexicographic order (resp. componentwise order), the algorithm runs in polynomial time (resp. exponential time). Moreover, if the number of dimensions is fixed, the computation of the Pareto frontier can be done in pseudo-polynomial time (polynomial if the weights of the game graph are encoded in unary). As a second contribution, in Sect. 3.3, based on the fixpoint algorithm, we synthesize the optimal strategies (one per order considered). In particular, we show that positional strategies suffice when considering the lexicographic order, although memory is needed in the componentwise case. Finally, in Sect. 4, we focus on the natural decision problem associated with our model: the constrained existence problem. Given a two-player multi-weighted reachability game and a cost profile $\mathbf{x}$, the answer to the constrained existence problem is positive when there exists a strategy of $\mathcal{P}_1$ that ensures $\mathbf{x}$. In the lexicographic case, we show that the problem belongs to PTIME; although it turns to be PSPACE-complete in the componentwise case.

The detailed proofs are provided in the full version of the article [2].

Related Work. Up to our knowledge, and quite surprisingly, two-player multi-weighted reachability games, as defined in this paper, were not studied before. Nevertheless, a one-player variant known as multi-constrained routing is known to be NP-complete and exact and approximate algorithms are, for example, provided in [11]. The time complexity of their exact algorithm matches our results since it runs in exponential time and they indicate that it is pseudo-polynomial if $d = 2$. The one-player setting is also studied in timed automata [10].

If we focus on two-player settings, another closely related model to multi-weighted reachability games is the one studied in [7]. The authors consider two-player generalized (qualitative) reachability games. In this setting $\mathcal{P}_1$ wants to reach several target sets in any order but does not take into account the cost of achieving that purpose. They prove that deciding the winner in such a game is PSPACE-complete. Moreover, they discuss the fact that winning strategies need memory. The memory is used in order to remember which target sets have already been reached. In our setting, we assume that there is only one target set but that the cost to reach it depends on the dimension. Memory is needed because we have to take into consideration the partial sum of weights up to now in order to make the proper choices in the future to ensure the required cost profile. Notice that if we would like to study the case where each dimension has its own target set, both types of memory would be needed.

If we consider other objectives than reachability, we can mention different works on multi-dimentional *energy* and *mean-payoff* objectives [3,5,8]. Moreover, in [1], they prove that the Pareto frontier in a multi-dimensional mean-payoff game is definable as a finite union of convex sets obtained from linear inequations. The authors also provide a Σ_2^P algorithm to decide if this set intersects a convex set defined by linear inequations.

Lexicographic preferences are used in stochastic games with lexicographic (qualitative) reachability-safety objectives [4]. The authors prove that lexico-optimal strategies exist but require finite-memory in order to know on which dimensions the corresponding objective is satisfied or not. They also provide an algorithm to compute the best ensured value and compute lexico-optimal strategies thanks to different computations of optimal strategies in single-dimensional games. Finally, they show that deciding if the best ensured value is greater than or equal to a tuple $\mathbf{x}$ is PSPACE-hard and in NEXPTIME $\cap$ CO-NEXPTIME.

2 Preliminaries

2.1 Two-Player Multi-weighted Reachability Games

Weighted Arena. We consider games that are played on an *(weighted) arena* by two players: $\mathcal{P}_1$ and $\mathcal{P}_2$. An arena $\mathcal{A}_d$ is a tuple $(V_1, V_2, E, \mathbf{w})$ where *(i)* $(V = V_1 \cup V_2, E)$ is a graph such that vertices V_i for $i \in \{1, 2\}$ are owned by $\mathcal{P}_i$ and $V_1 \cap V_2 = \emptyset$ and *(ii)* $\mathbf{w} : E \longrightarrow \mathbb{N}^d$ is a weight function which assigns d natural numbers to each edge of the graph. The variable d is called the number of *dimensions*. For all $1 \le i \le d$, we denote by w_i, with $\mathrm{w}_i : E \longrightarrow \mathbb{N}$, the projection of $\mathbf{w}$ on the ith component, *i.e.*, for all $e \in E$, if $\mathbf{w}(e) = (n_1, \dots, n_d)$ then, $\mathrm{w}_i(e) = n_i$. We define W as the largest weight that can appear in the values of the weight function, *i.e.*, $\mathrm{W} = \max\{\mathrm{w}_i(e) \mid 1 \le i \le d \text{ and } e \in E\}$.

Each time we consider a tuple $\mathbf{x} \in X^d$ for some set X, we write it in bold and we denote the ith component of this tuple by x_i. Moreover, we abbreviate the tuples $(0, \dots, 0)$ and $(\infty, \dots, \infty)$ by $\mathbf{0}$ and $\boldsymbol{\infty}$ respectively.

Plays and Histories. A *play* (resp. *history*) in $\mathcal{A}_d$ is an infinite (resp. finite) sequence of vertices consistent with the structure of the associated arena $\mathcal{A}_d$, *i.e.*, if $\rho = \rho_0\rho_1 \dots$ is a play then, for all $n \in \mathbb{N}$, $\rho_n \in V$ and $(\rho_n, \rho_{n+1}) \in E$. A history may be formally defined in the same way. The set of plays (resp. histories) are denoted by $\mathrm{Plays}_{\mathcal{A}_d}$ (resp. $\mathrm{Hist}_{\mathcal{A}_d}$). When the underlying arena is clear from the context we only write Plays (resp. Hist). We also denote by Hist_1 the set of histories which end in a vertex owned by $\mathcal{P}_1$, *i.e.*, $\mathrm{Hist}_1 = \{h = h_0h_1 \dots h_n \mid h \in \mathrm{Hist} \text{ and } h_n \in V_1\}$. For a given vertex $v \in V$, the sets $\mathrm{Plays}(v)$, $\mathrm{Hist}(v)$, $\mathrm{Hist}_1(v)$ denote the sets of plays or histories starting in v. Finally, for a history $h = h_0 \dots h_n$, the vertex h_n is denoted by $\mathrm{Last}(h)$ and $|h| = n$ is the length of h.

Multi-weighted Reachability Games. We consider *multi-weighted reachability games* such that $\mathcal{P}_1$ has a target set that he wants to reach from a given initial vertex. Moreover, crossing edges on the arena implies the increasing of the d cumulated costs for $\mathcal{P}_1$. While in 1-weighted reachability game $\mathcal{P}_1$ aims at reaching his target set as soon as possible (minimizing his cost), in the general d-weighted case he wants to find a *trade-off* between the different components.

More formally, $\mathrm{F} \subseteq V$ which is a subset of vertices that $\mathcal{P}_1$ wants to reach is called the *target set* of $\mathcal{P}_1$. The *cost function* $\mathbf{Cost} : \mathrm{Plays} \longrightarrow \overline{\mathbb{N}}^d$ of $\mathcal{P}_1$ provides, given a play ρ, the cost of $\mathcal{P}_1$ to reach his target set F along ρ.[1] This cost corresponds to the sum of the weight of the edges, component by component, until he reaches F or is equal to ∞ for all components if it is never the case. For all $1 \le i \le d$, we denote by $\mathrm{Cost}_i : \mathrm{Plays} \longrightarrow \overline{\mathbb{N}}$, the projection of $\mathbf{Cost}$ on the ith component. Formally, for all $\rho = \rho_0\rho_1 \dots \in \mathrm{Plays}$:

$$\mathrm{Cost}_i(\rho) = \begin{cases} \sum_{n=0}^{\ell-1} w_i(\rho_n, \rho_{n+1}) & \text{if } \ell \text{ is the least index such that } \rho_\ell \in \mathrm{F} \\ \infty & \text{otherwise} \end{cases}$$

[1] Where the following notation is used: $\overline{\mathbb{N}} = \mathbb{N} \cup \{\infty\}$.

and $\mathbf{Cost}(\rho) = (\mathrm{Cost}_1(\rho), \ldots, \mathrm{Cost}_d(\rho))$ is called a *cost profile*.

If $h = h_0 \ldots h_\ell$ is a history, $\mathbf{Cost}(h) = \sum_{n=0}^{\ell-1} \mathbf{w}(h_n, h_{n+1})$ is the accumulated costs, component by component, along the history. We assume that $\mathbf{Cost}(v) = \mathbf{0}$, for all $v \in V$.

Definition 1 (Multi-weighted reachability game). *Given a target set* $\mathrm{F} \subseteq V$*, the tuple* $\mathcal{G}_\mathrm{d} = (\mathcal{A}_\mathrm{d}, \mathrm{F}, \mathbf{Cost})$ *is called a d-weighted reachability game, or more generally a multi-weighted reachability game.*

In a d-weighted reachability game $\mathcal{G}_\mathrm{d} = (\mathcal{A}_\mathrm{d}, \mathrm{F}, \mathbf{Cost})$, an initial vertex $v_0 \in V$ is often fixed and the game $(\mathcal{G}_\mathrm{d}, \mathrm{v}_0)$ is called an *initialized multi-weighted reachability game*. A play (resp. history) of $(\mathcal{G}_\mathrm{d}, \mathrm{v}_0)$ is a play (resp. history) of $\mathcal{A}_\mathrm{d}$ starting in v_0.

In the rest of this document, for the sake of readability we write (initialized) game instead of (initialized) d-weighted reachability game.

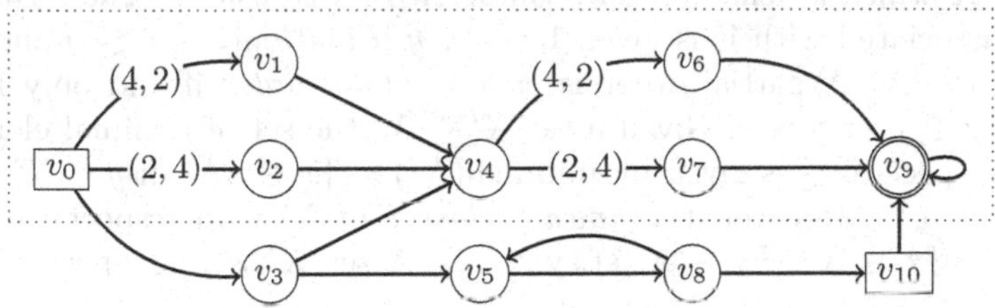

Fig. 1. Example of the arena $\mathcal{A}_2$ of a game $\mathcal{G}_2$. The target set is $\mathrm{F} = \{v_9\}$ and the weight function is given by the label of the edges. Edges without label have a weight of $(1, 1)$. The dotted rectangle is a restriction of the arena specifically used in Example 2.

Example 1. We consider as a running example the game $\mathcal{G}_2$ such that its arena $\mathcal{A}_2 = (V_1, V_2, E, \mathbf{w})$ is depicted in Fig. 1. In this example the set of vertices of $\mathcal{P}_1$ (resp. $\mathcal{P}_2$) are depicted by rounded (resp. rectangular) vertices and the vertices that are part of the target set are doubly circled/framed. The weight function $\mathbf{w}$ labels the corresponding edges. We follow those conventions all along this document. Here, $V_1 = \{v_1, v_2, v_3, v_4, v_5, v_6, v_7, v_8, v_9\}$, $V_2 = \{v_0, v_{10}\}$, $\mathrm{F} = \{v_9\}$ and, for example, $\mathbf{w}(v_0, v_2) = (2, 4)$. For all edges without label, we assume that the weight is $(1, 1)$, *e.g.*, $\mathbf{w}(v_3, v_4) = (1, 1)$. Do not pay attention to the dotted rectangle for the moment.

Let us now study the cost profiles of two different plays. First, the play $\rho = v_0 v_1 v_4 v_6 v_9^\omega$ has a cost profile of $\mathbf{Cost}(\rho) = (4, 2) + (1, 1) + (4, 2) + (1, 1) = (10, 6)$ since ρ visits F in v_9. Moreover, $\mathrm{Cost}_1(\rho) = 10$ and $\mathrm{Cost}_2(\rho) = 6$. Second, the play $\rho' = v_0 v_3 (v_5 v_8)^\omega$ has a cost profile of (∞, ∞) since it does not reach F.

Strategies. A *strategy* of player i, $i \in \{1, 2\}$, provides the next action of $\mathcal{P}_\mathrm{i}$. Formally, a strategy of $\mathcal{P}_\mathrm{i}$ from a vertex v is a function $\sigma_i : \mathrm{Hist}_i(v) \longrightarrow V$ such that for all $h \in \mathrm{Hist}_i(v)$, $(\mathrm{Last}(h), \sigma_i(h)) \in E$. We denote by Σ_i^v the set of strategies of $\mathcal{P}_i$ from $v \in V$. Notice that in an initialized game $(\mathcal{G}_d, v_0)$, unless we specify something else, we assume that the strategies are defined from v_0.

Moreover, given two strategies σ_1 of $\mathcal{P}_1$ and σ_2 of $\mathcal{P}_2$, there is only one play which is consistent with (σ_1, σ_2) from v_0. This play is called the *outcome* of (σ_1, σ_2) from v_0 and is denoted by $\langle \sigma_1, \sigma_2 \rangle_{v_0}$.

We differentiate two classes of strategies: *positional* strategies and *finite-memory* strategies. A positional strategy σ_i only depends on the last vertex of the history, *i.e.*, for all $h, h' \in \text{Hist}_i$, if $\text{Last}(h) = \text{Last}(h')$ then, $\sigma_i(h) = \sigma_i(h')$. It is finite-memory if it can be encoded by a finite-state machine.

Partial Orders. Given two cost profiles $\mathbf{x}$ and $\mathbf{y}$ in $\overline{\mathbb{N}}^d$, $\mathcal{P}_1$ should be able to decide which one is the most beneficial to him. In order to do so, we consider two *partial orders* in the rest of this document: the *componentwise* order and the *lexicographic* order.

We recall some related definitions. A *partial order* on X is a binary relation $\lesssim \subseteq X \times X$ which is reflexive, antisymmetric and transitive. The *strict partial* order $<$ associated with it is given by $x < y$ if and only if $x \lesssim y$ and $x \neq y$, for all $x, y \in X$. A partial order is called a *total order* if and only if for all $x, y \in X$, $x \lesssim y$ or $y \lesssim x$. Given a set $X' \subseteq X$, the set of minimal elements of X' with respect to $\lesssim$ is given by $\text{minimal}(X') = \{x \in X' \mid \text{ if } y \in X' \text{ and } y \lesssim x, \text{ then } x = y\}$. Moreover, the *upward closure* of X' with respect to $\lesssim$ is the set $\uparrow X' = \{\mathbf{x} \in X \mid \exists \mathbf{y} \in X' \text{ st. } \mathbf{y} \lesssim \mathbf{x}\}$. A set X' is said upward closed if $\uparrow X' = X'$.

In what follows we consider two partial orders on $\overline{\mathbb{N}}^d$. The *lexicographic* order, denoted by $\leq_L$, is defined as follows: for all $\mathbf{x}, \mathbf{y} \in \overline{\mathbb{N}}^d$, $\mathbf{x} \leq_L \mathbf{y}$ if and only if either *(i)* $x_i = y_i$ for all $i \in \{1, \ldots, d\}$ or *(ii)* there exists $i \in \{1, \ldots, d\}$ such that $x_i < y_i$ and for all $k < i$, $x_k = y_k$. The *componentwise* order, denoted by $\leq_C$, is defined as: for all $\mathbf{x}, \mathbf{y} \in \overline{\mathbb{N}}^d$, $\mathbf{x} \leq_C \mathbf{y}$ if and only if for all $i \in \{1, \ldots, d\}$, $x_i \leq y_i$. Although the lexicographic order is a total order, the componentwise order is not.

2.2 Studied Problems

We are now able to introduce the different problems that are studied in this paper: the *ensured values problem* and the *constrained existence problem.*

Ensured Values. Given a game $\mathcal{G}_d$ and a vertex v, we define the *ensured values* from v as the cost profiles that $\mathcal{P}_1$ can ensure from v whatever the behavior of $\mathcal{P}_2$. We denote the set of ensured values from v by $\text{Ensure}_{\lesssim}(v)$, *i.e.*, $\text{Ensure}_{\lesssim}(v) = \{\mathbf{x} \in \overline{\mathbb{N}}^d \mid \exists \sigma_1 \in \Sigma_1^v \text{ st. } \forall \sigma_2 \in \Sigma_2^v, \mathbf{Cost}(\langle \sigma_1, \sigma_2 \rangle_v) \lesssim \mathbf{x}\}$. Moreover, we say that a strategy σ_1 of $\mathcal{P}_1$ from v *ensures* the cost profile $\mathbf{x} \in \overline{\mathbb{N}}^d$ if for all strategies σ_2 of $\mathcal{P}_2$ from v, we have that $\mathbf{Cost}(\langle \sigma_1, \sigma_2 \rangle_v) \lesssim \mathbf{x}$.

We denote by $\text{minimal}(\text{Ensure}_{\lesssim}(v))$ the set of minimal elements of $\text{Ensure}_{\lesssim}(v)$ with respect to $\lesssim$. If $\lesssim$ is the lexicographic order, the set of minimal elements of $\text{Ensure}_{\leq_L}(v)$ with respect to $\leq_L$ is a singleton, as $\leq_L$ is a total order, and is called the *upper value* from v. We denote it by $\overline{\text{Val}}(v)$. On the other hand, if $\lesssim$ is the componentwise order, the set of minimal elements of $\text{Ensure}_{\leq_C}(v)$ with respect to $\leq_C$ is called the *Pareto frontier* from v and is denoted by $\text{Pareto}(v)$.

Definition 2 (Ensured Values Problems). *Let* $(\mathcal{G}_d, v_0)$ *be an initialized game. Depending on the partial order, we distinguish two problems:* (i) ***computation of the upper value,*** $\overline{\text{Val}}(v_0)$, *and* (ii) ***computation of the Pareto frontier,*** $\text{Pareto}(v_0)$.

Theorem 1. *Given an initialized game* $(\mathcal{G}_d, v_0)$,

1. *The upper value* $\overline{\text{Val}}(v_0)$ *can be computed in polynomial time.*
2. *The Pareto frontier can be computed in exponential time.*
3. *If* d *is fixed, the Pareto frontier can be computed in pseudo-polynomial time.*

Statement 1 is obtained by Theorem 3, Statements 2 and 3 are proved by Theorem 4.

A strategy σ_1 of $\mathcal{P}_1$ from v is said *Pareto-optimal* from v if σ_1 ensures $\mathbf{x}$ for some $\mathbf{x} \in \text{Pareto}(v)$. If we want to explicitly specify the element $\mathbf{x}$ of the Pareto frontier which is ensured by the Pareto-optimal strategy we say that the strategy σ_1 is $\mathbf{x}$-*Pareto-optimal* from v. Finally, a strategy σ_1 of $\mathcal{P}_1$ from v is said *lexico-optimal* if it ensures the only $\mathbf{x} \in \overline{\text{Val}}(v)$.

In Sect. 3.3, we show how to obtain *(i)* a $\mathbf{x}$-Pareto-optimal strategy from v_0 for each $\mathbf{x} \in \text{Pareto}(v_0)$ and *(ii)* a lexico-optimal strategy from v_0 which is positional. Notice that, as in Example 2, Pareto-optimal strategies sometimes require finite-memory.

Constrained Existence. We are also interested in deciding, given a cost profile $\mathbf{x}$, whether there exists a strategy σ_1 of $\mathcal{P}_1$ from v_0 that ensures $\mathbf{x}$. We call this decision problem the *constrained existence problem* (CE problem).

Definition 3 (Constrained Existence Problem – CE Problem). *Given an initialized game* $(\mathcal{G}_d, v_0)$ *and* $\mathbf{x} \in \mathbb{N}^d$, *does there exist a strategy* $\sigma_1 \in \Sigma_1^{v_0}$ *such that for all strategies* $\sigma_2 \in \Sigma_2^{v_0}$, $\mathbf{Cost}(\langle \sigma_1, \sigma_2 \rangle_{v_0}) \lesssim \mathbf{x}$?

The complexity results of this problem are summarized in the following theorem which is restated and discussed in Sect. 4.

Theorem 2. *If* $\lesssim$ *is the lexicographic order, the CE problem is solved in* PTIME. *If* $\lesssim$ *is the componentwise order, the CE problem is* PSPACE*-complete.*

We conclude this section by showing that memory may be required by $\mathcal{P}_1$ in order to ensure a given cost profile.

Example 2. We consider the game such that its arena is a restriction of the arena given in Fig. 1. This restricted arena is inside the dotted rectangle. For clarity, we assume that the arena is only composed by vertices $v_0, v_1, v_2, v_4, v_6, v_7$ and v_9 and their associated edges. We prove that with the componentwise order $\leq_C$, memory for $\mathcal{P}_1$ is required to ensure the cost profile $(8,8)$. There are only two positional strategies of $\mathcal{P}_1$: σ_1 defined such that $\sigma_1(v_4) = v_6$ and τ_1 defined such that $\tau_1(v_4) = v_7$. For all the other vertices, $\mathcal{P}_1$ has no choice. With σ_1, if $\mathcal{P}_2$ chooses v_1 from v_0, the resulting cost profile is $(10,6)$. In the same way, with τ_1, if $\mathcal{P}_2$ chooses v_2 from v_0, the resulting cost profile is $(6,10)$. This proves that $\mathcal{P}_1$ cannot ensure $(8,8)$ from v_0 with a positional strategy. This is nevertheless possible if $\mathcal{P}_1$ plays a finite-memory strategy. Indeed, by taking into account the past choice of $\mathcal{P}_2$, $\mathcal{P}_1$ is able to ensure $(8,8)$: if $\mathcal{P}_2$ chooses v_1 (resp. v_2) from v_0 then, $\mathcal{P}_1$ should choose v_7 (resp. v_6) from v_4 resulting in a cost profile of $(8,8)$ in both cases.

3 Ensured Values

This section is devoted to the computation of the sets $\text{minimal}(\text{Ensure}_{\lesssim}(v))$ for all $v \in V$. In Sect. 3.1, we provide a fixpoint algorithm which computes these sets. In Sect. 3.2, we study the time complexity of the algorithm both for the lexicographic and the componentwise orders. Finally, in Sect. 3.3, we synthesize lexico and Pareto-optimal strategies.

3.1 Fixpoint Algorithm

Our algorithm that computes the sets $\text{minimal}(\text{Ensure}_{\lesssim}(v))$ for all $v \in V$ shares the key idea of some classical shortest path algorithms. First, for each $v \in V$, we compute the set of cost profiles that $\mathcal{P}_1$ ensures from v in k steps. Then, once all these sets are computed, we compute the sets of cost profiles that can be ensured by $\mathcal{P}_1$ from each vertex but in $k+1$ steps. And so on, until the sets of cost profiles are no longer updated, meaning that we have reached a fixpoint.

For each $k \in \mathbb{N}$ and each $v \in V$, we define the set $\text{Ensure}^k(v)$ as the set of cost profiles that can be ensured by $\mathcal{P}_1$ within k steps. Formally, $\text{Ensure}^k(v) = \{\mathbf{x} \in \overline{\mathbb{N}}^d \mid \exists \sigma_1 \in \Sigma_1^v \text{ st. } \forall \sigma_2 \in \Sigma_2^v, \mathbf{Cost}(\langle\sigma_1,\sigma_2\rangle_v) \lesssim \mathbf{x} \wedge |\langle\sigma_1,\sigma_2\rangle_v|_{\mathrm{F}} \leq k\}$[2], where for all $\rho = \rho_0\rho_1\ldots \in \text{Plays}$, $|\rho|_{\mathrm{F}} = k$ if k is the least index such that $\rho_k \in \mathrm{F}$ and $|\rho|_{\mathrm{F}} = -\infty$ otherwise.

Notice that the sets $\text{Ensure}^k(v)$ are upward closed and that they are infinite sets except if $\text{Ensure}^k(v) = \{\infty\}$. This is the reason why, in the algorithm, we only store sets of minimal elements denoted by $\mathrm{I}^k(v)$. Thus, the correctness of the algorithm relies on the property that for all $k \in \mathbb{N}$ and all $v \in V$, $\text{minimal}(\text{Ensure}^k(v)) = \mathrm{I}^k(v)$.

The fixpoint algorithm is provided by Algorithm 1 in which, if X is a set of cost profiles, and $v, v' \in V$, $X + \mathbf{w}(v,v') = \{\mathbf{x} + \mathbf{w}(v,v') \mid \mathbf{x} \in X\}$. For the moment, do not pay attention to Lines 10 to 13, we come back to them later.

Example 3. We now explain how the fixpoint algorithm runs on Example 1. Table 1 represents the fixpoint of the fixpoint algorithm both for the lexicographic and componentwise orders. Remark that the fixpoint is reached with

[2] To lighten the notations, we omit the mention of $\lesssim$ in subscript.

Algorithm 1: Fixpoint algorithm

```
for v ∈ F do I^0(v) = {0}
for v ∉ F do I^0(v) = {∞}

repeat
    for v ∈ V do
        if v ∈ F then I^{k+1}(v) = {0}

        else if v ∈ V_1 then
            I^{k+1}(v) = minimal( ⋃_{v' ∈ Succ(v)} ↑ I^k(v') + w(v, v') )
            for x ∈ I^{k+1}(v) do
                if x ∈ I^k(v) then f_v^{k+1}(x) = f_v^k(x)
                else
                    f_v^{k+1}(x) = (v', x') where v' and x' are such that
                      v' ∈ Succ(v), x = x' + w(v, v') and x' ∈ I^k(v')

        else if v ∈ V_2 then
            I^{k+1}(v) = minimal( ⋂_{v' ∈ Succ(v)} ↑ I^k(v') + w(v, v') )
until I^{k+1}(v) = I^k(v) for all v ∈ V
```

Table 1. Fixpoint of the fixpoint algorithm reached at step $k^* = 4$.

	$\precsim$	v_0	v_1,v_2	v_3	v_4	v_5	v_6,v_7,v_{10}	v_8	v_9
$\mathrm{I}^*(\cdot)$	$\leq_L$	$\{(8,8)\}$	$\{(4,6)\}$	$\{(4,4)\}$	$\{(3,5)\}$	$\{(3,3)\}$	$\{(1,1)\}$	$\{(2,2)\}$	$\{(0,0)\}$
	$\leq_C$	$\{(8,8)\}$	$\{(6,4),(4,6)\}$	$\{(4,4)\}$	$\{(5,3),(3,5)\}$	$\{(3,3)\}$	$\{(1,1)\}$	$\{(2,2)\}$	$\{(0,0)\}$

$k^* = 4$, while the algorithm takes one more step in order to check that $\mathrm{I}^4(v) = \mathrm{I}^5(v)$ for all $v \in V$. We only focus on some relevant steps of the algorithm with the componentwise order $\leq_C$.

Let us first assume that the first step is computed and is such that $\mathrm{I}^1(v_9) = \{(0,0)\}$ since $v_9 \in \mathrm{F}$, $\mathrm{I}^1(v) = \{(1,1)\}$ if $v \in \{v_6, v_7, v_{10}\}$ and $\mathrm{I}^1(v) = \{(\infty,\infty)\}$ for all other vertices. We now focus on the computation of $\mathrm{I}^2(v_4)$. By Algorithm 1, $\mathrm{I}^2(v_4) = \mathrm{minimal}(\uparrow \mathrm{I}^1(v_6) + (4,2) \cup \uparrow \mathrm{I}^1(v_7) + (2,4)) = \mathrm{minimal}(\uparrow \{(5,3)\} \cup \uparrow \{(3,5)\}) = \{(5,3),(3,5)\}$.

We now assume: $\mathrm{I}^3(v_0) = \{(\infty,\infty)\}$, $\mathrm{I}^3(v_1) = \mathrm{I}^3(v_2) = \mathrm{I}^3(v_3) = \{(4,6),(6,4)\}$, $\mathrm{I}^3(v_4) = \{(5,3),(3,5)\}$ and $\mathrm{I}^3(v_5) = \{(3,3)\}$. We compute $\mathrm{I}^4(v_0)$ which is equal to $\mathrm{minimal}(\uparrow \{(4,6),(6,4)\} + (4,2) \cap \uparrow \{(4,6),(6,4)\} + (2,4) \cap \uparrow \{(4,6),(6,4)\} + (1,1)) = \mathrm{minimal}(\uparrow \{(8,8),(10,6)\} \cap \uparrow \{(6,10),(8,8)\} \cap \uparrow \{(5,7),(7,5)\}) = \mathrm{minimal}(\uparrow \{(8,8)\} \cap \uparrow \{(5,7),(7,5)\}) = \{(8,8)\}$. Finally, we compute $\mathrm{I}^4(v_3) = \mathrm{minimal}(\uparrow \{(6,4),(4,6)\} \cup \uparrow \{(4,4)\}) = \mathrm{minimal}(\{(6,4),(4,6),(4,4)\}) = \{(4,4)\}$.

Termination. We focus on the termination of the fixpoint algorithm.

Proposition 1. *The fixpoint algorithm terminates in less than* $|V| + 1$ *steps.*

The proof of this proposition relies on Propositions 2 and 3. Proposition 2 is interesting on its own. It states that if there exists a strategy σ_1 of $\mathcal{P}_1$ which ensures a cost profile $\mathbf{x} \in \mathbb{N}^d$ from $v \in V$ then, there exists another strategy σ_1' of $\mathcal{P}_1$ which also ensures $\mathbf{x}$ from v but such that the number of edges between v and the first occurrence of a vertex in F is less than or equal to $|V|$, and this regardless of the behavior of $\mathcal{P}_2$.

Proposition 2. *Given a game* $\mathcal{G}_{\mathrm{d}}$*, a vertex* $v \in V$ *and a cost profile* $\mathbf{x} \in \mathbb{N}^d$*, if there exists a strategy* σ_1 *of* $\mathcal{P}_1$ *such that for all strategies* σ_2 *of* $\mathcal{P}_2$ *we have that* $\mathbf{Cost}(\langle \sigma_1, \sigma_2 \rangle_v) \precsim \mathbf{x}$ *then, there exists* σ_1' *of* $\mathcal{P}_1$ *such that for all* σ_2 *of* $\mathcal{P}_2$ *we have:* (i) $\mathbf{Cost}(\langle \sigma_1', \sigma_2 \rangle_v) \precsim \mathbf{x}$ *and* (ii) $|\langle \sigma_1', \sigma_2 \rangle_v|_{\mathrm{F}} \leq |V|$.

The strategy σ_1' is obtained by adequately removing cycles formed by the strategy σ_1. Let us point out that Proposition 2 does not imply that σ_1' is positional. Indeed, in Example 2, the finite-memory strategy is the only strategy that ensures the cost profile $(8, 8)$, it satisfies conditions *(i)* and *(ii)* of Proposition 2 but requires memory.

Proposition 3. *We have:* (i) *for all* $k \in \mathbb{N}$ *and for all* $v \in V$*,* $\mathrm{Ensure}^k(v) \subseteq \mathrm{Ensure}^{k+1}(v)$*; and* (ii) *there exists* $k^* \leq |V|$*, such that for all* $v \in V$ *and for all* $\ell \in \mathbb{N}$*,* $\mathrm{Ensure}^{k^*+\ell}(v) = \mathrm{Ensure}^{k^*}(v)$*.*

Properties stated in Proposition 3 hold by definition of $\mathrm{Ensure}^k(v)$ and Proposition 2. Moreover, the step k^* is a particular step of the algorithm that we call the *fixpoint* of the algorithm. Notice that even if the fixpoint is reached at step k^*, the algorithm needs one more step in order to check that the fixpoint is reached. In the remaining part of this document, we write $\mathrm{Ensure}^*(v)$ (resp. $\mathrm{I}^*(v)$) instead of $\mathrm{Ensure}^{k^*}(v)$ (resp. $\mathrm{I}^{k^*}(v)$).

Correctness. The fixpoint algorithm (Algorithm 1) exactly computes the sets $\mathrm{minimal}(\mathrm{Ensure}_{\precsim}(v))$ for all $v \in V$, *i.e.*, for all $v \in V$, $\mathrm{minimal}(\mathrm{Ensure}_{\precsim}(v)) = \mathrm{I}^*(v)$. This is a direct consequence of Proposition 4.

Proposition 4. *For all* $k \in \mathbb{N}$ *and all* $v \in V$*,* $\mathrm{minimal}(\mathrm{Ensure}^k(v)) = \mathrm{I}^k(v)$*.*

3.2 Time Complexity

In this section we provide the time complexity of the fixpoint algorithm. The algorithm runs in polynomial time for the lexicographic order and in exponential time for the componentwise order. In this latter case, if d is fixed, the algorithm is pseudo-polynomial, *i.e.*, polynomial if the weights are encoded in unary.

Theorem 3. *If* $\precsim$ *is the lexicographic order, the fixpoint algorihtm runs in time polynomial in* $|V|$ *and* d*.*

Theorem 4. *If $\lesssim$ is the componentwise order, the fixpoint algorithm runs in time polynomial in* W *and* $|V|$ *and exponential in d.*

Theorem 3 relies on the fact that Line 9 and Line 16 can be performed in polynomial time. Indeed, in the lexicographic case, for all $k \in \mathbb{N}$ and all $v \in V$, $\mathrm{I}^k(v)$ is a singleton. Thus these operations amounts to computing a minimum or a maximum between at most $|V|$ values. Theorem 4 can be obtained thanks to representations of upward closed sets and operations on them provided in [6].

3.3 Synthesis of Lexico-Optimal and Pareto-Optimal Strategies

To this point, we have only explained the computation of the ensured values and we have not yet explained how lexico and Pareto-optimal strategies are recovered from the algorithm. This is the reason of the presence of Lines 10 to 13 in Algorithm 1. Notice that in Line 13, we are allowed to assume that $\mathbf{x}'$ is in $\mathrm{I}^k(v')$ instead of $\uparrow \mathrm{I}^k(v')$ because for all $k \in \mathbb{N}$, for all $v \in V_1 \backslash \mathrm{F}$, $\mathrm{I}^{k+1}(v) = \text{minimal}\left(\bigcup_{v' \in \mathsf{Succ}(v)} \mathrm{I}^k(v) + \mathbf{w}(v, v')\right)$.

Roughly speaking, the idea behind the functions f_v^k is the following. At each step $k \geq 1$ of the algorithm and for all vertices $v \in V_1 \backslash \mathrm{F}$, we have computed the set $\mathrm{I}^k(v)$. At that point, we know that given $\mathbf{x} \in \mathrm{I}^k(v)$, $\mathcal{P}_1$ can ensure a cost profile of $\mathbf{x}$ from v in at most k steps. The role of the function f_v^k is to keep in memory which next vertex, $v' \in \mathrm{Succ}(v)$, $\mathcal{P}_1$ should choose and what is the cost profile $\mathbf{x}' = \mathbf{x} - \mathbf{w}(v, v')$ which is ensured from v' in at most $k-1$ steps. If different such successors exist one of them is chosen arbitrarily.

In other words, f_v^k provides information about how $\mathcal{P}_1$ should behave locally in v if he wants to ensure one of the cost profile $\mathbf{x} \in \mathrm{I}^k(v)$ from v in at most k steps. In this section, we explain how, from this local information, we recover a global strategy which is $\mathbf{x}$-Pareto optimal from v (resp. lexico-optimal from v) for some $v \in V$ and some $\mathbf{x} \in \mathrm{I}^*(v) \backslash \{\infty\}$, if $\lesssim$ is the componentwise order (resp. the lexicographic order).

We introduce some additional notations. Since for all $k \in \mathbb{N}$ and all $v \in V$, $f_v^k : \mathrm{I}^k(v) \longrightarrow V \times \overline{\mathbb{N}}^d$, if $(v', \mathbf{x}') = f_v^k(\mathbf{x})$ for some $\mathbf{x} \in \mathrm{I}^k(v)$ then, we write $f_v^k(\mathbf{x})[1] = v'$ and $f_v^k(\mathbf{x})[2] = \mathbf{x}'$. Moreover, for all $v \in V$, we write f_v^* instead of $f_v^{k^*}$. Finally, if X is a set of cost profiles, $\min_{\leq_{\mathrm{L}}}(X) = \{\mathbf{x} \in X \mid \forall \mathbf{y} \in X, (\mathbf{y} \leq_{\mathrm{L}} \mathbf{x} \implies \mathbf{y} = \mathbf{x})\}$.

For all $u \in V$ and all $\mathbf{c} \in \mathrm{I}^*(u) \backslash \{\infty\}$, we define a strategy $\sigma_1^* \in \Sigma_1^u$. The aim of this strategy is to ensure $\mathbf{c}$ from u by exploiting the functions f_v^*. The intuition is as follows. If the past history is hv with $v \in V_1$, $\mathcal{P}_1$ has to take into account the accumulated partial costs $\mathbf{Cost}(hv)$ up to v in order the make adequately his next choice to ensure $\mathbf{c}$ at the end of the play. For this reason, he selects some $\mathbf{x} \in \mathrm{I}^*(v)$ such that $\mathbf{x} \lesssim \mathbf{c} - \mathbf{Cost}(hv)$ and follows the next vertex dictated by $f_v^*(\mathbf{x})[1]$.

Definition 4. *Given $u \in V$ and $\mathbf{c} \in \mathrm{I}^*(u)\backslash\{\infty\}$, we define a strategy $\sigma_1^* \in \Sigma_1^u$ such that for all $hv \in \mathrm{Hist}_1(u)$, let $\mathcal{C}(hv) = \{\mathbf{x}' \in \mathrm{I}^*(v) \mid \mathbf{x}' \lesssim \mathbf{c} - \mathbf{Cost}(hv) \wedge \mathbf{x}' \leq_{\mathrm{L}} \mathbf{c} - \mathbf{Cost}(hv)\}$,*

$$\sigma_1^*(hv) = \begin{cases} v' & \text{for some } v' \in \mathrm{Succ}(v), \text{ if } \mathcal{C}(hv) = \emptyset \\ f_v^*(\mathbf{x})[1] & \text{where } \mathbf{x} = \min_{\leq_{\mathrm{L}}} \mathcal{C}(hv), \text{ if } \mathcal{C}(hv) \neq \emptyset \end{cases}.$$

Remark 1. For some technical issues, when we have to select a representative in a set of incomparable elements, the $\leq_{\mathrm{L}}$ order is used in the definitions of $\mathcal{C}(hv)$ and of the strategy. Nevertheless, Definition 4 holds both for the lexicographic and the componentwise orders.

For all $u \in V$ and $\mathbf{c} \in \mathrm{I}^*(u)\backslash\{\infty\}$, the strategy σ_1^* defined in Definition 4 ensures $\mathbf{c}$ from u. In particular, σ_1^* is lexico-optimal and $\mathbf{c}$-Pareto-optimal from u.

Theorem 5. *Given $u \in V$ and $\mathbf{c} \in \mathrm{I}^*(u)\backslash\{\infty\}$, the strategy $\sigma_1^* \in \Sigma_1^u$ defined in Definition 4 is such that for all $\sigma_2 \in \Sigma_2^u$, $\mathbf{Cost}(\langle\sigma_1^*, \sigma_2\rangle_u) \lesssim \mathbf{c}$.*

Although the strategy defined in Definition 4 is a lexico-optimal strategy from u, it requires finite-memory. However, for the lexicographic order, positional strategies are sufficient.

Proposition 5. *If $\lesssim$ is the lexicographic order, for $u \in V$ and $\mathbf{c} \in \mathrm{I}^*(u)\backslash\{\infty\}$, the strategy ϑ_1^* defined as: for all $hv \in \mathrm{Hist}_1(u)$, $\vartheta_1^*(hv) = f_v^*(\mathbf{x})[1]$ where $\mathbf{x}$ is the unique cost profile in $\mathrm{I}^*(v)$, is a positional lexico-optimal strategy from u.*

4 Constrained Existence

Finally, we focus on the constrained existence problem (CE problem).

Theorem 6. *If $\lesssim$ is the lexicographic order, the CE problem is solved in* PTIME.

Theorem 6 is immediate since, in the lexicographic case, we can compute the upper value $\overline{\mathrm{Val}}(v_0)$ in polynomial time (Theorem 3).

Theorem 7. *If $\lesssim$ is the componentwise order, the CE problem is* PSPACE*-complete.*

PSpace-easiness. Proposition 2 allows us to prove that the CE problem with the componentwise order is in APTIME. The alternating Turing machine works as follows: all vertices of the game owned by $\mathcal{P}_1$ (resp. $\mathcal{P}_2$) correspond to disjunctive states (resp. conjunctive states). A path of length $|V|$ is accepted if and only if, *(i)* the target set is reached along that path and *(ii)* the sum of the weights until an element of the target set is $\leq_{\mathrm{C}} \mathbf{x}$. If such a path exists, there exists a strategy of $\mathcal{P}_1$ that ensures the cost profile $\mathbf{x}$. This procedure is done in polynomial time and since APTIME = PSPACE, we get the result.

PSpace-hardness. The hardness part of Theorem 7 is based on a polynomial reduction from the QUANTIFIED SUBSET-SUM problem, proved PSPACE-complete [12, Lemma 4]. This problem is defined as follows. Given a set of natural numbers $N = \{a_1, \dots, a_n\}$ and a threshold $T \in \mathbb{N}$, we ask if the formula $\Psi = \exists x_1 \in \{0,1\} \forall x_2 \in \{0,1\} \exists x_3 \in \{0,1\} \dots \exists x_n \in \{0,1\}, \sum_{1 \leq i \leq n} x_i a_i = T$ is true.

References

1. Brenguier, R., Raskin, J.-F.: Pareto curves of multidimensional mean-payoff games. In: Kroening, D., Păsăreanu, C.S. (eds.) CAV 2015. LNCS, vol. 9207, pp. 251–267. Springer, Cham (2015). https://doi.org/10.1007/978-3-319-21668-3_15
2. Brihaye, T., Goeminne, A.: Multi-weighted reachability games (2023). https://arxiv.org/abs/2308.09625
3. Chatterjee, K., Doyen, L., Henzinger, T.A., Raskin, J.: Generalized mean-payoff and energy games. In: Lodaya, K., Mahajan, M. (eds.) IARCS Annual Conference on Foundations of Software Technology and Theoretical Computer Science, FSTTCS 2010, Chennai, India, 15–18 December 2010, LIPIcs, vol. 8, pp. 505–516. Schloss Dagstuhl - Leibniz-Zentrum für Informatik (2010). https://doi.org/10.4230/LIPIcs.FSTTCS.2010.505
4. Chatterjee, K., Katoen, J.-P., Weininger, M., Winkler, T.: Stochastic games with lexicographic reachability-safety objectives. In: Lahiri, S.K., Wang, C. (eds.) CAV 2020. LNCS, vol. 12225, pp. 398–420. Springer, Cham (2020). https://doi.org/10.1007/978-3-030-53291-8_21
5. Chatterjee, K., Randour, M., Raskin, J.-F.: Strategy synthesis for multi-dimensional quantitative objectives. In: Koutny, M., Ulidowski, I. (eds.) CONCUR 2012. LNCS, vol. 7454, pp. 115–131. Springer, Heidelberg (2012). https://doi.org/10.1007/978-3-642-32940-1_10
6. Delzanno, G., Raskin, J.-F.: Symbolic representation of upward-closed sets. In: Graf, S., Schwartzbach, M. (eds.) TACAS 2000. LNCS, vol. 1785, pp. 426–441. Springer, Heidelberg (2000). https://doi.org/10.1007/3-540-46419-0_29
7. Fijalkow, N., Horn, F.: Les jeux d'accessibilité généralisée. Tech. Sci. Informatiques **32**(9-10), 931–949 (2013). https://doi.org/10.3166/tsi.32.931-949
8. Juhl, L., Guldstrand Larsen, K., Raskin, J.-F.: Optimal bounds for multiweighted and parametrised energy games. In: Liu, Z., Woodcock, J., Zhu, H. (eds.) Theories of Programming and Formal Methods. LNCS, vol. 8051, pp. 244–255. Springer, Heidelberg (2013). https://doi.org/10.1007/978-3-642-39698-4_15
9. Laroussinie, F., Markey, N., Oreiby, G.: Model-checking timed ATL for durational concurrent game structures. In: Asarin, E., Bouyer, P. (eds.) FORMATS 2006. LNCS, vol. 4202, pp. 245–259. Springer, Heidelberg (2006). https://doi.org/10.1007/11867340_18
10. Larsen, K.G., Rasmussen, J.I.: Optimal conditional reachability for multi-priced timed automata. In: Sassone, V. (ed.) FoSSaCS 2005. LNCS, vol. 3441, pp. 234–249. Springer, Heidelberg (2005). https://doi.org/10.1007/978-3-540-31982-5_15
11. Puri, A., Tripakis, S.: Algorithms for the multi-constrained routing problem. In: Penttonen, M., Schmidt, E.M. (eds.) SWAT 2002. LNCS, vol. 2368, pp. 338–347. Springer, Heidelberg (2002). https://doi.org/10.1007/3-540-45471-3_35
12. Travers, S.D.: The complexity of membership problems for circuits over sets of integers. Theor. Comput. Sci. **369**(1-3), 211–229 (2006). https://doi.org/10.1016/j.tcs.2006.08.017

On the Complexity of Robust Eventual Inequality Testing for C-Finite Functions

Eike Neumann(✉)

Swansea University, Swansea, UK
neumaef1@gmail.com

Abstract. We study the computational complexity of a robust version of the problem of testing two univariate C-finite functions for eventual inequality at large times. Specifically, working in the bit-model of real computation, we consider the eventual inequality testing problem for real functions that are specified by homogeneous linear Cauchy problems with arbitrary real coefficients and initial values. In order to assign to this problem a well-defined computational complexity, we develop a natural notion of polynomial-time decidability of subsets of computable metric spaces which extends our recently introduced notion of maximal partial decidability. We show that eventual inequality of C-finite functions is polynomial-time decidable in this sense.

1 Introduction

Linear dynamical systems capture the evolution of a wide range of real-world systems of interest in physics, biology, economics, and other natural sciences. In computing, they arise in connection with questions such as loop termination, model checking, and probabilistic or quantum computation. Unsurprisingly, questions surrounding decidability of liveness and safety properties of such systems have a long history in theoretical computer science, dating back at least to the mid-1970s with the work of Berstel and Mignotte [6] on decidable properties of the zero sets of linear recurrence sequences – a line of research that itself goes back at least to the beginning of the 20th century, as exhibited by Skolem's famous structural result of 1934. A simple-looking but notoriously difficult open problem in this area is the so-called *Positivity Problem* [23]: decide whether a given linear recurrence sequence or C-finite function is positive. The Positivity Problem is known to be decidable for linear recurrences of order up to five, but known to likely fall outside the scope of current techniques for recurrences of order six or higher [21]. The situation is better for simple linear recurrences, where Positivity is known to be decidable up to order nine [20] and Ultimate Positivity – to decide if a recurrence sequence is eventually positive – is known to be decidable for all orders [22]. Similar results have been obtained for the continuous-time analogues of these problems for C-finite functions [5,9,10].

Recently, there has been increased interest in the study of robust reachability questions for linear dynamical systems: in [2], point-to-point reachability for

Full version of submitted manuscript available as [18].

O. Bournez et al. (Eds.): RP 2023, LNCS 14235, pp. 98–112, 2023.
https://doi.org/10.1007/978-3-031-45286-4_8

linear systems with rounded orbits is studied. In [11], a version of the Skolem Problem (to decide whether a recurrence sequence has a zero), where arbitrarily small perturbations are performed after each step of the iteration, is shown to be decidable. Most notably, in [1] two robust versions of the Positivity Problem are studied: in the first version, one is given a fixed rational linear recurrence equation and a rational ball of initial values and asked to decide whether the linear recurrence is positive for all initial values in the ball. In the second version, one is given a linear recurrence equation and a single vector of initial values and asked if there exists a ball about the initial values such that all sequences satisfying the equation with initial values in the ball are positive. The second version is shown to be decidable in polynomial space, while the first version is shown to be at least as hard as the standard Positivity Problem. Both results are shown to hold true also for analogous versions of the Skolem Problem.

We recently proposed [17] another natural robust variant of the Positivity Problem based on computable analysis. Computable analysis [7,29] is the study of computation on data that is in general not known exactly but can be measured to arbitrary accuracy. Algorithms are provided infinitary objects as inputs via streams of increasingly accurate approximations. This allows one to provide arbitrary real numbers (as opposed to "efficiently discrete" countable subsets of real numbers such as rational or algebraic numbers) as inputs to algorithms. Here, robustness is built into the definition of computation, for any decision that an algorithm makes can only depend on a finite approximation to its input. See [17] for further motivation.

We showed that the Positivity Problem and the Ultimate Positivity Problem for linear recurrences with arbitrary real parameters are maximally partially decidable (see below for a definition). In this paper we improve one of these results by showing that the Ultimate Inequality Problem for C-finite functions which are specified by arbitrary real parameters is maximally partially decidable in polynomial time. This is a significant improvement over our previous decidability result, which requires exponentially many queries to the theory of the reals. The focus on the continuous-time setting is mainly for the sake of variety. Our results apply, *mutatis mutandis*, to linear recurrence sequences as well. More concretely, we consider functions that are specified by homogeneous linear differential equations

$$f^{(n)}(t) + c_{n-1}f^{(n-1)}(t) + \cdots + c_1 f^{(1)}(t) + c_0 f(t) = 0 \tag{1}$$

with constant coefficients $c_0, \ldots, c_{n-1} \in \mathbb{R}$ and initial values $f^{(k)}(0) = u_k \in \mathbb{R}$, $k = 0, \ldots, n-1$. Functions of this type are also called *C-finite*. The Ultimate Inequality Problem[1] is to decide, given $(c, u, d, v) \in \mathbb{R}^{2n} \times \mathbb{R}^{2m}$ if there exists $t_0 >$

[1] Over discrete inputs, the Ultimate Inequality Problem reduces to the Ultimate Positivity Problem, *i.e.*, the special case of Ultimate Inequality where the second function is identically equal to zero. However, the standard reduction maps robust instances of Ultimate Inequality to non-robust instances of Ultimate Positivity, which is why we consider Ultimate Inequality instead. For the same reason, we cannot assume without loss of generality that $n = m$ in the definition of the Ultimate Inequality Problem.

0 such that the function f satisfying the differential Eq. (1) with coefficients c and initial values $f^{(k)}(0) = u_k$ is greater than or equal to the function g satisfying an equation analogous to (1) with coefficient vector d and initial values $g^{(k)}(0) = v_k$ for all $t \geq t_0$. Thus, we admit functions as inputs that are specified by Cauchy problems with arbitrary real coefficients and initial values. Computation on such objects can be defined using computable metric spaces [8]. A *computable metric space* is a separable metric space (X, d) together with a dense sequence $(x_k)_k$ and a computable map $\delta\colon \mathbb{N}^3 \to \mathbb{Q}$ such that $|\delta(k, \ell, n) - d(x_k, x_\ell)| < 2^{-n}$. A point $x \in X$ can be provided as an input to an algorithm in the form of an infinite sequence $(k_n)_n$ of integers such that $|x - x_{k_n}| < 2^{-n}$. A sequence $(k_n)_n$ with this property is called a *name*[2] of x.

The prototypical example of a computable metric space is the space $\mathbb{R}$ of real numbers with the usual Euclidean distance, where $(x_k)_k$ is a suitable enumeration of the rational numbers. The computability of the map δ ensures that the distance function $d\colon X \times X \to \mathbb{R}$ is computable when $\mathbb{R}$ is given this structure of computable metric space.

Since an algorithm can only read a finite approximation to its input before committing to a decision, connected computable metric spaces do not have any non-trivial decidable subsets. For this reason, decision problems have arguably been somewhat neglected by the computable analysis community in the past. In order to extend the definition of decidability from $\mathbb{N}$ to arbitrary computable metric spaces in a more meaningful way, we proposed [17] the following notion:

Let $A \subseteq X$ be a subset of a computable metric space X. A *partial algorithm* for deciding A is an algorithm that takes as input (a name of) a point $x \in X$ and either halts and outputs a boolean value or runs indefinitely. Should the algorithm halt on an input x, it is required to correctly report whether x belongs to A. A *maximal partial algorithm* for deciding A is a partial algorithm for deciding A which halts on all names of all points $x \in X \setminus \partial A$. If there exists a maximal partial algorithm for deciding A, then A is called *maximally partially decidable*. A point in $X \setminus \partial A$ is also called a *robust instance* of A, while a point in ∂A is called a *boundary instance* of A. A maximal partial algorithm for deciding a set A can be alternatively defined as a partial algorithm whose halting set is maximal among the halting sets of all partial algorithms for deciding A. This motivates the name "maximal partial algorithm". For metric spaces X whose closed balls are compact, maximal partial decidability can be further characterised using rational balls as inputs [17, Proposition 2.2]. This helps clarify how maximal partial decidability compares with notions of robust decidability proposed by other authors.

[2] Note that we do not require names to be computable. In our computational model, the algorithm is given access to the name of its input as a black box. As a consequence, algorithms may operate on all points of X, not just on the computable points. This should be distinguished from the related notion of *Markov computability* (see [29, Chapter 9.6] and references therein), where algorithms operate on computable points which are presented as Gödel numbers of Turing machines.

In this paper, we extend the definition of maximal partial decidability to polynomial-time decidability. For general background on computational complexity in analysis, see [15]. Using second-order polynomials [14] and parametrised spaces [19], the notion of "maximal partial decidability in polynomial time" can be defined in great generality. For our present purpose, an ad-hoc definition will suffice. We restrict ourselves to computable metric spaces X such that there exists a function size: $X \to \mathbb{N}$ with the property that every point $x \in X$ is the limit of a sequence $(x_{k_n})_n$ with $|x_{k_n} - x| < 2^{-n}$ and $k_n \leq \text{size}(x) + O(n)$. This property is closely related to σ-compactness of X. See [25] and [16] for discussions on spaces that admit size-functions of this type. We tacitly assume in the sequel that all inputs to our algorithms are guaranteed to satisfy a size bound of this form.

The only computable metric spaces we consider in this paper are subspaces of countable sums of finite products of $\mathbb{R}$ or $\mathbb{C}$. Size functions for these spaces can easily be constructed from size functions for the spaces $\mathbb{K}^n$ where $\mathbb{K} \in \{\mathbb{R}, \mathbb{C}\}$. For the latter, we may put $\text{size}((x_1, \dots, x_n)) = O\left(n + \sum_{j=1}^{n} \lceil \log_2(|x_j| + 1) \rceil\right)$.

Let X be a computable metric space admitting a size function as above. Let $A \subseteq X$ be a subset of X. We say that A is maximally partially decidable in polynomial time if there exist a polynomial $P \in \mathbb{N}[x]$ and a maximal partial algorithm for deciding A such that given a name of a point $x \in X \setminus \partial A$, the algorithm halts within at most $P(\text{size}(x) - \log(\min\{d(x, \partial A), 1\}))$ steps (with the convention that $\min\{d(x, \partial A), 1\} = 1$ if $\partial A = \emptyset$). Observe that we recover the definition of polynomial-time decidability of subsets of $\mathbb{N}$ by interpreting $\mathbb{N}$ as a computable metric space with the discrete metric.

For $\mathbb{K} \in \{\mathbb{R}, \mathbb{C}\}$ and $n \geq 0$, write $C_n(\mathbb{K}) = \mathbb{K}^n \times \mathbb{K}^n$. Let

$$[\![\cdot]\!] : \sum_{n \in \mathbb{N}} C_n(\mathbb{K}) \to C(\mathbb{R}, \mathbb{K})$$

be the function that maps $(c, u) \in C_n(\mathbb{K})$ to the unique solution $f : \mathbb{R} \to \mathbb{K}$ of the Cauchy problem (1) with coefficient vector c and initial values u. By convention, if $n = 0$, then the single element of $C_n(\mathbb{K})$ represents the constant zero function. The following is our main result:

Theorem 1. *The Ultimate Inequality Problem is maximally partially decidable in polynomial time. More precisely, given $(c, u, d, v) \in C_n(\mathbb{R}) \times C_m(\mathbb{R})$ we can maximally partially decide in polynomial time if $[\![(c, u)]\!](t) \geq [\![(d, v)]\!](t)$ for all sufficiently large t.*

As part of the proof of Theorem 1, we establish the following result of independent interest:

Theorem 2. *Equality comparison is maximally partially decidable in polynomial time. More precisely, given $(c, u, d, v) \in C_n(\mathbb{C}) \times C_m(\mathbb{C})$ we can maximally partially decide in polynomial time if $[\![(c, u)]\!] \neq [\![(d, v)]\!]$.*

It can be proved essentially as in [17, Proposition 10.1] that the sets of boundary instances of both problems above have measure zero, so that our algorithms halt on almost every input.

2 Proof Outline

Let us briefly outline the proof of Theorem 1. For the purpose of exposition, consider the special case of the Ultimate Positivity Problem, *i.e.*, the special case of Ultimate Inequality where the second function is identically equal to zero. Formally speaking, this is the special case where the second input is an element of $C_0(\mathbb{R})$, so that we know for certain that the input represents the constant zero function.

The polynomial $\chi_c(z) = z^n + c_{n-1}z^{n-1} + \cdots + c_1 z + c_0$ is called the characteristic polynomial of (the differential equation with coefficient vector) c. The roots of χ_c are also called the *characteristic roots* or *eigenvalues* of (the differential equation with coefficient vector) c. Write $\sigma_c \subseteq \mathbb{C}$ for the set of all roots of χ_c. The Cauchy problem (1) has a unique solution $[\![(c, u)]\!] : \mathbb{R} \to \mathbb{C}$, which has the shape

$$[\![(c,u)]\!](t) = \sum_{\lambda \in \sigma_c} P_\lambda(c,u,t)e^{\lambda t} = \sum_{j=1}^{N} \sum_{k=0}^{m_j - 1} a_{j,k} t^k e^{\lambda_j t}. \tag{2}$$

The representation (2) is also called the *exponential polynomial solution* of (1).

A result by Bell and Gerhold [4, Theorem 2] asserts that non-zero linear recurrence sequences without positive real characteristic roots admit positive and negative values infinitely often. As an immediate consequence, one obtains an analogous result for linear differential equations (see also the proof of [5, Theorem 12]): if f satisfies a linear differential Eq. (1) where χ_c has no real roots, then either f is identically zero or there exist unbounded sequences $(t_j)_j$ and $(s_j)_j$ of positive real numbers with $f(t_j) < 0$ and $f(s_j) > 0$.

It is well-known [15, p. 117] that the roots of a polynomial can be computed in polynomial time in the sense described above: there exists a polynomial-time algorithm which takes as input a vector $(c_0, \ldots, c_{n-1}) \in \mathbb{C}^n$ of complex numbers and returns as output a vector $(z_0, \ldots, z_{n-1}) \in \mathbb{C}^n$ of complex numbers such that $(z_0, \ldots, z_{n-1})$ contains all roots of the polynomial $\chi_c(z)$, counted with multiplicity. Any such algorithm is necessarily multi-valued, which means that it is allowed to return different outputs $(z_0, \ldots, z_{n-1})$ for different names of the same input $(c_0, \ldots, c_{n-1})$. Of course, any two valid outputs for the same input agree up to permutation.

It is relatively easy to see that $[\![(c,u)]\!]$ is robustly eventually positive if and only if χ_c has a simple real root ρ with $\rho > \mathrm{Re}(\lambda)$ for all other roots such that the coefficient of ρ in the exponential polynomial solution (2) is strictly positive. We will show that we can detect this situation and in this case compute the sign of the coefficient in polynomial time.

The case where $[\![(c,u)]\!]$ robustly fails to be eventually positive is much more difficult. Intuitively, this can happen for two (potentially overlapping) reasons:

A The polynomial χ_c has a root λ with non-zero imaginary part such that $\mathrm{Re}(\lambda) > \rho$ for all real roots ρ of χ_c and the coefficient $P_\lambda(c, u, t)$ in (2) is non-zero. In this case, eventual inequality fails due to [4, Theorem 2] or the proof of [5, Theorem 12].

B The polynomial χ_c has real roots and the leading coefficient of the polynomial $P_\rho(c, u, t)$ in (2) corresponding to the largest real root ρ is strictly negative, robustly under small perturbations of (c, u).

There is no obvious reason why either of these properties should be semi-decidable, since the polynomials $P_\lambda(c, u, t)$ do not depend continuously on c. Indeed, the degree of $P_\lambda(c, u, t)$ depends on the multiplicity of the roots of χ_c, which is unstable under small perturbations. When the multiplicities of the roots are fixed, the polynomials P_λ depend computably on the input data. For generic initial values, the function which sends input data with fixed eigenvalue multiplicities to the leading coefficient of P_λ is unbounded as eigenvalues $\mu \neq \lambda$ are moved towards λ. This implies that if λ has multiplicity > 2, then the leading coefficient of P_λ will "jump discontinuously" from a finite value to an arbitrarily large value under arbitrarily small perturbations of the input data.

More precisely, consider a vector $(m_1, \dots, m_N)$ of positive integers with $m_1 + \dots + m_N = n$. Let $F_{m_1,\dots,m_N}$ denote the function which maps a complex vector $(\lambda_1, \dots, \lambda_N, u_0, \dots, u_{n-1})$ to the leading coefficient of the polynomial $P_{\lambda_1}(c, u, t)$, where c is the unique coefficient vector whose characteristic polynomial has roots $\lambda_1, \dots, \lambda_N$ with respective multiplicities $m_1, \dots, m_N$. The precise behaviour of the functions $F_{m_1,\dots,m_N}$ is captured by the following result:

Theorem 3.

1. *Let $n \geq 1$. Let $1 \leq m_1 \leq n$. We have*

$$\begin{aligned} &F_{m_1,1,\dots,1}(\Lambda_1, \Lambda_{m_1+1}, \dots \Lambda_n, U_0, \dots, U_{n-1}) \\ &\quad = \frac{G_{m_1,n}(\Lambda_1, \Lambda_{m_1+1}, \dots, \Lambda_n, U_0, \dots, U_{n-1})}{(m_1 - 1)! \prod_{j=m_1+1}^{n} (\Lambda_j - \Lambda_1)}, \end{aligned}$$

where

$$\begin{aligned} &G_{m_1,n}(\lambda_1, \lambda_{m_1+1}, \dots, \lambda_n, u_0, \dots, u_{n-1}) \\ &\quad = \sum_{j=1}^{n} (-1)^{j+m_1} A_{n,m_1,j}(\lambda_1, \lambda_{m_1+1}, \dots \lambda_n) u_{j-1} \end{aligned}$$

for integer polynomials $A_{n,m_1,j} \in \mathbb{Z}[\Lambda_1, \Lambda_{m_1+1}, \dots \Lambda_n]$ with $A_{n,m_1,n} = 1$. Moreover, the polynomials $A_{n,m_1,j}$ can be evaluated on a complex vector in polynomial time, uniformly in n, m_1, and j (when these integers are given in unary).

2. *There exists a polynomial $\Omega \in \mathbb{N}[X, Y, Z]$ with the following property: Let $(\lambda_1, \lambda_{m_1+1}, \dots, \lambda_n, u_0, \dots, u_{n-1})$ and $(\mu_1, \mu_{m_1+1}, \dots, \mu_n, v_0, \dots, v_{n-1})$ be complex vectors. Let C be a positive integer with*

$C \geq \log(\max\{||u||_\infty, ||v||_\infty, ||\lambda||_\infty, ||\mu||_\infty\} + 1)$.
If $|\lambda_j - \mu_j| < 2^{-\Omega(n,C,p)}$ *and* $|u_j - v_j| < 2^{-\Omega(n,C,p)}$ *for all* j, *then*
$G_{m_1,n}(\lambda_1, \lambda_{m_1+1}, \ldots, \lambda_n, u_0, \ldots, u_{n-1})$ *and*
$G_{m_1,n}(\mu_1, \mu_{m_1+1}, \ldots, \mu_n, v_0, \ldots, v_{n-1})$ *have distance at most* 2^{-p}.

3. *Let* $n \geq 1$. *Let* $m_1, \ldots, m_N$ *be positive integers with* $m_1 + \cdots + m_N = n$. *Then* $F_{m_1,m_2,\ldots,m_N}(\Lambda_1, \ldots, \Lambda_N, U_0, \ldots, U_{n-1})$ *is equal to*
$F_{m_1,\underbrace{1,\ldots,1}_{m_2\ times},\ldots,\underbrace{1,\ldots,1}_{m_N\ times}}(\Lambda_1, \underbrace{\Lambda_2, \ldots, \Lambda_2}_{m_2\ times}, \ldots, \underbrace{\Lambda_N, \ldots, \Lambda_N}_{m_N\ times}, U_0, \ldots, U_{n-1})$.

The proof of Theorem 3 can be roughly outlined as follows: It is easy to see that the coefficients in (2) satisfy a linear equation of the form

$$\widetilde{V}_{m_1,\ldots,m_N}(\lambda_1, \ldots, \lambda_N) \cdot \left[[a_{j,k}]_{k=0,\ldots,m_1-1} \right]_{j=1,\ldots,N} = [u_j]_{j=0,\ldots,n-1}. \quad (3)$$

where $\widetilde{V}_{m_1,\ldots,m_N}(\lambda_1, \ldots, \lambda_N)$ is a *modified generalised Vandermonde matrix*. The functions $F_{m_1,\ldots,m_N}$ can be computed explicitly from the cofactor expansion of $\widetilde{V}_{m_1,\ldots,m_N}(\lambda_1, \ldots, \lambda_N)$. Full details are given in [18, Section 3].

In order to verify Condition (A) we approximate the roots of χ_c to finite error 2^{-N}. We can then identify roots $\lambda_1, \ldots, \lambda_m$ that are guaranteed to have non-zero imaginary part and real part larger than any real root. To verify that one of the coefficients $P_{\lambda_j}(c, u, t)$ does not vanish, we construct the differential equation c' with $\chi_{c'}(z) = \chi_c / \prod_{j=1}^m (z - \lambda_j)$ and check that the functions $[\![(c, u)]\!]$ and $[\![(c', u)]\!]$ are different (to some finite accuracy). In case this check does not succeed, we start over with increased accuracy.

Let us now sketch how to verify Condition B. We first approximate the characteristic roots to error 2^{-N} for some integer N. We identify the largest "potentially real" root ρ, *i.e.* the root with largest real part among those roots whose imaginary part has absolute value less than 2^{-N}. We compute an upper bound on the multiplicity m_1 of this root. We then evaluate all "potential numerators" $\operatorname{Re} G_{\ell,n}(\rho, \rho, \ldots, \rho, \lambda_{m_1+1}, \ldots, \lambda_n, u)$ for $\ell = 1, \ldots, m_1$ as far as the accuracy of the root approximations allows, to check if the sign of the real part of the leading coefficient of $P_\rho(c, u, t)$ is guaranteed to be negative[3]. It is possible for this check to succeed while ρ is not a real root. However, in that case $\operatorname{Re} \rho$ is larger than all real roots and $P_\rho(c, u, t)$ is non-zero, so that Condition (A) is met. If the check does not succeed we start over, increasing the accuracy to $N + 1$.

In the full algorithm, we run all three of the above searches simultaneously while increasing the accuracy to which we approximate the roots. To establish polynomial running time, we show that if our searches do not succeed after "many" steps, then there exists a "small" perturbation of the given problem instance which is eventually positive, and another "small" perturbation which is not eventually positive. This is made possible mainly by the special shape of the polynomials $G_{m_1,n}$. In particular, since the coefficient of u_{n-1} in $G_{m_1,n}$ is equal to ± 1, it is possible to perturb the value of $G_{m_1,n}$ by perturbing u_{n-1}, which does

[3] If ρ is real with multiplicity $\ell \leq m_1$, then the leading coefficient of $P_\rho(c, u, t)$ has the same sign as $(-1)^{n-\ell} \operatorname{Re} G_{\ell,n}(\rho, \lambda_2, \ldots, \lambda_{n+1-\ell}, u_0, \ldots, u_{n-1})$.

not interfere with simultaneous perturbations of the eigenvalues. If the search to verify Condition (A) does not halt after "many" steps, the coefficients $P_\lambda(c, u, t)$ of all roots with "large" imaginary part whose real part is larger than the real part of the largest root with "small" imaginary part can be made zero by a "small" perturbation of[4] (c, u). The precise result is as follows:

Lemma 1. *Let* $(c, u, d, v) \in C_n(\mathbb{C}) \times C_m(\mathbb{C})$. *Let* $(u_j)_j$ *be the linear recurrence sequence with initial values* u *and characteristic polynomial* χ_c. *Let* $(v_j)_j$ *be the linear recurrence sequence with initial values* v *and characteristic polynomial* χ_d. *Assume that* $|u_j - v_j| < \varepsilon < 1$ *for* $j = \min\{n, m\}, \ldots, n + m$. *Then there exists a perturbation* $(\widetilde{c}, \widetilde{u}, \widetilde{d}, \widetilde{v}) \in C_n(\mathbb{C}) \times C_m(\mathbb{C})$ *of* (c, u, d, v) *by at most*

$$O\left((n + m + ||c||_\infty + ||d||_\infty + ||u||_\infty)^{O((n+m)\log(n+m))}\right) \varepsilon^{\frac{1}{4(n+m)-2}}$$

with respect to the spectral distance, such that $[\![(\widetilde{c}, \widetilde{u})]\!] = \left[\!\!\left[\left(\widetilde{d}, \widetilde{v}\right)\right]\!\!\right]$.

The proof to Lemma 1 is elementary, but less obvious than one might expect. It can be extracted from a sufficiently constructive proof of the fact that if two linear recurrences of order at most n agree in their first n terms, then they agree everywhere. Theorem 2 follows immediately from Lemma 1. A full proof of both results is given in [18, Section 4].

In perturbation arguments we will almost always perturb the roots of the characteristic polynomial χ_c, rather than perturbing its coefficients directly. It will therefore be useful to work with a suitable distance function. Let (c, u), $(c', u') \in \mathbb{K}^{2n}$. Let $\lambda_1, \ldots, \lambda_n \in \mathbb{C}$ denote the roots of χ_c, listed with multiplicity. Let $\mu_1, \ldots, \mu_n \in \mathbb{C}$ denote the roots of $\chi_{c'}$, listed with multiplicity. Define the *spectral distance* $d_\sigma(c, c')$ of c and c' as

$$d_\sigma(c, c') = \inf_{\pi \in S_n} \max_{j=1,\ldots,n} \{|\lambda_j - \mu_{\pi(j)}|\}.$$

The spectral distance $d_\sigma((c, u), (c', u'))$ of $(c, u) \in C_n(\mathbb{K})$ and $(c', u') \in C_n(\mathbb{K})$ is then defined as $d_\sigma((c, u), (c', u')) = ||u - u'||_\infty + d_\sigma(c, c')$.

The next proposition, proved in [18, Appendix A], shows that the running time of an algorithm is bounded polynomially in the spectral distance to the set of boundary instances, then it is bounded polynomially in the ordinary distance.

Proposition 1. *Let* $c, c' \in \mathbb{K}^n$ *with* $c \neq c'$. *Then we have:*

$$-\log d_\sigma(c, c') \leq -\log d(c, c') + 2n\log(n) + 2n\log(\max\{||c||_\infty, ||c'||_\infty\}).$$

Theorem 3 has the following straightforward corollary, which is useful for constructing perturbations. A proof can be found in [18, Appendix E].

Proposition 2. *Let* $m, n_1, n_2 \in \mathbb{N}$ *with* $m \leq n_2 \leq n_1$. *Let* $\lambda_1, \ldots, \lambda_{n_1}$, $u_0, \ldots,$ $u_{n_1-1} \in \mathbb{C}$. *Assume that the finite sequence* $(u_k)_k$ *satisfies the linear recurrence*

[4] Observe that the coefficients $P_\lambda(c, u, t)$ themselves are not necessarily small, since large coefficients for close characteristic roots could cancel each other out.

with initial values $u_0, \ldots, u_{n_2-1}$ *and characteristic polynomial* $\prod_{j=1}^{n_2}(z-\lambda_j)$. *Then the number* $G_{m,n_1}(\lambda_1, \lambda_{m+1}, \ldots, \lambda_{n_1}, u_0, \ldots, u_{n_1-1})$ *is equal to*

$$\left(\prod_{j=n_2+1}^{n_1}(\lambda_j-\lambda_1)\right) G_{m,n_2}(\lambda_1, \lambda_{m+1}, \ldots, \lambda_{n_2}, u_0, \ldots, u_{n_2-1}).$$

3 Proof of Theorem 1

We now give the full algorithm for inequality testing:

Algorithm 1.

- ***Input.*** *Two C-finite functions, specified by Cauchy problems* $(c, u, d, v) \in C_n(\mathbb{R}) \times C_m(\mathbb{R})$.
- ***Behaviour.*** *The algorithm may halt and return a truth value or run indefinitely. If the algorithm halts, it returns "true" if and only if* $[\![(c,u)]\!](t) \geq [\![(d,v)]\!](t)$ *for all sufficiently large* $t \geq 0$.
- ***Procedure.***
  1. *Compute* $\lambda_1, \ldots, \lambda_{n+m} \in \mathbb{C}$ *such that the list* $\lambda_1, \ldots, \lambda_n$ *contains all roots of* χ_c, *and the list* $\lambda_{n+1}, \ldots, \lambda_{n+m} \in \mathbb{C}$ *contains all roots of* χ_d, *listed with multiplicity.*
  2. *Compute an integer* $B > \log\left(\max\left\{||u||_\infty, ||v||_\infty, |\lambda_1|, \ldots, |\lambda_{n+m}|\right\} + 1\right)$.
  3. *For* $N \in \mathbb{N}$:

     3.1. *Let* $M = \max\{\Omega(n, B, N+1), \Omega(m, B, N+1)\}$, *where* Ω *is the polynomial from the second item of Theorem 3.*

     3.2. *Query the numbers* λ_j *for approximations* $\widetilde{\lambda}_j \in \mathbb{Q}[i]$ *with* $|\operatorname{Re}(\widetilde{\lambda}_j) - \operatorname{Re}(\lambda_j)| < 2^{-M}$ *and* $|\operatorname{Im}(\widetilde{\lambda}_j) - \operatorname{Im}(\lambda_j)| < 2^{-M}$ *such that the polynomials* $\prod_{j=1}^{n}\left(z-\widetilde{\lambda}_j\right)$ *and* $\prod_{j=n+1}^{n+m}\left(z-\widetilde{\lambda}_j\right)$ *have real coefficients.*

     3.3. *Compute an* $(n+m)\times(n+m)$*-matrix encoding the relation* $\preceq_M \subseteq \{1, \ldots, n+m\}$ *which is defined as follows:* $j \preceq_M k$ *if and only if* $\operatorname{Re}(\widetilde{\lambda}_j) - 2^{-M} < \operatorname{Re}(\widetilde{\lambda}_k) + 2^{-M}$.

     3.4. *Compute the sets*

     $$\mathcal{M}_1 = \{k \in \{1, \ldots, n\} \mid j \preceq_M k \text{ for all } j \in \{1, \ldots, n+m\}\}$$

     and

     $$\mathcal{M}_2 = \{k \in \{n+1, \ldots, n+m\} \mid j \preceq_M k \text{ for all } j \in \{1, \ldots, n+m\}\}.$$

     3.5. *Initialise two Kleeneans*[5] `c-positive?` *and* `d-positive?` *with value "unknown".*

[5] *i.e.*, variables that can assume three values: `true`, `false`, and `unknown`.

3.6. If $|\mathcal{M}_1| = 1$:
Writing $\mathcal{M}_1 = \{j_1\}$ and $\{j_2, \ldots, j_n\} = \{1, \ldots, n\} \setminus \{j_1\}$, compute an approximation to $(-1)^{n-1} G_{1,n}(\lambda_{j_1}, \ldots, \lambda_{j_n}, u_0, \ldots, u_{n-1})$ to error 2^{-N-1}. If the result is greater than or equal to 2^{-N}, assign the value ***true*** *to* **c-positive?**. *If the result is less than or equal to -2^{-N}, assign the value* ***false*** *to* **c-positive?**.

3.7. If $|\mathcal{M}_2| = 1$:
Writing $\mathcal{M}_2 = \{k_1\}$ and $\{k_2, \ldots, k_n\} = \{n+1, \ldots, n+m\} \setminus \{k_1\}$, compute an approximation to $(-1)^{m-1} G_{1,m}(\lambda_{k_1}, \ldots, \lambda_{k_n}, v_0, \ldots, v_{m-1})$ to error 2^{-N-1}. If the result is greater than or equal to 2^{-N}, assign the value ***true*** *to* **d-positive?**. *If the result is less than or equal to -2^{-N}, assign the value* ***false*** *to* **d-positive?**.

3.8. If $|\mathcal{M}_1| = 1$, $|\mathcal{M}_2| = 0$, and **c-positive?** $\neq$ ***unknown****: Halt and output the value of* **c-positive?**.

3.9. If $|\mathcal{M}_1| = 0$ and $|\mathcal{M}_2| = 1$, and **d-positive?** $\neq$ ***unknown****: Halt and output the negated value of* **d-positive?**.

3.10. If $|\mathcal{M}_1| = 1$, $|\mathcal{M}_2| = 1$, and **c-positive?** $= \neg$**d-positive?**: *Halt and output the value of* **c-positive?**.

3.11. Compute the set $\mathcal{R}$ of indexes $j \in \{1, \ldots, n+m\}$ such that $\mathrm{Im}(\widetilde{\lambda}_j) < 2^{-M}$.

3.12. Compute the sets
$\mathcal{MR}_1 = \{k \in \mathcal{R} \cap \{1, \ldots, n\} \mid j \preceq_M k$ for all $j \in \mathcal{R}\}$ and
$\mathcal{MR}_2 = \{k \in \mathcal{R} \cap \{n+1, \ldots, n+m\} \mid j \preceq_M k$ for all $j \in \mathcal{R}\}$.

3.13. Compute the set $\mathcal{C} = \{j \in \{1, \ldots, n+m\} \mid j \not\preceq_M k$ for all $k \in \mathcal{R}\}$.

3.14. If $\mathcal{C}$ is non-empty:

3.14.1. Compute the coefficients $e \in \mathbb{R}^{n+m}$ of the polynomial $\chi_e = \prod_{j \in \{1, \ldots, n+m\}} (z - \lambda_j)$.

3.14.2. Compute the first $n+m$ terms of the sequence $w_j = u_j - v_j$.

3.14.3. Compute the coefficients $e' \in \mathbb{C}^{\ell}$ of the polynomial $\chi_{e'} = \prod_{j \in \{1, \ldots, n+m\} \setminus \mathcal{C}} (z - \lambda_j)$.

3.14.4. Let $(w'_j)_j$ be the recurrence sequence with initial values $w'_j = w_j$ for $j = 0, \ldots, \ell - 1$ and characteristic polynomial $\chi_{e'}$.

3.14.5. Compute rational approximations $\widetilde{\varepsilon}_j$ to $\varepsilon_j = |w_j - w'_j|$ for $j = \ell, \ldots, n+m-1$ to error 2^{-M}.

3.14.6. If $\widetilde{\varepsilon}_j > 2^{-M}$ for some j, halt and output ***false***.

3.15. Initialise an empty list $L = \langle\rangle$.

3.16. Let $m_1 = |\mathcal{MR}_1|$. Let $m_2 = |\mathcal{MR}_2|$.

3.17. If $m_1 > 0$:

3.17.1 Pick an arbitrary index $j_1 \in \mathcal{MR}_1$.

3.17.2 Let $\{j_{m_1+1}, \ldots, j_n\} = \{1, \ldots, n\} \setminus \{j_1\}$

3.17.3 For $\ell = 1, \ldots, m_1$, Compute a rational approximation α_ℓ to

$$(-1)^{n-\ell} \operatorname{Re} G_{\ell,n}\left(\lambda_{j_1}, \underbrace{\lambda_{j_1}, \ldots, \lambda_{j_1}}_{m_1-\ell \text{ times}}, \lambda_{j_{m_1+1}}, \ldots, \lambda_{j_n}, u_0, \ldots, u_{n-1}\right)$$

to error 2^{-N-1}. Add α_ℓ to L.

3.18. If $m_2 > 0$:

3.18.1 Pick an arbitrary index $k_1 \in \mathcal{MR}_2$.

3.18.2 Let $\{k_{m_2+1}, \ldots, k_m\} = \{n+1, \ldots, n+m\} \setminus \{k_2\}$

3.18.3 For $\ell = 1, \ldots, m_2$, Compute a rational approximation α_ℓ to

$$(-1)^{m-\ell+1} \operatorname{Re} G_{\ell,m}\left(\lambda_{k_1}, \underbrace{\lambda_{k_1}, \ldots, \lambda_{k_1}}_{m_2-\ell \text{ times}}, \lambda_{k_{m_2+1}}, \ldots, \lambda_{k_m}, \boldsymbol{v}\right)$$

to error 2^{-N-1}. Add α_ℓ to L.

*3.19. If all elements of L are strictly smaller than -2^{-N}: halt and output **false**.*

Lemma 2. *Algorithm 1 is correct and runs in polynomial time.*

Proof. It is relatively easy to see that if Algorithm 1 halts, then it outputs the correct result. A full proof is given in [18, Lemma 17].

Let us now show that Algorithm 1 halts in polynomial time on all robust instances. The roots of χ_c and χ_d can be computed in polynomial time [15, p. 117]. It is relatively easy to see that each iteration of the For-loop takes polynomially many steps. It therefore suffices to show that the number of iterations of the For-loop is bounded polynomially in the negative logarithm of the spectral distance of (c, u, d, v) to the set of boundary instances of Ultimate Inequality.

Assume that the algorithm does not halt within the first N iterations of the For-loop. We will show that there exist polynomially controlled perturbations of (c, u, d, v) such that one of the perturbed instances is a "Yes"-instance and the other is a "No"-instance. Write $f = [\![(c, u)]\!]$ and $g = [\![(d, v)]\!]$.

First consider the case where we have $|\mathcal{M}_1| = 1$ and $|\mathcal{M}_2| = 0$. In this case we have $f(t) - g(t) = F_{1,n}(\lambda_{j_1}, \ldots, \lambda_{j_n}, u_0, \ldots, u_{n-1})e^{\lambda_{j_1}t} + o(e^{\lambda_{j_1}t})$. Since the algorithm does not halt in Step 3.8, $|G_{1,n}(\lambda_{j_1}, \ldots, \lambda_{j_n}, u_0, \ldots, u_{n-1})|$ is less than 2^{-N}.

It follows from the definition of $G_{1,n}$ (Theorem 3) that there exists a perturbation of u_{n-1} by at most 2^{-N} such that $(-1)^{n-1}G_{1,n}(\lambda_{j_1}, \ldots, \lambda_{j_n}, u_0, \ldots, u_{n-1})$ is strictly positive, and another perturbation by at most 2^{-N} such that it is strictly negative. It is easy to see that the sign of $F_{1,n}(\lambda_{j_1}, \ldots, \lambda_{j_n}, u_0, \ldots, u_{n-1})$ is equal to that of $(-1)^{n-1}G_{1,n}(\lambda_{j_1}, \ldots, \lambda_{j_n}, u_0, \ldots, u_{n-1})$. Hence, these perturbations yield a "Yes"-instance and a "No"-instance of Ultimate Inequality. The cases where $|\mathcal{M}_1| = 0$ and $|\mathcal{M}_2| = 1$ or $|\mathcal{M}_1| = 1$ and $|\mathcal{M}_2| = 1$ are treated analogously.

It remains to consider the case where $|\mathcal{M}_1| \geq 2$ or $|\mathcal{M}_2| \geq 2$. Let us assume without loss of generality that $|\mathcal{M}_1| \geq 2$. Let us first construct the "No"-instance. Consider the set of all λ_j with $j \in \mathcal{M}_1$. If this set contains only real numbers, then the ball of radius 2^{-M} about the largest real root of $\chi_c \cdot \chi_d$ contains at least two real roots of χ_c, counted with multiplicity. It follows that there exist perturbations $\widetilde{c}$ and $\widetilde{d}$ of c and d by at most 2^{-M} such that all roots of $\chi_{\widetilde{c}} \cdot \chi_{\widetilde{d}}$ have non-zero imaginary part. We can further ensure that all roots of $\chi_{\widetilde{c}} \cdot \chi_{\widetilde{d}}$ with maximal real part are roots of $\chi_{\widetilde{c}}$. Let λ be a root of $\chi_{\widetilde{c}}$ with maximal real part. Let $P_\lambda \in \mathbb{C}[t]$ denote its coefficient in the exponential polynomial solution of $(\widetilde{c}, u)$. Up to an arbitrarily small perturbation, P_λ is non-zero. This induces an arbitrarily small perturbation of u by (3). It then follows from [4, Theorem 2] or the proof of [5, Theorem 12] that $[\![(\widetilde{c}, \widetilde{u})]\!] - [\![(\widetilde{d}, v)]\!]$ assumes negative values at arbitrarily large times, so that $(\widetilde{c}, \widetilde{u}, \widetilde{d}, v)$ is a "No"-instance of Ultimate Inequality.

Let us now construct the "Yes"-instance. Assume that the set $\mathcal{C}$ computed in Step 3.14 is non-empty. Since the algorithm does not halt in Step 3.14.6, it follows exactly like in the proof of Lemma 1 that there exists a perturbation $(\widetilde{c}, \widetilde{u}, \widetilde{d}, \widetilde{v})$ of (c, u, d, v) by at most

$$\delta := O\left((n + m + ||c||_\infty + ||d||_\infty + ||u||_\infty)^{O((n+m)\log(n+m))} \right) 2^{\frac{-M}{4(n+m)-2}}$$

such that we have $[\![(\widetilde{c}, \widetilde{u})]\!] = \sum_{j=1}^{n} P_j(t)e^{\widetilde{\lambda}_j t}$ and $[\![(\widetilde{d}, \widetilde{v})]\!] = \sum_{j=n+1}^{n+m} P_j(t)e^{\widetilde{\lambda}_j t}$ with $|\operatorname{Re}(\lambda_j) - \operatorname{Re}(\widetilde{\lambda}_j)| < \delta$, $|\operatorname{Im}(\lambda_j) - \operatorname{Im}(\widetilde{\lambda}_j)| < \delta$, and $P_j = 0$ for all $j \in \mathcal{C}$. In particular, the real part of the dominant characteristic roots of $[\![(\widetilde{c}, \widetilde{u})]\!] - [\![(\widetilde{d}, \widetilde{v})]\!]$ is δ-close to the real part of a characteristic root whose imaginary part is at most 2δ. Further, we can ensure that $\gcd(\chi_c, \chi_{\widetilde{c}}) = \left(\prod_{j\in\{1,\dots,n\}\setminus\mathcal{C}} (z - \lambda_j)\right)$ and $\gcd(\chi_d, \chi_{\widetilde{d}}) = \left(\prod_{j\in\{n+1,\dots,n+m\}\setminus\mathcal{C}} (z - \lambda_j)\right)$. If $\mathcal{C}$ is empty, then these properties already hold true for $(\widetilde{c}, \widetilde{u}, \widetilde{d}, \widetilde{v}) = (c, u, d, v)$.

Now, let $c' \in \mathbb{R}^{n'}$ and $d' \in \mathbb{R}^{m'}$ be defined by letting their characteristic polynomials be $\chi_{c'} = \prod_{j\in\{1,\dots,n\}\setminus\mathcal{C}} \left(z - \widetilde{\lambda}_j\right)$ and $\chi_{d'} = \prod_{j\in\{n+1,\dots,n+m\}\setminus\mathcal{C}} \left(z - \widetilde{\lambda}_j\right)$. Define initial values $u' = (\widetilde{u}_0, \dots, \widetilde{u}_{n'-1})$ and $v' = (\widetilde{v}_0, \dots, \widetilde{v}_{m'-1})$. Then we have $[\![(c', u')]\!] = [\![(\widetilde{c}, \widetilde{u})]\!]$ and $[\![(d', v')]\!] = [\![\left(\widetilde{d}, \widetilde{v}\right)]\!]$.

Let N' be the largest integer with $2^{-\max\{\Omega(n,B,N'+1),\Omega(m,B,N'+1)\}} > \delta$. Then, since Ω is a polynomial and $-\log\delta$ depends polynomially on N, n, and B we have $N' \geq \alpha(n+m+B)N^{1/\beta(n+m+B)} - \gamma(n+m+B)$ for polynomials $\alpha, \beta, \gamma \in \mathbb{N}[x]$.

Observe that $\mathcal{MR}_1 \cup \mathcal{MR}_2 \neq \emptyset$. Since the algorithm does not halt, there exists – without loss of generality – an $\ell \in \{1, \dots, m_1\}$ such that

$$(-1)^{n-\ell} \operatorname{Re} G_{\ell,n}\left(\lambda_{j_1}, \underbrace{\lambda_{j_1}, \dots, \lambda_{j_1}}_{m_1-\ell \text{ times}}, \lambda_{j_{m_1+1}}, \dots, \lambda_{j_n}, u_0, \dots, u_{n-1}\right) \geq -2^{-N'}.$$

By construction, there exist $\widetilde{\lambda}_{j_1}, \widetilde{\lambda}_{j_{\ell+1}}, \dots, \widetilde{\lambda}_{j_{m_1}}$, whose real and imaginary parts are 2^{-M}-close to those of λ_{j_k}, such that $\widetilde{\lambda}_{j_1}$ is real and $\operatorname{Re}(\widetilde{\lambda}_{j_1}) > \operatorname{Re}(\nu)$ for all $\nu \in \{\widetilde{\lambda}_{j_{\ell+1}}, \dots, \widetilde{\lambda}_{j_{m_1}}, \lambda_{j_{m_1+1}}, \dots, \lambda_{j_n}\}$. We then have that

$$G_{\ell,n}\left(\widetilde{\lambda}_{j_1}, \widetilde{\lambda}_{j_{\ell+1}}, \dots, \widetilde{\lambda}_{j_{m_1}}, \lambda_{j_{m_1+1}}, \dots, \lambda_{j_{n'}}, \lambda_{j_{n'+1}} \dots, \lambda_{j_n}, \widetilde{u}_0, \dots, \widetilde{u}_{n-1}\right)$$

is greater than or equal to $-2^{-N'+1}$.

Further, up to perturbing the characteristic roots of $\chi_{d'}$ by at most 2^{-M}, we can ensure that $\operatorname{Re}(\widetilde{\lambda}_{j_1}) > \operatorname{Re}(\nu)$ for all characteristic roots ν of $\chi_{d'}$.

Up to relabelling, we may assume that $\mathcal{C} \cap \{1, \dots, n\} = \{j_{n'+1}, \dots, j_n\}$. By Proposition 2 we have that

$G_{\ell,n}\left(\widetilde{\lambda}_{j_1},\widetilde{\lambda}_{j_{\ell+1}},\ldots,\widetilde{\lambda}_{j_{m_1}},\lambda_{j_{m_1+1}},\ldots,\lambda_{j_{n'}},\lambda_{j_{n'+1}}\ldots,\lambda_{j_n},\widetilde{u}_0,\ldots,\widetilde{u}_{n-1}\right)$
is equal to

$$\left(\prod_{k=n'+1}^{n}\left(\lambda_{j_k}-\widetilde{\lambda}_{j_1}\right)\right)G_{\ell,n'}\left(\widetilde{\lambda}_{j_1},\widetilde{\lambda}_{j_{\ell+1}},\ldots,\widetilde{\lambda}_{j_{m_1}},\lambda_{j_{m_1+1}},\ldots,\lambda_{j_{n'}},\widetilde{u}_0,\ldots,\widetilde{u}_{n'-1}\right)$$

so that the number

$$(-1)^{n-\ell}G_{\ell,n'}\left(\widetilde{\lambda}_{j_1},\widetilde{\lambda}_{j_{\ell+1}},\ldots,\widetilde{\lambda}_{j_{m_1}},\lambda_{j_{m_1+1}},\ldots,\lambda_{j_{n'}},\widetilde{u}_0,\ldots,\widetilde{u}_{n'-1}\right)$$

is greater than or equal to $\left(\prod_{k=n'+1}^{n}\left(\lambda_{j_k}-\widetilde{\lambda}_{j_1}\right)\right)^{-1}-2^{-N'+1}$.

Now, $\left(\prod_{k=n'+1}^{n}\left(\lambda_{j_k}-\widetilde{\lambda}_{j_1}\right)\right)$ is a positive real number, since the numbers λ_{j_k} with $k\geq n'+1$ come in complex conjugate pairs. Further, the difference $n-n'$ is even, so that $(-1)^{n-\ell}=(-1)^{n'-\ell}$. Let C be an upper bound on $\left|\lambda_{j_k}-\widetilde{\lambda}_{j_1}\right|$. We obtain that

$$(-1)^{n'-\ell}G_{\ell,n'}\left(\widetilde{\lambda}_{j_1},\widetilde{\lambda}_{j_{\ell+1}},\ldots,\widetilde{\lambda}_{j_{m_1}},\lambda_{j_{m_1+1}},\ldots,\lambda_{j_{n'}},\widetilde{u}_0,\ldots,\widetilde{u}_{n'-1}\right)$$

is greater than or equal to $-C^n2^{-N'+1}$.

By Proposition 2, the left-hand side of this inequality can be written as

$$\widetilde{u}_{n'-1}+\sum_{j=1}^{n'-1}(-1)^{n'+j}A_{n',\ell,j}\left(\widetilde{\lambda}_{j_1},\widetilde{\lambda}_{j_{\ell+1}},\ldots,\widetilde{\lambda}_{j_{m_1}},\lambda_{j_{m_1+1}},\ldots,\lambda_{j_{n'}}\right)\widetilde{u}_{j-1},$$

so that there exists a perturbation $\widetilde{\widetilde{u}}_{n'-1}$ of $\widetilde{u}_{n'-1}$ by at most $C^n2^{-N'+1}$ such that the number

$$(-1)^{n'-\ell}G_{\ell,n'}\left(\widetilde{\lambda}_{j_1},\widetilde{\lambda}_{j_{\ell+1}},\ldots,\widetilde{\lambda}_{j_{m_1}},\lambda_{j_{m_1+1}},\ldots,\lambda_{j_{n'}},\widetilde{u}_0,\ldots,\widetilde{u}_{n'-2},\widetilde{\widetilde{u}}_{n'-1}\right)$$

is strictly positive. Observing that N' is controlled polynomially in the input data and that the logarithm of C is bounded polynomially in the input data, we obtain that the perturbation is controlled polynomially. We have thus constructed polynomially controlled perturbations $(\widetilde{c'},\widetilde{u'})\in C_{n'}(\mathbb{R})$ and $(\widetilde{d'},v')\in C_{m'}(\mathbb{R})$ of (c',u') and (d',v') with $\left[\!\left[\left(\widetilde{c'},\widetilde{u'}\right)\right]\!\right](t)\geq\left[\!\left[\left(\widetilde{d'},v'\right)\right]\!\right](t)$ for all large t. We can construct polynomially controlled perturbations of (c,u) and (d,v) by adding back the characteristic roots we have removed by passing from c to c' and from d to d' and by extending $\widetilde{u'}$ and v' using the linear recurrence equations with characteristic polynomial $\chi_{\widetilde{c'}}$ and $\chi_{\widetilde{d'}}$ respectively. By a calculation similar to the proof of Lemma 1, this yields polynomially controlled perturbations of (c,u) and (d,v).

References

1. Akshay, S., Bazille, H., Genest, B., Vahanwala, M.: On robustness for the skolem and positivity problems. In: Berenbrink, P., Monmege, B. (eds.) 39th International Symposium on Theoretical Aspects of Computer Science, STACS 2022, Marseille, France (Virtual Conference), 15–18 March 2022, LIPIcs, vol. 219, pp. 5:1–5:20. Schloss Dagstuhl - Leibniz-Zentrum für Informatik (2022). https://doi.org/10.4230/LIPIcs.STACS.2022.5
2. Baier, C., et al.: Reachability in dynamical systems with rounding. In: Saxena, N., Simon, S. (eds.) 40th IARCS Annual Conference on Foundations of Software Technology and Theoretical Computer Science, FSTTCS 2020, BITS Pilani, K K Birla Goa Campus, Goa, India (Virtual Conference), 14–18 December 2020, LIPIcs, vol. 182, pp. 36:1–36:17. Schloss Dagstuhl - Leibniz-Zentrum für Informatik (2020). https://doi.org/10.4230/LIPIcs.FSTTCS.2020.36
3. Basu, S., Pollack, R., Roy, M.F.: Algorithms in Real Algebraic Geometry. Springer-Verlag, Heidelberg (2006). https://doi.org/10.1007/3-540-33099-2
4. Bell, J.P., Gerhold, S.: On the positivity set of a linear recurrence. Israel J. Math. **157**, 333–345 (2007). https://doi.org/10.1007/s11856-006-0015-1
5. Bell, P.C., Delvenne, J.C., Jungers, R.M., Blondel, V.D.: The continuous Skolem-Pisot problem. Theor. Comput. Sci. **411**(40), 3625–3634 (2010). https://doi.org/10.1016/j.tcs.2010.06.005
6. Berstel, J., Mignotte, M.: Deux propriétés décidables des suites récurrentes linéaires. Bull. de la Société Mathématique de France **104**, 175–184 (1976). https://doi.org/10.24033/bsmf.1823
7. Brattka, V., Hertling, P. (eds.): Handbook of Computability and Complexity in Analysis. Theory and Applications of Computability, Springer, Cham (2021). https://doi.org/10.1007/978-3-030-59234-9
8. Brattka, V., Presser, G.: Computability on subsets of metric spaces. Theor. Comput. Sci. **305**(1–3), 43–76 (2003)
9. Chonev, V., Ouaknine, J., Worrell, J.: On recurrent reachability for continuous linear dynamical systems. In: Proceedings of the 31st Annual ACM/IEEE Symposium on Logic in Computer Science, LICS 2016, pp. 515–524. Association for Computing Machinery, New York (2016). https://doi.org/10.1145/2933575.2934548
10. Chonev, V., Ouaknine, J., Worrell, J.: On the skolem problem for continuous linear dynamical systems. In: Chatzigiannakis, I., Mitzenmacher, M., Rabani, Y., Sangiorgi, D. (eds.) 43rd International Colloquium on Automata, Languages, and Programming (ICALP 2016). Leibniz International Proceedings in Informatics (LIPIcs), vol. 55, pp. 100:1–100:13. Schloss Dagstuhl-Leibniz-Zentrum fuer Informatik, Dagstuhl (2016). https://doi.org/10.4230/LIPIcs.ICALP.2016.100
11. D'Costa, J., et al.: The pseudo-skolem problem is decidable. In: Bonchi, F., Puglisi, S.J. (eds.) 46th International Symposium on Mathematical Foundations of Computer Science, MFCS 2021, Tallinn, Estonia, 23–27 August 2021, LIPIcs, vol. 202, pp. 34:1–34:21. Schloss Dagstuhl - Leibniz-Zentrum für Informatik (2021). https://doi.org/10.4230/LIPIcs.MFCS.2021.34
12. Everest, G., van der Poorten, A., Shparlinski, I., Ward, T.: Recurrence Sequences. American Mathematical Society (2003)
13. Kauers, M., Pillwein, V.: When can we detect that a P-finite sequence is positive? In: Koepf, W. (ed.) Symbolic and Algebraic Computation, International Symposium, ISSAC 2010, Munich, Germany, 25–28 July 2010, Proceedings, pp. 195–201. ACM (2010)

14. Kawamura, A., Cook, S.: Complexity theory for operators in analysis. ACM Trans. Comput. Theory **4**(2), 5:1–5:24 (2012). https://doi.org/10.1145/2189778.2189780
15. Ko, K.I.: Complexity theory of real functions. In: Progress in Theoretical Computer Science, Birkhäuser Boston Inc., Boston (1991). https://doi.org/10.1007/978-1-4684-6802-1
16. Kunkle, D., Schröder, M.: Some examples of non-metrizable spaces allowing a simple type-2 complexity theory. Electron. Notes Theor. Comput. Sci. **120**, 111–123 (2005). https://doi.org/10.1016/j.entcs.2004.06.038
17. Neumann, E.: Decision problems for linear recurrences involving arbitrary real numbers. Log. Methods Comput. Sci. **17**(3) (2021). https://doi.org/10.46298/lmcs-17(3:16)2021
18. Neumann, E.: On the complexity of robust eventual inequality testing for C-finite functions. preprint https://arxiv.org/abs/2307.00363 (2023)
19. Neumann, E., Steinberg, F.: Parametrised second-order complexity theory with applications to the study of interval computation. Theor. Comput. Sci. **806**, 281–304 (2020). https://doi.org/10.1016/j.tcs.2019.05.009
20. Ouaknine, J., Worrell, J.: On the positivity problem for simple linear recurrence sequences. In: Esparza, J., Fraigniaud, P., Husfeldt, T., Koutsoupias, E. (eds.) ICALP 2014. LNCS, vol. 8573, pp. 318–329. Springer, Heidelberg (2014). https://doi.org/10.1007/978-3-662-43951-7_27
21. Ouaknine, J., Worrell, J.: Positivity problems for low-order linear recurrence sequences. In: Chekuri, C. (ed.) Proceedings of the Twenty-Fifth Annual ACM-SIAM Symposium on Discrete Algorithms, SODA 2014, Portland, Oregon, USA, 5–7 January 2014, pp. 366–379. SIAM (2014). https://doi.org/10.1137/1.9781611973402.27
22. Ouaknine, J., Worrell, J.: Ultimate positivity is decidable for simple linear recurrence sequences. In: Esparza, J., Fraigniaud, P., Husfeldt, T., Koutsoupias, E. (eds.) ICALP 2014. LNCS, vol. 8573, pp. 330–341. Springer, Heidelberg (2014). https://doi.org/10.1007/978-3-662-43951-7_28
23. Ouaknine, J., Worrell, J.: On linear recurrence sequences and loop termination. ACM SIGLOG News **2**(2), 4–13 (2015). https://dl.acm.org/citation.cfm?id=2766191
24. Pillwein, V., Schussler, M.: An efficient procedure deciding positivity for a class of holonomic sequences. ACM Commun. Comput. Algebra **49**(3), 90–93 (2015)
25. Schröder, M.: Spaces allowing type-2 complexity theory revisited. Math. Logic Q. **50**(45), 443–459 (2004). https://doi.org/10.1002/malq.200310111
26. Schönhage, A.: The fundamental theorem of algebra in terms of computational complexity. Preliminary Report, Mathematisches Institut der Universität Tübingen (1982)
27. Tijdeman, R., Mignotte, M., Shorey, T.: The distance between terms of an algebraic recurrence sequence. Journal für die reine und angewandte Mathematik **349**, 63–76 (1984). http://eudml.org/doc/152622
28. Vereshchagin, N.: Occurrence of zero in a linear recursive sequence. Mat. Zametki **38**(2), 177–189 (1985)
29. Weihrauch, K.: Computable Analysis. Springer, Heidelberg (2000). https://doi.org/10.1007/978-3-642-56999-9
30. Ziegler, M., Brattka, V.: Computability in linear algebra. Theor. Comput. Sci. **326**(1–3), 187–211 (2004)

Adaptive Directions for Bernstein-Based Polynomial Set Evolution

Alberto Casagrande[1(✉)] and Carla Piazza[2]

[1] Dip. di Matematica e Geoscienze, Università di Trieste, Trieste, Italy
acasagrande@units.it

[2] Dip. di Matematica, Informatica e Fisica, Università di Udine, Udine, Italy
carla.piazza@uniud.it

Abstract. Dynamical systems are systems in which states evolve according to some laws. Their simple definition hides a powerful tool successfully adopted in many domains from physics to economy and medicine. Many techniques have been proposed so far to study properties, forecast behaviors, and synthesize controllers for dynamical systems, in particular, for the continuous-time case. Recently, methods based on Bernstein polynomials emerged as tools to investigate non-linear evolutions for sets of states in discrete-time dynamical systems. These approaches represent sets as parallelotopes having fixed axis/directions, and, during the evolution, they update the parallelotope boundaries to over-approximate the reached set.

This work suggests a heuristic to identify a new set of axis/directions to reduce over-approximation. The heuristic has been implemented and successfully tested in some examples.

1 Introduction

Dynamical systems are mathematical models in which a function or a set of functions, named *dynamic laws* or *dynamics*, rule the evolution of a state (e.g., see [4,12]). In the last century, they have been successfully used to model and analyze many Natural and artificial phenomena in various domains.

Dynamical systems can be classified as either discrete or continuous time depending on how their time elapses: systems consisting of synchronized components, or having coarse time granularity, represent time in a discrete non-dense domain; Natural time-driven events usually model time as a continuous dimension. To a certain extent, discrete-time dynamical systems can approximate continuous-time ones. Numerical integration methods such as Euler or

This work is partially supported by PRIN project NiRvAna CUP G23C22000400005 and National Recovery and Resilience Plan (NRRP), Mission 4 Component 2 Investment 1.4 - Call for tender No. 3138 of 16 December 202, rectified by Decree n.3175 of 18 December 2021 of Italian Ministry of University and Research funded by the European Union - NextGenerationEU; Project code CN_00000033, Concession Decree No. 1034 of 17 June 2022 adopted by the Italian Ministry of University and Research, CUP G23C22001110007, Project title "National Biodiversity Future Center - NBFC".

O. Bournez et al. (Eds.): RP 2023, LNCS 14235, pp. 113–126, 2023.
https://doi.org/10.1007/978-3-031-45286-4_9

Runge-Kutta are commonly employed to investigate continuous-time models. This approach involves discretization to calculate discrete step dynamics, which are then used to approximate the continuous-time flow pipe (e.g., see [1,2]).

This paper focuses on discrete-time dynamical systems. We are interested in over-approximating the image of polytopes in the case of polynomial dynamics. This problem is at the core of `sapo` engine [5,8,11], which deals with the non-linear dynamics using Bernstein coefficients. Such an approach has proved effective in analyzing complex systems [14].

Even though the Bernstein-based approach allows for the analysis of high-dimensional polynomial systems over a long time horizon, the surge of over-approximation errors along the computation remains the main issue. One could observe the dramatic effect even on a single time step when the dynamical laws are *normal* to the faces of the starting polytope. The methods presented in [5,11] control the computational complexity by preserving the initial polytope directions at each step. In order to mitigate the problem, [11] introduced two different ways of exploiting such *static* directions, which, however, do not solve the issue even in the simplest case of a parallelotope.

This work aims to prove an efficient yet possibly effective way to improve the over-approximation in the above-described framework by changing polytope directions during the computation. First, we generalize the reachability algorithm and parametrize it with respect to the image directions. Then we propose a method for choosing the adaptive directions depending on the dynamics and half of the polytope faces. We also provide two main improvements by keeping the initial static directions and exploiting the remaining half faces. This approach does not affect the asymptotic complexity of the original method. We implemented the method in `sapo` and tested it on two toy, yet critical, examples.

Many approaches have been proposed so far for reachability analysis over non-linear dynamical systems. For instance, Flow* [6] and HyPro [20] deal with non-linear systems by exploiting Taylor model arithmetic techniques. They also handle hybrid systems, but not parametric ones. More detailed comparison can be found in [18] where a methodology that integrates different tools for the design of cyber-physical systems is described. ARCH-COMP (International Workshop on Applied Verification of Continuous and Hybrid Systems - Competition) reports the state of the art in the field, see, e.g., [13,14]. Moreover, we mention here two works that look strictly related to our approach. In [3], the authors consider linear systems and focus on hybrid zonotopes, a data structure for zonotopes that effectively handle unions of 2^n zonotopes. The idea is extended to non-linear systems in [22]. However, the emptiness test on hybrid zonotopes is NP-complete, and the number of constraints increases at each time step. In our case, instead, the emptiness test is polynomial, and the number of constraints in the set representation does not change.

The paper is organized as follows: Sect. 2 introduces the notation and terminology; Sect. 3 deals with the representations of convex polytopes at the basis of our work; Sect. 4 describes a procedure for over-approximating the image of a polytope subject to a discrete-time dynamical system parametric with respect to the directions; Sect. 5 presents our proposal on the use of adaptive directions for improving such over-approximations; we implemented the proposed approach in the tool sapo, and Sect. 6 uses the enhanced version of sapo to provide evidence of the effectiveness of the proposal; finally, Sect. 7 draw some conclusions.

2 Notation and Basics

Let $\mathbf{x} = < x_1, \ldots, x_n >$ and $\mathbf{y} = < y_1, \ldots, y_n >$ be the two vectors in $\mathbb{R}^n$, let $\mathbf{i} = < i_1, \ldots, i_n >$ and $\mathbf{j} = < j_1, \ldots, j_n >$ be two vectors in $\mathbb{N}^n$, and let c be a value in $\mathbb{R}$. We write $\mathbf{x} \leq \mathbf{y}$ to state that $x_k \leq y_k$ for any $k \in [1, n]$, $\mathbf{x}/\mathbf{y}$ to denote the vector $< x_1/y_1, \ldots, x_n/y_n >$, and $c{*}\mathbf{x}$ in place of $< c * x_1, \ldots c * x_n >$. Moreover, $\binom{\mathbf{i}}{\mathbf{j}} \stackrel{\text{def}}{=} \prod_{k=1}^{n} \binom{i_k}{j_k}$ and $\mathbf{x}^{\mathbf{i}}$ stands for $\prod_{k=1}^{n} x_k^{i_k}$. If $\mathbf{A}$ is a matrix, we write $\mathbf{A}_i$ to denote the i-th row of $\mathbf{A}$. We may write $< \mathbf{x_1}, \ldots, \mathbf{x_m} >$ to represent the $m \times n$-matrix whose rows are the n-dimensional vectors $\mathbf{x_1}, \ldots, \mathbf{x_m}$.

If $f : \mathbb{R}^n \to \mathbb{R}^m$ is a function and $S \subseteq \mathbb{R}^n$, as it is standard in the literature, we write $f(S)$ to denote the image of S through f, i.e., $f(S) \stackrel{\text{def}}{=} \{f(\mathbf{v} \mid \mathbf{v} \in S\}$.

A *term* is the multiplication of some variables (potentially, none or in multiple copies) and one single constant, e.g., $-3 * x * y * x$, 5, and $0 * x$ are terms, while $0 * x + 5$ and $-3 * x * 4$ are not. Any term involving a subset of the variables $x_1, \ldots, x_n$ can be expressed as $c{*}\mathbf{x}^{\mathbf{i}}$ where c is a constant value, $\mathbf{x}$ is the variable vector $< x_1, \ldots, x_n >$, and $\mathbf{i}$ is an opportune vector $< i_1, \ldots, i_n >$ of natural values.

Example 1. Let us consider the variable vector $\mathbf{x} = < x, y, z >$. The term $7x^2z$ can be expressed as $7 * \mathbf{x}^{\mathbf{i}}$ where $\mathbf{i} = < 2, 0, 1 >$.

Two functions $f_1(x_1, \ldots, x_n)$ and $f_2(x_1, \ldots, x_n)$ are *equivalent* when their values are the same for every interpretation of the variables $x_1, \ldots, x_n$, i.e., $f_1(x_1, \ldots, x_n) = f_2(x_1, \ldots, x_n)$ for every value of $x_1, \ldots, x_n$.

Polynomials are expressions exclusively built by variables, constants, addictions, and multiplications. A polynomial is in *normal form* if it has the form $\sum_{\mathbf{i} \in I} a_{\mathbf{i}} * \mathbf{x}^{\mathbf{i}}$ where I is a finite set of natural power vectors and the $a_{\mathbf{i}}$s are constant values depending on $\mathbf{i}$. Every polynomial has an equivalent polynomial in normal form.

Bernstein basis of degree $\mathbf{d} \in \mathbb{N}^n$ are basis for the space of polynomials over $\mathbb{R}^n$ having degree at most $\mathbf{d}$. For $\mathbf{i} = < i_1, \ldots, i_n > \in \mathbb{N}^n$, the $\mathbf{i}$-th Bernstein basis of degree $\mathbf{d} = < d_1, \ldots, d_n >$ is defined as: $\mathcal{B}_{\mathbf{d},\mathbf{i}}(< x_1, \ldots, x_n >) \stackrel{\text{def}}{=} \prod_{k=1}^{n} \beta_{d_k, i_k}(x_k)$ where $\beta_{d,i}(x) \stackrel{\text{def}}{=} \binom{d}{i} x^i (1 - x)^{d-i}$.

Any polynomial $\pi(\mathbf{x}) = \sum_{\mathbf{i} \in I} a_{\mathbf{i}} \mathbf{x}^{\mathbf{i}}$ can be represented by using Bernstein basis as: $\pi(\mathbf{x}) = \sum_{\mathbf{i} \leq \mathbf{d}} b_{\mathbf{d},\mathbf{i}} \mathcal{B}_{\mathbf{d},\mathbf{i}}(\mathbf{x})$ where $b_{\mathbf{d},\mathbf{i}} = \sum_{\mathbf{j} \leq \mathbf{i}} \frac{\binom{\mathbf{i}}{\mathbf{j}}}{\binom{\mathbf{d}}{\mathbf{j}}} a_{\mathbf{j}}$ are the *Bernstein coefficients*. See [5] for examples.

We are interested in Bernstein polynomial representation and, in particular, in Bernstein coefficients because of the following theorem.

Theorem 1 ([21]). *Let $\pi(\mathbf{x})$ be a polynomial having degree $\mathbf{d} \in \mathbb{R}^n$ and let $b_{\mathbf{d},\mathbf{i}}$ be the Bernstein coefficients of $\pi(\mathbf{x})$. For any $\mathbf{v} \in [0,1]^n$, it holds that*

$$\min_{\mathbf{i} \leq \mathbf{d}} \{b_{\mathbf{d},\mathbf{i}}\} \leq \pi(\mathbf{v}) \leq \max_{\mathbf{i} \leq \mathbf{d}} \{b_{\mathbf{d},\mathbf{i}}\}$$

Theorem 1 provides upper and lower bounds for the image of the hypercube $[0,1]^n$ through $\pi(\mathbf{x})$ in terms of $\pi(\mathbf{x})$'s Bernstein coefficients.

2.1 Dynamical Systems

A *dynamical system* is a system equipped with a *state* that evolves in time according to a function called *dynamic law.* The state denotes the system condition and changes according to the dynamic laws as time elapses. The space of all the system states is called *state space.*

Depending on the nature of the underlying time structure, we distinguish *continuous-time dynamical systems,* in which time elapses continuously, from *discrete-time dynamical systems* whose state changes by successive applications of the dynamic law called *epochs.* This work focuses on discrete-time dynamical systems and leaves continuous-time dynamical systems for future works.

Definition 1 (Discrete-Time Dynamical System). *A* discrete-time dynamical system *is a tuple $D = (X, f)$ where:*
- *$X \subseteq \mathbb{R}^n$ is the state space;*
- *$f : X \to X$ is the dynamic law.*

A *trajectory* of a discrete-time dynamical system $D = (X, f)$ is a succession of values $\xi_0, \ldots, \xi_i, \ldots$, being either finite or infinite, such that $\xi_i \in \mathbb{X}$ and $\xi_{i+1} = f(\xi_i)$ for any ξ_i and ξ_{i+1} in the succession. A *trajectory of D from p* is a trajectory $\xi_0, \ldots, \xi_i, \ldots$ such that $\xi_0 = p$.

A state r *is reachable from p in n epochs by D* when there exists a trajectory $\xi_0, \ldots, \xi_n$ for D from p and $\xi_n = r$. In this case, we may equivalently say that *D reaches r from p in n epochs.*

Example 2. Let us consider the discrete-time dynamical system $D = (X, f)$ where $X = \{< x, y > \in \mathbb{R}^2 \mid y \leq 50 \wedge x > 0\}$ and $f(< x, y >) = < x^2, y + x >$. The trajectory of D from $< 2, 0 >$ is $< 2, 0 >, < 4, 2 >, < 16, 6 >$. Thus, $< 2, 0 >$, $< 4, 2 >$, and $< 16, 6 >$ are reachable from $< 2, 0 >$ by D.

3 Representing Sets

Dynamical systems provide an elegant way to model natural phenomena, and reachability analysis is a powerful tool to reason about them automatically. However, it is generally unlikely to precisely identify one single system state, for instance, because of Heisenberg's uncertainty principle or due to measurement errors. It is, instead, relatively common to deal with approximate values, which can be represented as sets, or sets themselves.

Convex polytopes are intersections of half-spaces in $\mathbb{R}^n$ that can effectively approximate convex bounded sets.

Definition 2. *A* hyperplane *in* $\mathbb{R}^n$ *is a subset of* $\mathbb{R}^n$ *having the form* $h = \{\mathbf{v} \in \mathbb{R}^n \,|\, \mathbf{d}^T \cdot \mathbf{v} = c\}$ *where* $c \in \mathbb{R}$ *and* $\mathbf{d} \in \mathbb{R}^n$ *is a non-null vector.*

A half-space *of* $\mathbb{R}^n$ *is a set having the form* $H = \{\mathbf{v} \in \mathbb{R}^n \,|\, \mathbf{d}^T \cdot \mathbf{v} \leq c\}$ *where* $c \in \mathbb{R}$ *and* $\mathbf{d} \in \mathbb{R}^n$ *is a non-null vector, known as* direction *of* h*. Two half-spaces* $H_1 = \{\mathbf{v} \in \mathbb{R}^n \,|\, \mathbf{d_1}^T \cdot \mathbf{v} \leq c_1\}$ *and* $H_2 = \{\mathbf{v} \in \mathbb{R}^n \,|\, \mathbf{d_2}^T \cdot \mathbf{v} \leq c_2\}$ *are* opposite *when* $\mathbf{d_2} = -\mathbf{d_1}$*.*

A convex polytope P *is a finite intersection of half-spaces, i.e., there exist a vector* $\mathbf{c} \in \mathbb{R}^m$ *and a matrix* $\mathbf{D} \in \mathbb{R}^{n \times m}$*, known as* direction matrix*, such that* $P = \{\mathbf{v} \in \mathbb{R}^n \,|\, \mathbf{D} \cdot \mathbf{v} \leq \mathbf{c}\}$*.*

Let $P = \{\mathbf{v} \in \mathbb{R}^n \,|\, \mathbf{D} \cdot \mathbf{v} \leq < c_1, \ldots c_m >\}$ be a convex polytope and let $\mathbf{D}_i$ be the i-th row of the direction matrix $\mathbf{D}$, i.e., $\mathbf{D}_i$ is the direction of the i-th half-space. A *face* of P is a subset of P having the form $F_i = \{\mathbf{v} \in P \,|\, \mathbf{D}_i \cdot \mathbf{v} = c_i\}$ for some row i in $\mathbf{D}$. A *vertex* of P is the only value in a non-null intersection of n distinct faces of P, and an *edge* of P is the non-null intersection of $n-1$ distinct faces of P.

Parallelotopes in $\mathbb{R}^n$ are bounded convex polytopes consisting in the intersection of $2n$ pairwise opposite half-spaces and such that their direction matrices have rank n. A *bundle* is an intersection of parallelotopes. Bundles are polytopes, and any polytope can be represented as a bundle [5].

Beyond the standard representation for polytopes, any parallelotope $P = \{\mathbf{v} \in \mathbb{R}^n \,|\, \mathbf{D} \cdot \mathbf{v} \leq < c_1, \ldots c_m >\}$ can be represented in two other ways: the *min-max representation* and the *generator representation.*

In the min-max representation, each pair of opposite half-spaces $h_i = \{\mathbf{v} \in R^n \,|\, \mathbf{D}_i \cdot \mathbf{v} \leq c_i\}$ and $h_j = \{\mathbf{v} \in R^n \,|\, -\mathbf{D}_i \cdot \mathbf{v} \leq c_j\}$ is denoted by one single direction $\mathbf{D}_i$ and the associated upper and lower bounds, c_i and $-c_j$, respectively. Since P consists in the intersection of $2n$ pairwise opposite half-spaces, all the pairs of opposite half-spaces correspond to a matrix $\mathbf{M} \in \mathbb{R}^{n \times n}$ of directions and two vectors $\mathbf{l} = < l_1, \ldots, l_n >, \mathbf{u} = < u_1, \ldots, u_n > \in \mathbb{R}^n$: the lower and upper bound vectors, respectively. Thus, the set $\{\mathbf{v} \in \mathbb{R}^n \,|\, \mathbf{l} \leq \mathbf{M} \cdot \mathbf{v} \leq \mathbf{u}\}$ denotes the parallelotope P too.

The standard polytope representation of P can be obtained from the min-max representation as:

$$P = \left\{ \mathbf{v} \in \mathbb{R}^n \,|\, \begin{pmatrix} \mathbf{M} \\ -\mathbf{M} \end{pmatrix} \cdot \mathbf{v} \leq \begin{pmatrix} \mathbf{u} \\ -\mathbf{l} \end{pmatrix} \right\}. \tag{1}$$

The min-max representation allows computing the vertices of a parallelotope in $\mathbb{R}^n$. A vector $\mathbf{v}$ is a vertex if and only if $\mathbf{M} \cdot \mathbf{v} = \mathbf{w}$ where $\mathbf{w} = < w_1, \ldots, w_n >$ and $w_i \in \{l_i, u_i\}$. As a matter of the facts, if $\mathbf{M} \cdot \mathbf{v} = \mathbf{w}$, $\mathbf{w} = < w_1, \ldots, w_n >$, and $w_i \in \{l_i, u_i\}$, then $\mathbf{v}$ belongs to exactly n different faces of P, i.e., $\{\mathbf{z} \in P \,|\, \mathbf{M}_i \cdot \mathbf{z} = w_i\}$ for $i \in [1, n]$, and it is a vertex. If, instead, $w_j \notin \{l_j, u_j\}$ for some $j \in [1, n]$, neither $\{\mathbf{z} \in P \,|\, \mathbf{M}_j \cdot \mathbf{z} = l_j\}$ nor $\{\mathbf{z} \in P \,|\, \mathbf{M}_j \cdot \mathbf{z} = u_j\}$ contains $\mathbf{v}$, $\mathbf{v}$ itself belongs to $n-1$ distinct faces at most, and it is not a vertex. Regarding vertex computation, $\mathbf{M}$ is invertible because it is full rank by definition. Thus, all the parallelotope vertices have the form $\mathbf{M}^{-1} \cdot \mathbf{w}$ where $\mathbf{w} = < w_1, \ldots, w_n >$ and $w_i \in \{l_i, u_i\}$. Since we can build 2^n vector having the form $\mathbf{w} = < w_1, \ldots, w_n >$

with $w_i \in \{l_i, u_i\}$ at most (it may be the case that $l_i = u_i$ for some i), any parallelotope in $\mathbb{R}^n$ can have 2^n vertices edges at most.

Intriguingly, E is a parallelotope edge if and only if it has the form $E = \left\{\mathbf{M}^{-1} \cdot < w_1, \ldots, w_n > | \, w_j \in [l_j, u_j] \wedge \bigwedge_{i \neq j} w_i \in \{l_i, u_i\}\right\}$. Hence, every vertex of a n-dimensional parallelotope belongs to n edges at most.

The generator representation characterizes a parallelotope in $\mathbb{R}^n$ by using one *base vertex*, $\mathbf{b}$, and a set, $\mathcal{G} = \{\mathbf{g}_1, \ldots, \mathbf{g}_n\}$, of n linearly independent vectors, called *generators*. In this case, any parallelotope P is denoted by a set having the form $P = \{\mathbf{b} + \sum_{i=1}^{n} c_i \mathbf{g}_i \, | \, c_i \in [0, 1]\}$.

Let us notice that the above parallelotope generator representation is similar to the zonotope generator representation in [15]. However, the latter depicts zonotopes through their center, whereas the former focuses on one of the set vertices. While moving from one representation to the other is not tricky, later on, we will clarify the advantages of using the former in place of the latter in the context investigated by this work.

Switching from min-max representation to generator representation can be done in three steps: first, evaluate $\mathbf{M}^{-1}$; second, choose a base-vertex, e.g., we can set $\mathbf{b} = \mathbf{M}^{-1} \cdot \mathbf{l}$ by convention; third and final step, compute the generators as the differences between the vertices on the edges containing $\mathbf{b}$ and $\mathbf{b}$ itself, i.e., $\mathbf{g}_i = \mathbf{M}^{-1} \cdot \mathbf{w}_i - \mathbf{b}$, where $\mathbf{w}_i = < l_1, \ldots, l_{i-1}, u_i, l_{i+1}, \ldots, l_n >$. Hence, $\mathbf{g}_i = \mathbf{M}^{-1} \cdot (\mathbf{w}_i - \mathbf{l}) = \mathbf{M}^{-1} \cdot ((u_i - l_i)\mathbf{e}_i)$ where $\mathbf{e}_i$ is the i-th vector in the canonical base of $\mathbb{R}^n$, i.e., $\mathbf{e}_i[i] = 1$ and $\mathbf{e}_i[j] = 0$ for all $j \neq i$.

On the other hand, to get the min-max representation of a parallelotope P from the generator representation, we need to compute the direction matrix $\mathbf{M}$: $\mathbf{l}$ and $\mathbf{u}$ can, then, be obtained as $\mathbf{M} \cdot \mathbf{b}$ and $\mathbf{M} \cdot (\mathbf{b} + \sum_i g_i)$, respectively. A possible way to compute $\mathbf{M}$ is to notice that any direction $\mathbf{M}_i$ is normal to the face $F_i = \{\mathbf{v} \in P \, | \, \mathbf{M}_i \cdot \mathbf{v} = u_i\}$. Since F_i contains the vertices $\mathbf{b}$ and $\mathbf{b} + \mathbf{g}_j$ with $j \neq i$, we can evaluate a normal $\mathbf{n}_i$ to h_i as a vector that is normal to all the generators in $\mathcal{G}$, but $\mathbf{g}_i$. This can be achieved be setting $\mathbf{n}_i = < a_1, \ldots, a_n >$ where $\sum_i a_i x_i$ is the determinant of the matrix $<< x_1, \ldots x_n >, \mathbf{g}_1, \ldots, \mathbf{g}_{i-1}, \mathbf{g}_{i+1}, \ldots, \mathbf{g}_n >$. The direction $\mathbf{M}_i$ will be equal to $\mathbf{M}_i = \frac{\mathbf{u}_i}{\mathbf{n}_i \cdot \mathbf{b}} \mathbf{n}_i$.

4 Over-Approximating Parallelotope Images

A reachability algorithm to over-approximate the image of a parallelotope by using Bernstein coefficients was presented in [11]. In this section, we generalize it to be parametric with respect to the image directions.

Let $\pi_i : \mathbb{R}^n \rightarrow \mathbb{R}$ be a polynomial function for all $i \in [1, n]$ and let $\mathbf{M}^*$ be a $n \times n$-matrix having rank n. It follows that $\mu_i(\mathbf{x}) \stackrel{\text{def}}{=} \mathbf{M}_i^* \cdot \pi(\mathbf{x})$, where $\pi(\mathbf{x}) \stackrel{\text{def}}{=} < \pi_1(\mathbf{x}), \ldots, \pi_n(\mathbf{x}) >$, is polynomial too. If $\mathbf{v} \in [0, 1]^n$ and c_i and C_i are the minimum and the maximum among all the Bernstein coefficients of $\mu_i(\mathbf{x})$, respectively, then $c_i \leq \mu_i(\mathbf{v}) \leq C_i$ by Theorem 1 and the parallelotope

$$\{\mathbf{v} \in \mathbb{R}^n \, | < c_1, \ldots, c_n > \leq \mathbf{M}^* \cdot \mathbf{v} \leq < C_1, \ldots, C_n >\} \tag{2}$$

over-approximate the image of $[0,1]^n$ through $\pi(\mathbf{x})$, i.e., $\pi([0,1]^n)$.

This approach can be extended to compute the image of any generic parallelotope P by using its generator representation. If $< b_1, \ldots, b_n >$ and $\mathbf{g}_1, \ldots, \mathbf{g}_n$ are the base vertex and the generators of P, respectively, then, $\eta(< x_1, \ldots, x_n >) \stackrel{\text{def}}{=} \mathbf{b} + \sum_{i=1}^{n} x_i \mathbf{g}_i$ is a polynomial function mapping the hypercube $[0,1]^n$ into the parallelotope P. Thus, $\varrho_i(\mathbf{x}) \stackrel{\text{def}}{=} (\mu_i \circ \eta)(\mathbf{x})$ is a polynomial function and, analogously to the hypercube case, the parallelotope

$$\Pi_{\pi,\mathbf{M}^*}(P) \stackrel{\text{def}}{=} \{\mathbf{v} \in \mathbb{R}^n \,|\, < c_1^*, \ldots, c_n^* > \leq \mathbf{M}^* \cdot \mathbf{v} \leq < C_1^*, \ldots, C_n* >\}, \quad (3)$$

where c_i^* and C_i^* are the minimum and maximum among all the Bernstein coefficients of $\varrho_i(\mathbf{x})$, respectively, is a superset for the π image of P, i.e., $\Pi_{\pi,\mathbf{M}^*}(P) \subseteq \pi(P)$. Algorithm 1 details the described procedure.

Algorithm 1. Polynomial image of a parallelotope

Require: A parallelotope $P = \{\mathbf{v} \in \mathbb{R}^n \,|\, \mathbf{l} \leq \mathbf{M} \cdot \mathbf{v} \leq \mathbf{u}\}$, a polynomial function $\pi(\mathbf{x}) : \mathbb{R}^n \to \mathbb{R}^n$, and a full-rank $n \times n$-matrix $\mathbf{M}$.
Ensure: $\mathbf{l}^*, \mathbf{u}^* \in \mathbb{R}^n$ such that $\{\mathbf{v} \in \mathbb{R}^n \,|\, \mathbf{l}^* \leq \mathbf{M}^* \cdot \mathbf{v} \leq \mathbf{u}^*\} \supseteq \{\pi(\mathbf{v}) \,|\, \mathbf{v} \in P\}$

```
procedure POLYNOMIAL_IMAGE(P, π(x), M*)
    b, < g_1, ..., g_n > ← COMPUTE_GENERATOR_REPRESENTATION(P)
    ρ(x) ← π (b + Σ_{i=1}^n g_i · x)
    ALLOCATE_VECTORS({l*, u*}, n)
    for i ← 1 ... n do
        ϱ_i(x) ← M*_i · ρ(x)
        B_i ← COMPUTE_BERNSTEIN_COEFFICIENTS(ϱ_i(x))
        l*[i], u*[i] ← FIND_MIN(B_i), FIND_MAX(B_i)
    end for
    return l*, u*
end procedure
```

Since bundles are parallelotope intersections and polytopes can be exactly represented as bundles, Algorithm 1 can also be used to over-approximate the image of any polytope through polynomial functions by observing that $\pi\left(\bigcap_i P_i\right) \subseteq \bigcap_i \Pi_{\pi,\mathbf{M}}(P_i)$. Moreover, if P is defined by the min-max representation $\{\mathbf{v} \in \mathbb{R}^n \,|\, \mathbf{l} \leq \mathbf{M} \cdot \mathbf{v} \leq \mathbf{u}\}$, then $\mathbf{M}$ is a full-rank matrix and the set $\Pi_{\pi,\mathbf{M}}(P)$, whose min-max representation is characterized by $\mathbf{M}$ once more, can be computed as detailed above. By iterating the parallelotope image computation, we can over-approximate the set reachable from a polytope by a discrete-time polynomial dynamical system in a finite number of epochs.

4.1 Directions and Approximation

Algorithm 1 has effectively over-approximated the reachable set of many discrete-time polynomial dynamical systems (e.g., see [5,7,9,11]). The approximation accuracy depends on the directions of the original parallelotope, on the polynomial function $\pi(\mathbf{x})$, and on the directions of the reached set. All these parameters contribute to $\varrho_i(\mathbf{x})$'s definition, influence their Bernstein coefficients, and, consequently, determine the bounds of the final approximation. However, while the initial set of states and the function $\pi(\mathbf{x})$ directly depend on the specific instance of the considered reachability problem, the direction matrix $\mathbf{M}^*$ is arbitrary and may significantly impact the approximation. The following example shows a linear dynamical system whose reachability computation is severely affected when we choose $\mathbf{M}^*$ once for all before the evaluation.

Example 3. Let $D_1 = (\mathbb{R}^2, f_1)$ be a discrete-time dynamical system where

$$f_1(\mathbf{x}) \stackrel{\text{def}}{=} \begin{pmatrix} \cos\frac{\pi}{4} & -\sin\frac{\pi}{4} \\ \sin\frac{\pi}{4} & \cos\frac{\pi}{4} \end{pmatrix} \cdot \mathbf{x} = \frac{1}{\sqrt{2}} \begin{pmatrix} 1 & -1 \\ 1 & 1 \end{pmatrix} \cdot \mathbf{x},$$

i.e., $f_1(\mathbf{x})$ rotates $\mathbf{x}$ of an angle $\frac{\pi}{4}$ around the axis origin $< 0, 0 >$. Moreover, let P_1 be the parallelotope $P_1 \stackrel{\text{def}}{=} \{\mathbf{v} \in \mathbb{R}^2 \,|\, \mathbf{l} \leq \mathbf{M}\cdot\mathbf{v} \leq \mathbf{u}\}$ where $\mathbf{l} = < -0.5, -0.5 >$, $\mathbf{u} = < 0.5, 0.5 >$, and $\mathbf{M}_1 = < 1, 0 >$ and $\mathbf{M}_2 = < 0, 1 >$. Namely, P_1 is the square $[-0.5, 0.5]^2$.

The parallelotope $\Pi_{f_1,\mathbf{M}}(P_1)$ is a superset for $f_1(P_1)$ and, thus, it includes the $\frac{\pi}{4}$ rotation around $< 0, 0 >$ of the square $[-0.5, 0.5]^2$ which is the square whose vertices are $< 0, -1/\sqrt{2} >$, $< 1/\sqrt{2}, 0 >$, $< 0, 1/\sqrt{2} >$, and $< -1/\sqrt{2}, 0 >$ (see Fig. 1a). However, $\Pi_{f_1,\mathbf{M}}(P_1)$ and P_1 have the same directions, thus, $\Pi_{f_1,\mathbf{M}}(P_1)$ must include $[-1/\sqrt{2}, 1/\sqrt{2}]^2$ whose area is twice the area of $f_1(P_1)$ (actually, $\Pi_{f_1,\mathbf{M}}(P_1)$ is exactly $[-1/\sqrt{2}, 1/\sqrt{2}]^2$). The over-approximation escalates as we iterate the application of f and the computed over-approximation of the set $f_1^k(P_1)$ has an area 2^k time greater than that of $f_1^k(P_1)$ itself. On the other hand, $\Pi_{f_1,\mathbf{M}^*}(P_1)$, where $\mathbf{M}_1^* = < 1, 1 >^T$ and $\mathbf{M}_2^* = < -1, 1 >^T$, equals $f_1(P_1)$ and $f_1(f_1(P_1)) = P_1 = \Pi_{f_1,\mathbf{M}}(\Pi_{f_1,\mathbf{M}^*}(P_1))$. Thus, up to the arithmetic approximations, the set reachable from P_1 can be exactly approximated.

5 Adaptive Directions

Example 3 highlights the importance of selecting the appropriate directions at each image computation. Ideally, we would like to identify the matrix $\mathbf{M}^*$ that minimizes the volume of $\Pi_{\pi,\mathbf{M}^*}(P)$. However, this goal appears to be too ambitious, in particular, when $\pi(\mathbf{x})$ is non-linear.

This work proposes a heuristic for computing from $\pi(\mathbf{x})$ and $P = \{\mathbf{v} \in \mathbb{R}^n \,|\, \mathbf{l} \leq \mathbf{M}\cdot\mathbf{v} \leq \mathbf{u}\}$ a matrix $\mathbf{M}^*$ that *may* minimize the volume of $\Pi_{\pi,\mathbf{M}^*}(P)$. We suggest evaluating a linear approximation of $\pi(\mathbf{x})$ in a neighborhood of P and gauging its effects on the directions of P itself to get $\mathbf{M}^*$.

Let $\tilde{\pi}(\mathbf{x}) \stackrel{\text{def}}{=} \mathbf{T}\cdot\mathbf{x}+\mathbf{d}$ be the aimed approximation of $\pi(\mathbf{x})$. Thus, $\pi(\mathbf{q})-\pi(\mathbf{p}) \approx \tilde{\pi}(\mathbf{q}) - \tilde{\pi}(\mathbf{p}) = \mathbf{T}\cdot(\mathbf{q}-\mathbf{p})$ for any distinct points $\mathbf{p}, \mathbf{q} \in P$. If $\mathbf{b}$ is the base vertex and $\mathbf{g}_1, \ldots, \mathbf{g}_n$ are the generators of P, $\mathbf{b}$, $\mathbf{b}+\mathbf{g}_1$, ..., $\mathbf{b}+\mathbf{g}_n$ are vertices of P, they are pairwise different, and the following equation holds

$$\begin{pmatrix}\pi(\mathbf{b}+\mathbf{g}_1)-\pi(\mathbf{b})\\ \vdots \\ \pi(\mathbf{b}+\mathbf{g}_n)-\pi(\mathbf{b})\end{pmatrix}^T \approx \mathbf{T}\cdot\begin{pmatrix}(\mathbf{b}+\mathbf{g}_1)-\mathbf{b}\\ \vdots \\ (\mathbf{b}+\mathbf{g}_n)-\mathbf{b}\end{pmatrix}^T = \mathbf{T}\cdot\begin{pmatrix}\mathbf{g}_1\\ \vdots \\ \mathbf{g}_n\end{pmatrix}^T. \tag{4}$$

Let $\mathbf{M}_{\mathbf{b},\pi}$ be $< \pi(\mathbf{b}+\mathbf{g}_1)-\pi(\mathbf{b}), \ldots, \pi(\mathbf{b}+\mathbf{g}_n)-\pi(\mathbf{b}) >^T$. Since the generators are linearly independent, $< \mathbf{g}_1, \ldots, \mathbf{g}_n >^T$ is invertible, and $\mathbf{T}$ and $\mathbf{d}$ can be evaluated as $\mathbf{T} \approx \mathbf{M}_{\mathbf{b},\pi} \cdot \left(\mathbf{g}_1 \ldots \mathbf{g}_n\right)^{-1}$ and $\mathbf{d} \approx \pi(\mathbf{b}) - \mathbf{T}\cdot\mathbf{b}$, respectively.

By construction, $\tilde{\pi}(\mathbf{x}) = \pi(\mathbf{x})$ for any $\mathbf{x} \in \{\mathbf{b}, \mathbf{b}+\mathbf{g}_1, \ldots, \mathbf{b}+\mathbf{g}_n\}$. If $\pi(\mathbf{x})$ is linear, then $\tilde{\pi}(\mathbf{x})$ and $\pi(\mathbf{x})$ are the same function, but, in the general case, there may exist points $\mathbf{x}'$ in P such that $\tilde{\pi}(\mathbf{x}') \neq \pi(\mathbf{x}')$. However, $\tilde{\pi}(\mathbf{x})$ still approximates $\pi(\mathbf{x})$ in a neighborhood of the points $\mathbf{b}$, $\mathbf{b}+\mathbf{g}_1$, ..., $\mathbf{b}+\mathbf{g}_n$.

Each direction of P is normal to some of its faces F. When $\tilde{\pi}(F)$ is good approximation for $\pi(F)$, we use a normal vector to $\tilde{\pi}(F)$ as direction for the parallelotope over-approximating $\pi(P)$. Every face F_i is such that $F_i = \{\mathbf{v} \in P \,|\, \mathbf{M_i}\cdot\mathbf{v} = c_i\}$ for some direction $\mathbf{M_i}$ and bound c_i. If $\mathbf{M}_{\mathbf{b},\pi}$ is invertible, then $\mathbf{T}$, which equals $\mathbf{M}_{\mathbf{b},\pi}\cdot\left(\mathbf{g}_1 \cdots \mathbf{g}_n\right)^{-1}$, is invertible too, and $\mathbf{M_i}\cdot\mathbf{v} = \mathbf{M_i}\cdot\mathbf{T}^{-1}\cdot\mathbf{T}\cdot\mathbf{v} = c_i$. The vector $\mathbf{T}\cdot\mathbf{v}$ is a generic point in F_i transformed by $\mathbf{T}$ and, thus, $\mathbf{M_i}\cdot\mathbf{T}^{-1}$ is normal to both $\tilde{\pi}(F_i) - \mathbf{d}$ and $\tilde{\pi}(F_i)$. This last property holds for any direction $\mathbf{M_i}$ of P and we can define $\mathbf{M}^*$ as

$$\mathbf{M}^* = \begin{cases}\mathbf{M}\cdot\mathbf{T}^{-1} = \mathbf{M}\cdot\left(\mathbf{g}_1 \ldots \mathbf{g}_n\right)\cdot\mathbf{M}_{\mathbf{b},\pi}^{-1} & \text{if } \mathbf{M}_{\mathbf{b},\pi} \text{ is invertible}\\ \mathbf{M} & \text{if } \mathbf{M}_{\mathbf{b},\pi} \text{ is } \textit{not} \text{ invertible}\end{cases} \tag{5}$$

Figure 1b shows how adaptive direction impacts over-approximation accuracy. We want to remark that Eq. 5 is only used to compute new directions, while the image over-approximation is performed in the standard way by Algorithm 1.

Our technique is similar to the approaches used to update the directions in sets of states evolving according to linear dynamics (e.g., see [15–17,19]). Linear dynamics map parallelotopes, zonotopes, and ellipsoids into parallelotopes, zonotopes, and ellipsoids. Thus, computing the image of one of these sets requires the evaluation of the new generators/axis, which can be achieved by applying the linear function to the original generators/axis. When instead the dynamics are non-linear, the images of linear sets, such as parallelotopes, zonotopes, and ellipsoids, may be non-linear, and non-fully symbolic techniques can exclusively aim to over-approximate (or under-approximate) them. In the considered framework, the direction update is not mandatory, and we suggest it exclusively aiming for a tighter approximation.

Moreover, the approach considered in this work deals with two different representations of the same set: the min-max representation and the generator representation. The approximation function $\tilde{\pi}(\mathbf{x})$, i.e., $\mathbf{T}\cdot\mathbf{x}+\mathbf{d}$, depends on the dynamics and the generator representation. Then, we compute $\mathbf{M}^*$ considering $\mathbf{T}$ and the min-max representation of the same set. This two-step evaluation involving different representations of the same set is novel up to our knowledge.

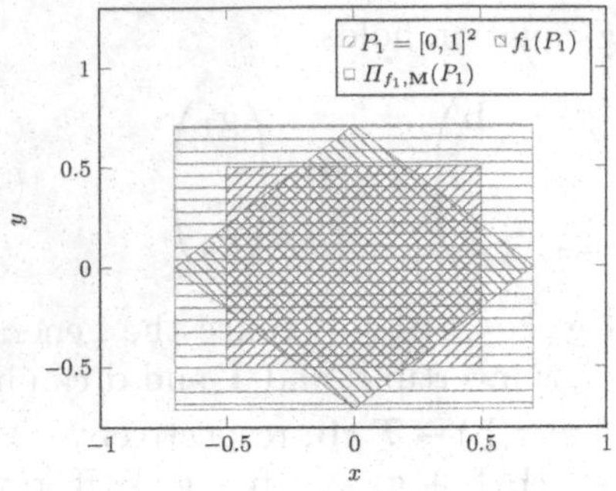

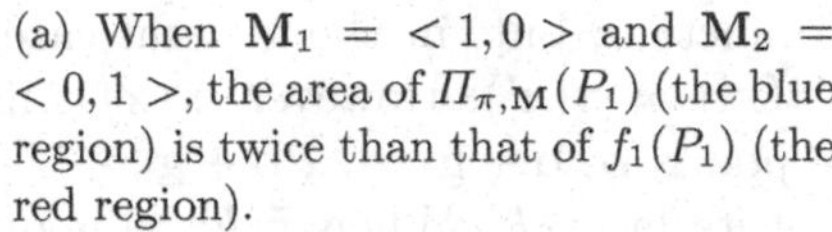

(a) When $\mathbf{M}_1 = <1, 0>$ and $\mathbf{M}_2 = <0, 1>$, the area of $\Pi_{\pi,\mathbf{M}}(P_1)$ (the blue region) is twice than that of $f_1(P_1)$ (the red region).

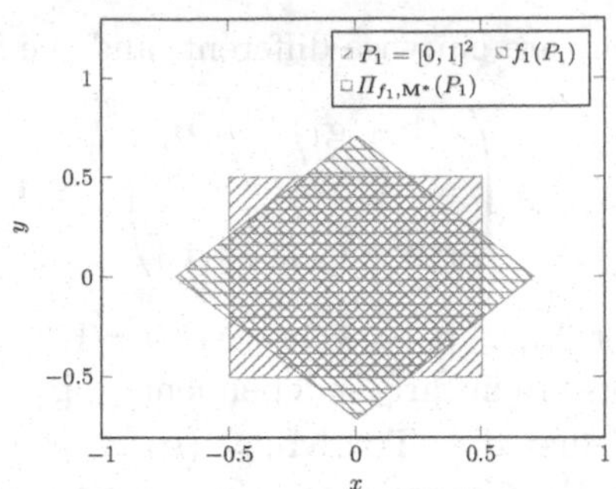

(b) When $\mathbf{M}^*$ is computed as in Eq. 5, $\mathbf{M}_1^* = <1, 1>$ and $\mathbf{M}_2^* = <1, -1>$ and $\Pi_{f_1,\mathbf{M}^*}(P_1)$ (the blue region) equals $f_1(P_1)$ (the red region).

Fig. 1. A representation of Example 3. The function f_1, that rotates the space around $<0, 0>$ of an angle $\frac{\pi}{4}$, is applied to the set $S = [0, 1]^2$ (the green region). (Color figure online)

We need to stress two relevant aspects of our proposal: first of all, there are no guarantees for $\tilde{\pi}(\mathbf{x})$ to accurately approximate the image of the faces in P through $\pi(\mathbf{x})$ unless $\pi(\mathbf{x})$ itself is linear. Because of this, we do not know whether $\Pi_{\pi,\mathbf{M}^*}$ is tighter or larger than $\Pi_{\pi,\mathbf{M}}$ when $\pi(\mathbf{x})$ is non-linear.

In the second place, we only use $n + 1$ point images to estimate $\tilde{\pi}(\mathbf{x})$: the images of $\mathbf{b}, \mathbf{b} + \mathbf{g}_1, \ldots, \mathbf{b} + \mathbf{g}_n$. These points belong to n of the $2n$ faces in P: those containing $\mathbf{b}$. Thus, even if $\tilde{\pi}(\mathbf{x})$ accurately approximates $\pi(\mathbf{x})$ on these faces, it does not consider the other n parallelotope faces.

Example 4. Let P_2 be the parallelotope $P_2 = \{\mathbf{v} \in \mathbb{R}^2 \mid \mathbf{l} \leq \mathbf{M} \cdot \mathbf{v} \leq \mathbf{u}\}$ where $\mathbf{l} \stackrel{\text{def}}{=} <-0.5, 0>$, $\mathbf{u} \stackrel{\text{def}}{=} <-0.5, 1>$, and $\mathbf{M}$ is the 2×2 identity matrix. i.e., P_2 is the square $[-0.5, 0.5] \times [0, 1]$. Moreover, let D_2 be the dynamical system $D_2 \stackrel{\text{def}}{=} (\mathbb{R}^2, f_2)$ where $f_2(<x_1, x_2>) = <x_1(1 + x_2), x_2>$. It is easy to see that $\mathbf{b} = \mathbf{l}$, $\mathbf{g}_1 = <1, 0>$, and $\mathbf{g}_2 = <0, 1>$. Thus, $f_2(\mathbf{b}) = <-0.5, 0>$, $f_2(\mathbf{b} + \mathbf{g}_1) = <0.5, 0>$, $f_2(\mathbf{b} + \mathbf{g}_2) = <-1, 1>$, and $\mathbf{M}_{\mathbf{b},f_2} = \left(1\ -0.5\ ;\ 0\ 1\right)$. It follows that $\mathbf{M}^* = \mathbf{M} \cdot (\mathbf{g}_1\ \mathbf{g}_2) \cdot \mathbf{M}_{\mathbf{b},f_2}^{-1} = \mathbf{I} \cdot \mathbf{I} \cdot \mathbf{M}_{\mathbf{b},f_2}^{-1} = <<1, 0.5>, <0, 1>>$, where $\mathbf{I}$ is the 2×2 identity matrix, and $\Pi_{f_2,\mathbf{M}^*}(R) = \{\mathbf{v} \in \mathbb{R}^2 \mid \mathbf{l}^* \leq \mathbf{M}^* \cdot \mathbf{v} \leq \mathbf{u}^*\}$ where $\mathbf{l}^* = <-0.5, 0>$ and $\mathbf{u}^* = <1.5, 1>$.

5.1 Avoiding Coarser Approximations

The concurrent use of both adaptive and non-adaptive directions can prevent the approximation from soaring when non-favorable directions are selected.

Let P_0 and P_0' be two alternative versions of the same set P such that the over-approximations of the reachable sets from P_0 are always represented by using the same direction matrix $\mathbf{M}$, while the direction matrices of over-approximation of the reachable sets from P'_0 changes at each evolution step. Since P_0 and P'_0 represent the same set, the two parallelotopes $P_1 = \Pi_{\pi,\mathbf{M}}(P_0)$ and $P'_1 = \Pi_{\pi,\mathbf{M}^*_0}(P'_0)$, where $\mathbf{M}^*_0$ is the direction matrix evaluated as in Eq. 5, include $\pi(P_0)$. As observed in Sect. 4, $\pi(P_0 \cap P'_0)$ can be over-approximated by the bundle $P_1 \cap P'_1$ and this bundle is a subset of P_1. If P_{i+1} and P'_{i+1} are the parallelotopes $\Pi_{\pi,\mathbf{M}}(P_i)$ and $\Pi_{\pi,\mathbf{M}^*_i}(P'_i)$, respectively, for any $i \in [0, k-1]$, then $\pi^k(P_0 \cap P'_0) \subseteq P_k \cap P'_k$ and $P_k \cap P'_k \subseteq P_k$. Thus, we can use bundle computation and parallelotope-dependent adaptive directions to avoid the possible approximation worsening due to the adaptive directions.

This approach impacts the computation time because it requires doubling the number of parallelotopes in each bundle, but it does not affect the time complexity of the procedure in terms of set operations, and the evolution of each set in the bundle can be independently computed in parallel. We should, however, notice that the complexity of some of the set operations (e.g., the emptiness test) polynomially depends on the number of directions in the bundle.

5.2 Fitting All the Parallelotope Faces

As highlighted by Example 4, even if $\tilde{\pi}(\mathbf{x})$ correctly approximates π on the faces containing the base vertex $\mathbf{b}$, it may diverge from π on the remaining faces which contain the opposite vertex $\mathbf{b}+\sum_{i=1}^{n} \mathbf{g_i}$. If $\{\mathbf{v} \in \mathbb{R}^n \mid \mathbf{l} \leq \mathbf{M}\cdot\mathbf{v} \leq \mathbf{u}\}$ is a min-max representation for the considered parallelotope P and $\mathbf{b}$ is conventionally set to be $\mathbf{M}^{-1} \cdot \mathbf{l}$ as suggested in Sect. 3, then

$$\mathbf{b} + \sum_{i=1}^{n} \mathbf{g_i} = \mathbf{M}^{-1} \cdot \left(\mathbf{l} + \sum_{i=1}^{n} (u_i - l_i)\mathbf{e_i} \right) = \mathbf{M}^{-1} \cdot (\mathbf{l} + (\mathbf{u} - \mathbf{l})) = \mathbf{M}^{-1} \cdot \mathbf{u}. \quad (6)$$

Since $\mathbf{l} \leq \mathbf{M} \cdot \mathbf{v} \leq \mathbf{u}$ holds if and only if $-\mathbf{u} \leq -\mathbf{M} \cdot \mathbf{v} \leq -\mathbf{l}$ holds too, P can also be represented as $\{\mathbf{v} \in \mathbb{R}^n \mid \overline{\mathbf{l}} \leq \overline{\mathbf{M}} \cdot \mathbf{v} \leq \overline{\mathbf{u}}\}$ where $\overline{\mathbf{l}} \stackrel{\text{def}}{=} -\mathbf{u}$, $\overline{\mathbf{u}} \stackrel{\text{def}}{=} -\mathbf{l}$, and $\overline{\mathbf{M}} \stackrel{\text{def}}{=} -\mathbf{M}$. This last min-max representation for P is the *opposite representation* with respect to the original one, as $\overline{\mathbf{M}}$ and $\mathbf{M}$ contain opposite directions. Following the established convention for base vertices, the base vertex $\overline{\mathbf{b}}$ of the min-max representation based on $\overline{\mathbf{l}}$, $\overline{\mathbf{u}}$, and $\overline{\mathbf{M}}$ is $\overline{\mathbf{b}} = \overline{\mathbf{M}}^{-1} \cdot \overline{\mathbf{l}} = -\mathbf{M}^{-1} \cdot -\mathbf{u} = \mathbf{M}^{-1} \cdot \mathbf{u}$ which is $\mathbf{b} + \sum_{i=1}^{n} \mathbf{g_i}$ by Eq. 6. Hence, from the two opposite min-max representations of P, we can build two alternative generator representations whose base vertices are $\mathbf{M}^{-1} \cdot \mathbf{l}$ and its opposite vertex $\mathbf{M}^{-1} \cdot \mathbf{u}$, respectively. Each of the two representations can approximate the effects of π on half of the faces of P: one on the faces containing $\mathbf{M}^{-1} \cdot \mathbf{l}$ and the other on the remaining faces, i.e., those containing the opposite vertex $\mathbf{M}^{-1} \cdot \mathbf{u}$.

The effectiveness of this proposal will be evaluated in Sect. 6 on an example similar to Example 4 (see Fig. 2).

The systematic application of the presented parallelotope double representation leads to a reachability algorithm exponential in the number of epochs. However, when, for each face F_i, the effects of $\pi(\mathbf{x})$ on F_i can be properly approximated by a linear function $\tilde{\pi}_i(\mathbf{x})$ (such as in the case presented by Example 4), there is no need to double the number of parallelotopes at each evolution step, and the approximation can be improved simply by including opposite representations of the same parallelotope in the initial bundle.

As in the previous case, this approach affects the execution time by doubling the number of parallelotopes, but does not change the asymptotic complexity of the reachability algorithm in terms of set operations.

Theorem 2. *Let $D = (\mathbb{R}^n, f)$ be a discrete-time dynamical system, let P be the parallelotope $\{\mathbf{v} \in \mathbb{R}^n \mid \mathbf{l} \leq \mathbf{M} \cdot \mathbf{v} \leq \mathbf{u}\}$, and let $\overline{P}$ be its opposite representation.*

The bundle $\mathcal{B} = \{\Pi_{f,\mathbf{M}}(P), \Pi_{f,\mathbf{M}^*}(P), \Pi_{f,\overline{\mathbf{M}}^*}(P)\}$, where $\mathbf{M}^*$ and $\overline{\mathbf{M}}^*$ are computed as in Eq. 5 from P and $\overline{P}$, over-approximates $f(P)$ not worse than $\Pi_{f,\mathbf{M}}(P)$. Moreover, both $\mathcal{B}$ and $\Pi_{f,\mathbf{M}}(P)$ can be computed with the same asymptotic computational complexity by using Algorithm 1 and Eq. 5.

6 Examples

We coded the techniques described in Sect. 5 in sapo, a software tool for analyzing polynomial dynamics based on Bernstein coefficients [5, 10], and we studied a few relevant examples. This section reports the analysis results and the execution time on a MacBook Pro M1 2020 with 16 GB of RAM.

First of all, we considered the dynamical system $D_1 = (\mathbb{R}^2, f_1)$ and the parallelotope P_1 defined in Example 3, and we computed the over-approximations of the set $f_1^{100}(P_1)$ reachable by D_1 from P_1 in 100 epochs using static and adaptive directions. In the former case, the approximation diverges as time elapses and sapo over-approximates $f_1^{100}(P_1)$ by $[-5.63\text{e}{+}14, 5.63\text{e}{+}14]^2$. On the contrary, using adaptive directions, sapo returns a quadrilateral R whose vertices are about $< 0.5, 0.5 >$, $< 0.5, -0.5 >$, $< -0.5, -0.5 >$, and $< -0.5, 0.5 >$. The two sets, P_1 and R, are equivalent up to approximation errors and this is the result expected by the theoretical analysis of the system. There are no appreciable differences in execution time; the analysis is instantaneous using both approaches. The approximation surge in static directions analysis avoided comparing the two strategies on a longer time horizon.

We also used sapo to over-approximate the set reachable from the square $P_2 = [-0.5, 0.5] \times [0, 1]$ by $D_3 = (\mathbb{R}^2, f_3)$, where $f_3 = < x_1(1 + 0.001x_2), x_2 >$, in exactly 1000 epochs. Figure 2 depicts the set $f_3^{1000}(P_2)$ and the over-approximations obtained using static, adaptive, and adaptive directions combined with opposite parallelotopes in the initial bundle. The static directions approach produces a rectangle loosely representing $f_3^{1000}(P_2)$ in 0.13 s. Adaptive directions alone do not help in improving the approximation, and the area of the obtained set does not change despite the computation time increasing to 0.18 s. When, instead, the original bundle consists of two opposite parallelotopes and the computation uses adaptive directions, sapo returns a much tighter approximation of the reached set in 0.51 s.

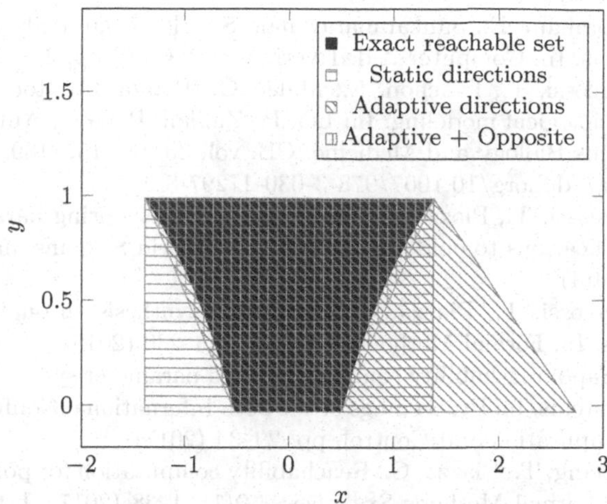

Fig. 2. Over-approximations of the set reachable by $(\mathbb{R}^2, < x(1 + 0.001y), y >)$ in 1000 epochs from P_2 (see Example 4) as computed by sapo. Both static and plain adaptive directions do not approximate the reachability set adequately. A tighter approximation can be obtained by including two opposite parallelotopes in the original bundle.

7 Conclusions

We presented a method for dealing with adaptive directions in the context of reachability computation over polynomial dynamical systems. The proposal aims to improve the over-approximation of the image of a polytope at the same asymptotic complexity of the algorithms presented in [8,11].

The method has already been implemented in sapo and has shown its effectiveness on critical examples.

Since sapo also deals with parametric systems and synthesis problems [5,9] we plan to analyze the parametric case in the future.

References

1. Althoff, M., Frehse, G., Girard, A.: Set propagation techniques for reachability analysis. Ann. Rev. Control Rob. Auton. Syst. **4**, 369–395 (2021)
2. Bak, S., Duggirala, P.S.: Simulation-equivalent reachability of large linear systems with inputs. In: Majumdar, R., Kunčak, V. (eds.) CAV 2017. LNCS, vol. 10426, pp. 401–420. Springer, Cham (2017). https://doi.org/10.1007/978-3-319-63387-9_20
3. Bird, T.J., Pangborn, H.C., Jain, N., Koeln, J.P.: Hybrid zonotopes: a new set representation for reachability analysis of mixed logical dynamical systems. Automatica **154**, 111107 (2023)
4. Brin, M., Stuck, G.: Introduction to Dynamical Systems. Cambridge University Press, Cambridge (2002)
5. Casagrande, A., Dang, T., Dorigo, L., Dreossi, T., Piazza, C., Pippia, E.: Parameter synthesis of polynomial dynamical systems. Inf. Comput. **289**, 104941 (2022)

6. Chen, X., Ábrahám, E., Sankaranarayanan, S.: Flow*: an analyzer for non-linear hybrid systems. In: Computer Aided Verification, CAV, pp. 258–263 (2013)
7. Dang, T., Dreossi, T., Fanchon, E., Maler, O., Piazza, C., Rocca, A.: Set-based analysis for biological modeling. In: Liò, P., Zuliani, P. (eds.) Automated Reasoning for Systems Biology and Medicine. CB, vol. 30, pp. 157–189. Springer, Cham (2019). https://doi.org/10.1007/978-3-030-17297-8_6
8. Dang, T., Dreossi, T., Piazza, C.: Parameter synthesis using parallelotopic enclosure and applications to epidemic models. In: Hybrid Systems and Biology, HSB, pp. 67–82 (2014)
9. Dang, T., Dreossi, T., Piazza, C.: Parameter synthesis through temporal logic specifications. In: Formal Methods, FM, pp. 213–230 (2015)
10. Dreossi, T.: Sapo: reachability computation and parameter synthesis of polynomial dynamical systems. In: Proceedings of the 20th International Conference on Hybrid Systems: Computation and Control, pp. 29–34 (2017)
11. Dreossi, T., Dang, T., Piazza, C.: Reachability computation for polynomial dynamical systems. Formal Methods Syst. Des. **50**(1), 1–38 (2017). https://doi.org/10.1007/s10703-016-0266-3
12. Galor, O.: Discrete Dynamical Systems. Springer, Heidelberg (2007)
13. Geretti, L., et al.: ARCH-COMP20 category report: continuous and hybrid systems with nonlinear dynamics. In: Frehse, G., Althoff, M. (eds.) ARCH20. 7th International Workshop on Applied Verification of Continuous and Hybrid Systems (ARCH20), vol. 74 of EPiC Series in Computing, pp. 49–75. EasyChair (2020)
14. Geretti, L., et al.: Arch-comp21 category report: continuous and hybrid systems with nonlinear dynamics. In: Frehse, G., Althoff, M. (eds.) 8th International Workshop on Applied Verification of Continuous and Hybrid Systems (ARCH21), vol. 80 of EPiC Series in Computing, pp. 32–54. EasyChair (2021)
15. Girard, A.: Reachability of uncertain linear systems using zonotopes. In: Morari, M., Thiele, L. (eds.) HSCC 2005. LNCS, vol. 3414, pp. 291–305. Springer, Heidelberg (2005). https://doi.org/10.1007/978-3-540-31954-2_19
16. Kurzhanski, A.B., Varaiya, P.: Ellipsoidal techniques for reachability analysis. In Hybrid Systems: Computation and Control, HSCC, pp. 202–214 (2000)
17. Le Guernic, C.: Reachability analysis of hybrid systems with linear continuous dynamics. PhD thesis, Université Joseph-Fourier-Grenoble I (2009)
18. Nuzzo, P., Sangiovanni-Vincentelli, A.L., Bresolin, D., Geretti, L., Villa, T.: A platform-based design methodology with contracts and related tools for the design of cyber-physical systems. Proc. IEEE **103**(11), 2104–2132 (2015)
19. Sassi, M.A.B., Testylier, R., Dang, T., Girard, A.: Reachability analysis of polynomial systems using linear programming relaxations. In Automated Technology for Verification and Analysis, ATVA, pp. 137–151 (2012)
20. Schupp, S., Ábrahám, E., Makhlouf, I.B., Kowalewski, S.: HyPro: a C++ library of state set representations for hybrid systems reachability Analysis. In: Barrett, C., Davies, M., Kahsai, T. (eds.) NFM 2017. LNCS, vol. 10227, pp. 288–294. Springer, Cham (2017). https://doi.org/10.1007/978-3-319-57288-8_20
21. Shisha, O.: The Bernstein form of a polynomial. J. Res. Natl. Bureau Stand. Math. Math. Phys. B **70**, 79 (1966)
22. Siefert, J.A., Bird, T.J., Koeln, J.P., Jain, N., Pangborn, H.C.: Successor sets of discrete-time nonlinear systems using hybrid zonotopes (2023)

Introducing Divergence for Infinite Probabilistic Models

Alain Finkel[1,2], Serge Haddad[1], and Lina Ye[1,3](✉)

[1] Université Paris-Saclay, CNRS, ENS Paris-Saclay, Laboratoire Méthodes Formelles, Gif-sur-Yvette, France
{alain.finkel,serge.haddad,lina.ye}@ens-paris-saclay.fr
[2] Institut Universitaire de France (IUF), Paris, France
[3] CentraleSupélec, Gif-sur-Yvette, France

Abstract. Computing the reachability probability in infinite state probabilistic models has been the topic of numerous works. Here we introduce a new property called *divergence* that when satisfied allows to compute reachability probabilities up to an arbitrary precision. One of the main interest of divergence is that our algorithm does not require the reachability problem to be decidable. Then we study the decidability of divergence for probabilistic versions of pushdown automata and Petri nets where the weights associated with transitions may also depend on the current state. This should be contrasted with most of the existing works that assume weights independent of the state. Such an extended framework is motivated by the modeling of real case studies. Moreover, we exhibit some divergent subclasses of channel systems and pushdown automata, particularly suited for specifying open distributed systems and networks prone to performance collapsing in order to compute the probabilities related to service requirements.

Keywords: Reachability probability · Infinite state probabilistic models · Divergence

1 Introduction

Probabilistic Models. In the 1980's, finite-state Markov chains have been considered for the modeling and analysis of probabilistic concurrent finite-state programs [19]. Since the 2000's, many works have been done to verify the infinite-state Markov chains obtained from probabilistic versions of automata extended with unbounded data (like stacks, channels, counters and clocks)[1]. The (qualitative and quantitative) model checking of *probabilistic pushdown automata* (pPDA) is studied in many papers, for example in [6,10–12,17] (see [5] for a

This work has been supported by ANR project BRAVAS (ANR-17-CE40-0028) with Alain Finkel and ANR project MAVeriQ (ANR-20-CE25-0012) with Serge Haddad.

[1] Surprisingly, in 1972, to the best of our knowledge, Santos gave the first definition of *probabilistic pushdown automata* [18] that did not open up a new field of research at the time.

O. Bournez et al. (Eds.): RP 2023, LNCS 14235, pp. 127–140, 2023.
https://doi.org/10.1007/978-3-031-45286-4_10

survey). In 1997, Iyer and Narasimha [15] started the study of *probabilistic lossy channel systems* (pLCS) and later both some qualitative and quantitative properties were shown decidable for pLCS [1]. *Probabilistic counter machines* (pCM) have also been studied [7–9].

Computing the Reachability Probability. In finite Markov chains, there is a well-known algorithm for computing exactly the reachability probabilities in polynomial time [3]. Here we focus on the problem of *Computing the Reachability Probability up to an arbitrary precision* (CRP) in *infinite* Markov chains. There are (at least) two possible research directions:

The first one is to consider the Markov chains associated with a particular class of probabilistic models (like pPDA or probabilistic Petri nets (pPN)) and some specific target sets and to exploit the properties of these models to design a CRP-algorithm. For instance in [5], the authors exhibit a PSPACE algorithm for pPDA and PTIME algorithms for single-state pPDA and for one-counter automata.

The second one consists in exhibiting a property of Markov chains that yields a generic algorithm for solving the CRP problem and then looking for models that generate Markov chains that fulfill this property. *Decisiveness* of Markov chains is such a property. Intuitively, decisiveness w.r.t. s_0 and A means that almost surely the random path σ starting from s_0 will reach A or some state s' from which A is unreachable. It has been shown that pPDA are not (in general) decisive but both pLCS and probabilistic Petri nets (pPN) are decisive (for pPN: when the target set is upward-closed [2]).

Two Limits of the Previous Approaches. The generic approach based on the decisiveness property has numerous applications but suffers the restriction that the reachability problem must be decidable in the corresponding non deterministic model. To the best of our knowledge, all generic approaches rely on a *decidable reachability problem.*

In most of the works, the probabilistic models associate a *constant* weight for transitions and get transition probabilities by normalizing these weights among the enabled transitions in the current state. This *forbids to model phenomena* like congestion in networks (resp. performance collapsing in distributed systems) when the number of messages (resp. processes) exceeds some threshold leading to an increasing probability of message arrivals (resp. process creations) before message departures (resp. process terminations).

Our Contributions

- In order to handle realistic phenomena (like congestion in networks), we consider *dynamic* weights i.e., weights depending on the current state.
- We introduce the new *divergence* property of Markov chains w.r.t. s_0 and A: given some precision θ, one can discard a set of states with either a small probability to be reached from s_0 or a small probability to reach A such that the remaining subset of states is finite and thus allows for an approximate computation of the reachability probability up to θ. For divergent Markov chains, we

provide a generic algorithm for the CRP-problem that *does not require* the decidability of the reachability problem. While decisiveness and divergence are not exclusive (both hold for finite Markov chains), they are complementary. In fact, divergence is somehow related to transience of Markov chains while decisiveness is somehow related to recurrence [13].
- In order to check divergence, we provide several simpler sufficient conditions based on existing and new results of martingale theory.
- We study for different models the decidability of divergence. Our first undecidability result implies that whatever the infinite models, one must restrict the kind of dynamics weights. Here we limit to polynomial weights, i.e. where a weight is defined by a polynomial whose variables are characteristics of the current state (e.g. the marking of a place in a Petri net).
- We prove, by a case study analysis, that divergence is decidable for a subclass of polynomial pPDA (i.e. pPDA with *polynomial* weights). We show that divergence is undecidable for polynomial pPNs w.r.t. an upward closed set.
- We provide two classes of divergent polynomial models. The first one is a probabilistic version of channel systems particularly suited for the modeling of open queuing networks. The second one is the probabilistic version of pushdown automata restricted to some typical behaviors of dynamic systems.

Organisation. Section. 2 recalls Markov chains, introduces divergent Markov chains, presents an algorithm for solving the CRP-problem. In Sect. 3, we study the decidability status of divergence for pPDA and pPN. Finally Sect. 4 presents two divergent subclasses of probabilistic channel systems and pPDA. All missing proofs and a second CRP-algorithm when reachability is decidable can be found in [14].

2 Divergence of Markov Chains

2.1 Markov Chains: Definitions and Properties

Notations. A set S is *countable* if there exists an injective function from S to the set of natural numbers: hence it could be finite or countably infinite. Let S be a countable set of elements called states. Then $Dist(S) = \{\Delta : S \to \mathbb{R}_{\geq 0} \mid \sum_{s \in S} \Delta(s) = 1\}$ is the set of *distributions* over S. Let $\Delta \in Dist(S)$, then the *support* of Δ is defined by $Supp(\Delta) = \Delta^{-1}(\mathbb{R}_{>0})$. Let $T \subseteq S$, then $S \setminus T$ will also be denoted $\overline{T}$.

Definition 1 (Effective Markov chain). *A* Markov chain $\mathcal{M} = (S, p)$ *is a tuple where:*

- *S is a countable set of states,*
- *p is the transition function from S to $Dist(S)$;*

When for all $s \in S$, $Supp(p(s))$ is finite and computable and the function p is computable, one says that $\mathcal{M}$ is effective.

Notations. The function p may be viewed as a $S \times S$ matrix defined by $p(s, s') = p(s)(s')$. Let $p^{(d)}$ denote the dth power of the transition matrix p. When S is countably infinite, we say that $\mathcal{M}$ is *infinite* and we sometimes identify S with $\mathbb{N}$. We also denote $p(s, s') > 0$ by $s \xrightarrow{p(s,s')} s'$. A Markov chain is also viewed as a transition system whose transition relation $\rightarrow$ is defined by $s \rightarrow s'$ if $p(s, s') > 0$. Let $A \subseteq S$, one denotes $Post^*_{\mathcal{M}}(A)$, the set of states that can be reached from some state of A and $Pre^*_{\mathcal{M}}(A)$, the set of states that can reach A. As usual, we denote $\rightarrow^*$, the transitive closure of $\rightarrow$ and we say that s' is *reachable from* s if $s \rightarrow^* s'$. We say that a subset $A \subseteq S$ is *reachable* from s if some $s' \in A$ is reachable from s. Note that every finite path of $\mathcal{M}$ can be extended into (at least) one infinite path.

Example 1. Let $\mathcal{M}_1$ be the Markov chain of Fig. 1. In any state $i > 0$, the probability for going to the "right", $p(i, i+1)$, is equal to $0 < p_i < 1$ and for going to the "left" $p(i, i-1)$ is equal to $1 - p_i$. In state 0, one goes to 1 with probability 1. $\mathcal{M}_1$ is effective if the function $n \mapsto p_n$ is computable.

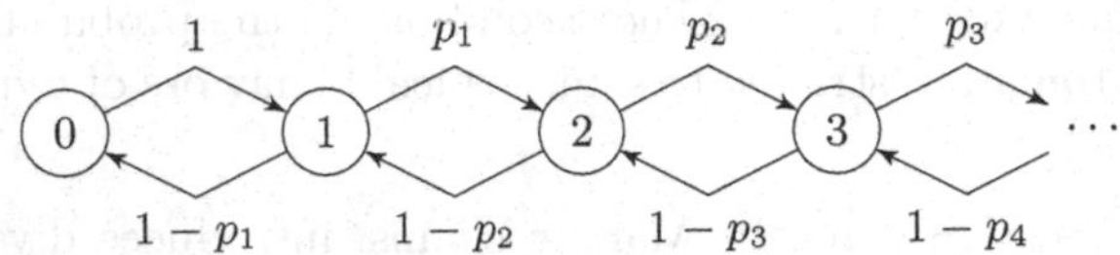

Fig. 1. A random walk $\mathcal{M}_1$

Given an initial state s_0, the *sampling* of a Markov chain $\mathcal{M}$ is an *infinite random sequence of states* (i.e., a path) $\sigma = s_0 s_1 \dots$ such that for all $i \geq 0$, $s_i \rightarrow s_{i+1}$. As usual, the corresponding σ-algebra whose items are called events is generated by the finite prefixes of infinite paths and the probability of an event Ev given an initial state s_0 is denoted $\mathbf{Pr}_{\mathcal{M},s_0}(Ev)$. In case of a finite path $s_0 \dots s_n$, $\mathbf{Pr}_{\mathcal{M},s_0}(s_0 \dots s_n) = \prod_{0 \leq i < n} p(s_i, s_{i+1})$.

Notations. From now on, $\mathbf{G}$ (resp. $\mathbf{F}$, $\mathbf{X}$) denotes the always (resp. eventual, next) operator of LTL.

Let $A \subseteq S$. We say that σ *reaches* A if $\exists i \in \mathbb{N}\ s_i \in A$ and that σ *visits* A if $\exists i > 0\ s_i \in A$. The probability that starting from s_0, the path σ reaches (resp. visits) A will be denoted by $\mathbf{Pr}_{\mathcal{M},s_0}(\mathbf{F}A)$ (resp. $\mathbf{Pr}_{\mathcal{M},s_0}(\mathbf{XF}A)$).

We now state qualitative and quantitative properties of a Markov chain.

Definition 2 (Irreducibility, recurrence, transience). *Let $\mathcal{M} = (S, p)$ be a Markov chain and $s \in S$. Then $\mathcal{M}$ is* irreducible *if for all $s, s' \in S$, $s \rightarrow^* s'$. s is* recurrent *if $\mathbf{Pr}_{\mathcal{M},s}(\mathbf{XF}\{s\}) = 1$ otherwise s is* transient.

In an irreducible Markov chain, all states are in the same category, either recurrent or transient [16]. Thus an irreducible Markov chain will be said transient or recurrent depending on the category of its states. In the remainder of this section, we will relate this category with techniques for computing reachability probabilities.

Example 2. Clearly $\mathcal{M}_1$ is irreducible. Moreover (see [14]), $\mathcal{M}_1$ is recurrent if and only if $\sum_{n\in\mathbb{N}}\prod_{1\leq m<n}\rho_m = \infty$ with $\rho_m = \frac{1-p_m}{p_m}$, and when transient, the probability that starting from i the random path visits 0 is equal to $\frac{\sum_{i\leq n}\prod_{1\leq m<n}\rho_m}{\sum_{n\in\mathbb{N}}\prod_{1\leq m<n}\rho_m}$.

One of our goals is to approximately compute reachability probabilities in infinite Markov chains. Let us formalize it. Given a finite representation of a subset $A \subseteq S$, one says that this representation is *effective* if one can decide the membership problem for A. With a slight abuse of language, we identify A with any effective representation of A. The Computing of Reachability Probability (CRP) problem is defined by:

- Input: an effective Markov chain $\mathcal{M}$, an (initial) state s_0, an effective subset of states A, and a rational number $\theta > 0$.
- Output: an interval $[low, up]$ such that $up - low \leq \theta$ and $\mathbf{Pr}_{\mathcal{M},s_0}(\mathbf{F}A) \in [low, up]$.

2.2 Divergent Markov Chains

Let us first discuss two examples before introducing the notion of *divergent* Markov chains.

Example 3. Consider again the Markov chain $\mathcal{M}_1$ of Fig. 1 with for all $n > 0$, $p_n = p > \frac{1}{2}$. In this case, for $m \geq 0$, $\mathbf{Pr}_{\mathcal{M}_1,m}(\mathbf{F}\{0\}) = \rho^m$ with $\rho = \frac{1-p}{p}$. Thus here the key point is that not only this reachability probability is less than 1 but it goes to 0 when m goes to ∞. This means that given some precision θ, one could "prune" states $n \geq n_0$ and compute the reachability probabilities of A in a finite Markov chain.
Consider the Markov chain of Fig. 2, where $\mathbf{Pr}_{\mathcal{M},0}(\mathbf{F}\{m, m+1, \ldots\}) = \prod_{n<m} p_n$ goes to 0 when m goes to ∞. As in the precedent example, computing the reachability probabilities of A can be also done in a finite Markov chain after pruning states $n \geq n_0$, given some precision θ.

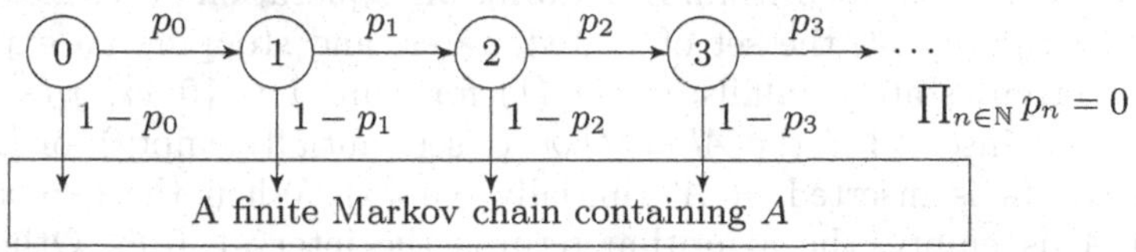

Fig. 2. An infinite (divergent) Markov chain

Intuitively, a divergent Markov chain w.r.t. s_0 and A generalizes these examples: given some precision θ, one can discard a set of states with either a small probability to be reached from s_0 ($f_0^{-1}([0,\theta])$ in the next definition) or a small probability to reach A (from any state of the set $f_1^{-1}([0,\theta])$ in the next definition), such that the remaining subset of states is finite and thus allows for an approximate computation of the reachability probability up to θ.

Definition 3 (divergent Markov chain). *Let $\mathcal{M}$ be a Markov chain, $s_0 \in S$ and $A \subseteq S$. We say that $\mathcal{M}$ is* divergent *w.r.t. s_0 and A if there exist two computable functions f_0 and f_1 from S to $\mathbb{R}_{\geq 0}$ such that:*

- *For all $0 < \theta < 1$, $\mathbf{Pr}_{\mathcal{M},s_0}(\mathbf{F} f_0^{-1}([0,\theta])) \leq \theta$;*
- *For all $s \in S$, $\mathbf{Pr}_{\mathcal{M},s}(\mathbf{F}A) \leq f_1(s)$;*
- *For all $0 < \theta < 1$, $\{s \mid f_0(s) \geq \theta \wedge f_1(s) \geq \theta\} \cap Post^*_{\mathcal{M}}(\{s_0\})$ is finite.*

Observation and Illustration. Let us remark that there may not exist, for general Markov chains, an algorithm to decide the existence of such functions f_0, f_1 and if there exist, to find them. Indeed as for decisiveness, divergence is a semantical property. But there exist some simpler sufficient conditions for divergence.

A finite Markov chain is divergent (letting $f_0 = f_1 = 1$) w.r.t. any s_0 and any A. In the first Markov chain of Example 3, $f_0 = 1$ and $f_1(m) = \rho^m$ and in the second Markov chain, $f_1 = 1$, $f_0(m) = \prod_{0 \leq n < m} p_n$ and $f_0(s) = 1$ for all s in the finite Markov chain containing A. Generalizing these two examples, the next proposition introduces a sufficient condition for divergence. Its proof is immediate by choosing ($f = f_0$ and $f_1 = 1$) or ($f = f_1$ and $f_0 = 1$).

Proposition 1. *Let $\mathcal{M}$ be a Markov chain, $s_0 \in S$, $A \subseteq S$, and a computable function f from S to $\mathbb{R}_{\geq 0}$ such that:*

- *For all $0 < \theta < 1$, $\mathbf{Pr}_{\mathcal{M},s_0}(\mathbf{F} f^{-1}([0,\theta])) \leq \theta$ or for all $s \in S$, $\mathbf{Pr}_{\mathcal{M},s}(\mathbf{F}A) \leq f(s)$;*
- *For all $0 < \theta < 1$, $\{s \mid f(s) \geq \theta\} \cap Post^*_{\mathcal{M}}(\{s_0\})$ is finite.*

Then $\mathcal{M}$ is divergent w.r.t. s_0 and A.

2.3 An Algorithm for Divergent Markov Chains

We now design an algorithm for accurately framing the reachability probability for a divergent (effective) Markov chain w.r.t. s_0 and an effective A.

Let us describe this algorithm. It performs an exploration of reachable states from s_0 maintaining S', the set of visited states, and stopping an exploration when the current state s fulfills: either (1) for some $i \in \{0,1\}$, $f_i(s) \leq \frac{\theta}{2}$ in which case s is inserted in the $AlmostLoose_i$ set (initially empty), or (2) $s \in A$ in which case s is inserted in A' (initially empty). When the exploration is ended, if A' is empty, the algorithm returns the interval $[0, \theta]$. Otherwise it builds $\mathcal{M}' = (S', p')$ a finite Markov chain over S' whose transition probabilities are the ones of $\mathcal{M}$ except for the states of $AlmostLoose_0 \cup AlmostLoose_1 \cup A'$, which are made absorbing. Finally it computes the vector of reachability probabilities starting from s_0 in $\mathcal{M}'$ (function `CompFinProb`) and returns the interval $[preach(A'), preach(A') + preach(AlmostLoose_0) + \frac{\theta}{2} \cdot preach(AlmostLoose_1)]$. The next proposition establishes the correctness of the algorithm.

Proposition 2. *Let $\mathcal{M}$ be a divergent Markov chain with $s_0 \in S$, $A \subseteq S$ and $\theta > 0$. Then Algorithm 1 solves the CRP problem.*

Algorithm 1: Framing the reachability probability

```
CompProb(M, s0, A, θ)
AlmostLoose0 ← ∅; AlmostLoose1 ← ∅; S' ← ∅
A' ← ∅; Front ← ∅; Insert(Front, s0)
while Front ≠ ∅ do
    s ← Extract(Front); S' ← S' ∪ {s}
    if f0(s) ≤ θ/2 then AlmostLoose0 ← AlmostLoose0 ∪ {s}
    else if f1(s) ≤ θ/2 then AlmostLoose1 ← AlmostLoose1 ∪ {s}
    else if s ∈ A then A' ← A' ∪ {s}
    else for s → s' ∧ s' ∉ S' do Insert(Front, s')
end
if A' = ∅ then return (0, θ)
Abs ← AlmostLoose0 ∪ AlmostLoose1 ∪ A'
for s ∈ Abs do p'(s, s) ← 1
for s ∈ S' \ Abs ∧ s' ∈ S' do p'(s, s') ← p(s, s')
preach ← CompFinProb(M', s0)  // M' = (S', p'): a finite Markov chain
return (preach(A'), preach(A') + preach(AlmostLoose0) + θ/2 · preach(AlmostLoose1))
```

We also provide an algorithm for models with a decidable reachability problem that returns $[0, 0]$ when A is unreachable and $[\ell, u]$ with $\ell > 0$ otherwise. This algorithm and the proof of its correctness are both presented in [14].

3 (Un)Decidability Results

We now study probabilistic versions of well-known models like Pushdown Automaton (PDA) and Petri nets (PN), for which we analyse the decidability of the divergence property.

3.1 Probabilistic Pushdown Automata

Let Γ be a finite alphabet. $\Gamma^{\leq k}$ is the set of words over Γ with length at most k. Let $w \in \Gamma^*$, then $|w|$ denotes its length. ε denotes the empty word.

Definition 4 (pPDA). *A (dynamic-)probabilistic pushdown automaton (pPDA) is a tuple $\mathcal{A} = (Q, \Gamma, \Delta, W)$ where:*

- *Q is a finite set of control states;*
- *Γ is a finite stack alphabet with $Q \cap \Gamma = \emptyset$;*
- *Δ is a subset of $Q \times \Gamma^{\leq 1} \times Q \times \Gamma^{\leq 2}$ such that for all $(q, \varepsilon, q', w) \in \Delta$, $|w| \leq 1$;*
- *W is a computable function from $\Delta \times \Sigma^*$ to $\mathbb{Q}_{>0}$.*

In the version of pPDA presented in [12], the weight function W goes from Δ to $\mathbb{Q}_{>0}$. In order to emphasize this restriction here and later we say that, in this case, the weight function is *static* and the corresponding models will be called static pPDA. In what follows, pPDA denotes the dynamic version.

An item (q, a, q', w) of Δ is also denoted $q \xrightarrow{?a!w} q'$ and $?a!\varepsilon$ is also simply denoted by $?a$. A *configuration* of $\mathcal{A}$ is a pair $(q, w) \in Q \times \Gamma^*$. We use the letters a, b, c, x, y for elements in Γ and w for a word in Γ^*.

Definition 5. *Let $\mathcal{A}$ be a pPDA. Then the Markov chain $\mathcal{M}_\mathcal{A} = (S_\mathcal{A}, p_\mathcal{A})$ is defined by:*

- $S_\mathcal{A} = Q \times \Gamma^*$ *is the set of configurations;*
- *For all* $(q, \varepsilon) \in S_\mathcal{A}$ *s.t.* $\{t = q \xrightarrow{?\varepsilon!w_t} q'\}_{t\in\Delta} = \emptyset$, $p_\mathcal{A}((q, \varepsilon), (q, \varepsilon)) = 1$;
- *For all* $(q, \varepsilon) \in S_\mathcal{A}$ *s.t.* $\{t = q \xrightarrow{?\varepsilon!w_t} q'\}_{t\in\Delta} \neq \emptyset$, *let* $W(q, \varepsilon) = \sum_{t=q \xrightarrow{?\varepsilon!w_t} q'} W(t, \varepsilon)$.
  Then: for all $t = q \xrightarrow{?\varepsilon!w_t} q' \in \Delta$, $p_\mathcal{A}((q, \varepsilon), (q', w_t)) = \frac{W(t,\varepsilon)}{W(q,\varepsilon)}$
- *For all* $(q, wa) \in S_\mathcal{A}$ *s.t.* $\{t = q \xrightarrow{?a!w_t} q'\}_{t\in\Delta} = \emptyset$, $p_\mathcal{A}((q, wa), (q, wa)) = 1$;
- *For all* $(q, wa) \in S_\mathcal{A}$ *s.t.* $\{t = q \xrightarrow{?a!w_t} q'\}_{t\in\Delta} \neq \emptyset$,
  let $W(q, wa) = \sum_{t=q \xrightarrow{?a!w_t} q'} W(t, wa)$. *Then:*
  for all $t = q \xrightarrow{?a!w_t} q' \in \Delta$, $p_\mathcal{A}((q, wa), (q', ww_t)) = \frac{W(t,wa)}{W(q,wa)}$

We now show that even for pPDA with a single state and with a stack alphabet reduced to a singleton, divergence is undecidable.

Theorem 1. *The divergence problem for pPDA is undecidable even with a single state and stack alphabet* $\{a\}$.

Due to this negative result on such a basic model, it is clear that one must restrict the possible weight functions. A pPDA $\mathcal{A}$ is said *polynomial* if for all $t \in \Delta$, $W(t, w)$ is a positive integer polynomial (i.e. whose coefficients are non negative and the constant one is positive) whose single variable is $|w|$.

Theorem 2. *The divergence problem w.r.t.* s_0 *and finite* A *for polynomial pPDA with a single state and stack alphabet* $\{a\}$ *is decidable (in linear time).*

3.2 Probabilistic Petri Nets

A probabilistic Petri net (resp. a probabilistic VASS) is a Petri net (resp. a VASS) with a computable weight function W. In previous works [2,4], the weight function W is a *static* one: i.e., a function from T, the finite set of transitions of the Petri net, to $\mathbb{N}_{>0}$. As above, we call these models *static* probabilistic Petri nets. We introduce here a more powerful function where the weight of a transition depends on the current marking.

Definition 6. *A* (dynamic-)probabilistic Petri net *(pPN)* $\mathcal{N} = (P, T, \mathbf{Pre}, \mathbf{Post}, W, \mathbf{m}_0)$ *is defined by:*

- P, *a finite set of places;*
- T, *a finite set of transitions;*
- $\mathbf{Pre}, \mathbf{Post} \in \mathbb{N}^{P\times T}$, *resp. the pre and post condition matrices;*
- W, *a computable function from* $T \times \mathbb{N}^P$ *to* $\mathbb{Q}_{>0}$ *the weight function;*

– $\mathbf{m}_0 \in \mathbb{N}^P$, *the initial marking.*

When for all $t \in T$, $W(t, -)$ is a positive polynomial whose variables are the place markings, we say that $\mathcal{N}$ is a *polynomial* pPN.

A marking $\mathbf{m}$ is an item of $\mathbb{N}^P$. Let t be a transition. Then t is *enabled* in $\mathbf{m}$ if for all $p \in P$, $\mathbf{m}(p) \geq \mathbf{Pre}(p,t)$. When enabled, the *firing* of t leads to marking $\mathbf{m}'$ defined for all $p \in P$ by $\mathbf{m}'(p) = \mathbf{m}(p) + \mathbf{Post}(p,t) - \mathbf{Pre}(p,t)$ which is denoted by $\mathbf{m} \xrightarrow{t} \mathbf{m}'$. Let $\sigma = t_1 \dots t_n$ be a sequence of transitions. We define the enabling and the firing of σ by induction. The empty sequence is always enabled in $\mathbf{m}$ and its firing leads to $\mathbf{m}$. When $n > 0$, σ is enabled if $\mathbf{m} \xrightarrow{t_1} \mathbf{m}_1$ and $t_2 \dots t_n$ is enabled in $\mathbf{m}_1$. The firing of σ leads to the marking reached by $t_2 \dots t_n$ from $\mathbf{m}_1$. A marking $\mathbf{m}$ is reachable from $\mathbf{m}_0$ if there is a firing sequence σ that reaches $\mathbf{m}$ from $\mathbf{m}_0$.

Definition 7. *Let $\mathcal{N}$ be a pPN. Then the* Markov chain $\mathcal{M}_\mathcal{N} = (S_\mathcal{N}, p_\mathcal{N})$ *associated with $\mathcal{N}$ is defined by:*

- *$S_\mathcal{N}$ is the set of reachable markings from $\mathbf{m}_0$;*
- *Let $\mathbf{m} \in S_\mathcal{N}$ and $T_\mathbf{m}$ be the set of transitions enabled in $\mathbf{m}$. If $T_\mathbf{m} = \emptyset$ then $p_\mathcal{N}(\mathbf{m},\mathbf{m}) = 1$. Otherwise let $W(\mathbf{m}) = \sum_{\mathbf{m} \xrightarrow{t} \mathbf{m}_t} W(t,\mathbf{m})$. Then for all $\mathbf{m} \xrightarrow{t} \mathbf{m}_t$, $p_\mathcal{N}(\mathbf{m},\mathbf{m}_t) = \frac{W(t,\mathbf{m})}{W(\mathbf{m})}$.*

Contrary to the previous result, restricting the weight functions to be polynomials does not yield decidability for pPNs.

Theorem 3. *The divergence problem of polynomial pPNs w.r.t. an upward closed set is undecidable.*

4 Illustration of Divergence

Due to the undecidability results, we propose syntactical restrictions for standard models like pushdown automata and channel systems that ensure divergence. Observing that function f_1 of Definition 3 is somewhat related to transience of Markov chains, we first establish a sufficient condition of transience from which we derive a sufficient condition of divergence for infinite Markov chains used for our two illustrations.

Theorem 4. *Let $\mathcal{M}$ be a Markov chain and f be a function from S to $\mathbb{R}$ with $B = \{s \mid f(s) \leq 0\}$ fulfilling $\emptyset \subsetneq B \subsetneq S$, $\varepsilon, K \in \mathbb{R}_{>0}$ and $d \in \mathbb{N}^*$ such that:*

$$\text{for all } s \in S \setminus B \qquad \sum_{s' \in S} p^{(d)}(s,s')f(s') \geq f(s) + \varepsilon \text{ and } \sum_{|f(s')-f(s)| \leq K} p(s,s') = 1 \tag{1}$$

Then for all $s \in S$ such that $f(s) > dK$,

$$\mathbf{Pr}_{\mathcal{M},s}(\mathbf{F}B) \leq c_1 e^{-c_2(f(s)-dK)}$$

where $c_1 = \sum_{n \geq 1} e^{-\frac{\varepsilon^2 n}{2(\varepsilon+K)^2}}$ and $c_2 = \frac{\varepsilon}{(\varepsilon+K)^2}$, which implies transience of $\mathcal{M}$ when it is irreducible.

Proposition 3. *Let $\mathcal{M}$ be a Markov chain and f be a computable function from S to $\mathbb{R}$ with $B = \{s \mid f(s) \leq 0\}$ fulfilling $\emptyset \subsetneq B \subsetneq S$, and for some $\varepsilon, K \in \mathbb{R}_{>0}$ and $d \in \mathbb{N}^*$, Equation (1). Assume in addition that for all $n \in \mathbb{N}$, $\{s \mid f(s) \leq n\}$ is finite. Then $\mathcal{M}$ is divergent w.r.t. any s_0 and any finite A.*

4.1 Probabilistic Channel Systems

Now we introduce a probabilistic variant of channel systems particularly appropriate for the modelling of open queuing networks. Here a special input channel c_{in} (that works as a counter) only receives the arrivals of anonymous clients all denoted by \$ (item 1 of the next definition). Then the service of a client corresponds to a message circulating between the other channels with possibly change of message identity until the message disappears (items 2 and 3).

Definition 8. *A* probabilistic open channel system (pOCS) $\mathcal{S} = (Q, Ch, \Sigma, \Delta, W)$ *is defined by:*

- *a finite set Q of states;*
- *a finite set Ch of channels, including c_{in};*
- *a finite alphabet Σ including \$;*
- *a transition relation $\Delta \subseteq Q \times Ch \times \Sigma_\varepsilon \times Ch \times \Sigma_\varepsilon \times Q$ that fulfills:*
  1. *For all $q \in Q$, $(q, c_{in}, \varepsilon, c_{in}, \$, q) \in \Delta$;*
  2. *For all $(q, c, a, c', a', q') \in \Delta$, $a = \varepsilon \Rightarrow a' = \$ \wedge c = c' = c_{in}$;*
  3. *For all $(q, c, a, c', a', q') \in \Delta$, $c \neq c_{in} \Rightarrow c' \neq c_{in}$;*
- *W is a function from $\Delta \times (\Sigma^*)^{Ch}$ to $\mathbb{Q}_{>0}$.*

Illustration. We consider three types of transitions between configurations: sending messages to the input channel (i.e., $(q, c_{in}, \varepsilon, c_{in}, \$, q)$) representing client arrivals; transferring messages between different channels (i.e., (q, c, a, c', a', q') with $\varepsilon \notin \{a, a'\}$ and $c' \neq c_{in}$) describing client services; and terminating message processing (i.e., $(q, c, a, c', \varepsilon, q')$ with $a \neq \varepsilon$) meaning client departures. All messages entering c_{in} are anonymous (i.e., denoted by \$). The left part of Fig. 3 is a schematic view of such systems. The left channel is c_{in}. All dashed lines represent message arrivals (to c_{in}) or departures. The solid lines model message transferrings.

The next definitions formalize the semantics of pOCS.

Definition 9. *Let $\mathcal{S}$ be a pOCS, $(q, \nu) \in Q \times (\Sigma^*)^{Ch}$ be a configuration and $t = (q, c, a, c', a', q') \in \Delta$. Then t is enabled in (q, ν) if $\nu(c) = aw$ for some w. The firing of t in (q, ν) leads to (q', ν') defined by:*

- *if $c = c'$ then $\nu'(c) = wa'$ and for all $c'' \neq c$, $\nu'(c'') = \nu(c'')$;*
- *if $c \neq c'$ then $\nu'(c) = w$, $\nu'(c') = \nu(c')a'$ and for all $c'' \notin \{c, c'\}$, $\nu'(c'') = \nu(c'')$.*

As usual one denotes the firing by $(q, \nu) \xrightarrow{t} (q', \nu')$. Observe that from any configuration at least one transition (a client arrival) is enabled.

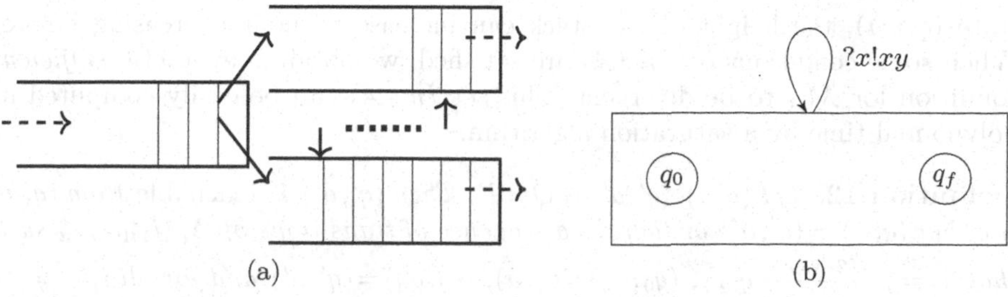

Fig. 3. A schematic view of pOCS (left) and a pPDA (right)

Definition 10. *Let $\mathcal{S}$ be a pOCS. Then the Markov chain $\mathcal{M}_\mathcal{S} = (S_\mathcal{S}, p_\mathcal{S})$ is defined by:*

- *$S_\mathcal{S} = Q \times (\Sigma^*)^{Ch}$ is the set of configurations;*
- *For all $(q,\nu) \in S_\mathcal{S}$ let $W(q,\nu) = \sum_{(q,\nu)\xrightarrow{t}(q',\nu')} W(t,\nu)$. Then: for all $(q,\nu) \xrightarrow{t} (q',\nu')$, $p_\mathcal{S}((q,\nu),(q',\nu')) = \frac{W(t,\nu)}{W(q,\nu)}$.*

The restrictions on pOCS w.r.t. standard CS do not change the status of the reachability problem.

Proposition 4. *The reachability problem of pOCS is undecidable.*

As discussed in the introduction, when the number of clients exceeds some threshold, the performances of the system drastically decrease and thus the ratio of arrivals w.r.t. the achievement of a task increase. We formalize it by introducing *uncontrolled* pOCS where the weights of transitions are constant except the ones of client arrivals which are specified by positive non constant polynomials. Let $\nu \in (\Sigma^*)^{Ch}$. Then $|\nu|$ denotes $\sum_{c \in Ch} |\nu(c)|$.

Definition 11. *Let $\mathcal{S}$ be a pOCS. Then $\mathcal{S}$ is* uncontrolled *if:*

- *For all $t = (q, c, a, c', a', q') \in \Delta$ with $a \neq \varepsilon$, $W(t,\nu)$ only depends on t and will be denoted $W(t)$;*
- *For all $t = (q, c_{in}, \varepsilon, c_{in}, \$, q)$, $W(t,\nu)$ is a positive non constant polynomial, whose single variable is $|\nu|$, and will be denoted $W_{in}(q, |\nu|)$.*

The next proposition establishes that an uncontrolled pOCS generates a divergent Markov chain. This model illustrates the interest of divergence: while reachability of a pOCS is undecidable, we can apply Algorithm 1.

Proposition 5. *Let $\mathcal{S}$ be a uncontrolled pOCS. Then $\mathcal{M}_\mathcal{S}$ is divergent.*

4.2 Probabilistic Pushdown Automata

Increasing pPDA. We introduce the subset of *increasing pairs*, denoted as $Inc(\mathcal{A})$, which is a subset of $Q \times \Gamma$ that contains pairs (q, a) such that from state (q, wa), the height of the stack can increase without decreasing before. When some conditions on $Inc(\mathcal{A})$ are satisfied, we obtain a syntactic *sufficient* condition for $\mathcal{M}_\mathcal{A}$ to be divergent. This set $Inc(\mathcal{A})$ can be easily computed in polynomial time by a saturation algorithm.

Definition 12. *Let* $(q, a), (q', a') \in Q \times \Gamma$. *Then* (q', a') *is* reachable *from* (q, a) *if either* $(q, a) = (q', a')$ *or there is a sequence of transitions of* Δ, $(t_i)_{0 \le i < d}$ *such that:* $t_i = q_i \xrightarrow{?a_i!a_{i+1}} q_{i+1}$, $(q_0, a_0) = (q, a)$, $(q_d, a_d) = (q', a')$ *and for all* i, $a_i \neq \varepsilon$. *The set of* increasing pairs $Inc(\mathcal{A}) \subseteq Q \times \Gamma$ *is the set of pairs* (q, a) *that can reach a pair* (q', a') *with some* $q' \xrightarrow{?a'!bc} q'' \in \Delta$.

Definition 13. *A pPDA* $\mathcal{A}$ *is* increasing *if:*

- $Inc(\mathcal{A}) = Q \times \Gamma$;
- *for all* $t = q \xrightarrow{?a!w} q' \in \Delta$ *such that* $|w| \le 1$, $W(t, -)$ *is an integer constant denoted* W_t;
- *for all* $t = q \xrightarrow{?a!bc} q' \in \Delta$, $W(t, -)$ *is a non constant integer polynomial where its single variable is the height of the stack denoted* W_t;
- *for all* $q \xrightarrow{?a} q' \in \Delta$, *there exists* $q \xrightarrow{?a!bc} q'' \in \Delta$.

Illustration. The right part of Fig. 3 is an abstract view of a pPDA modelling of a server simultaneously handling multiple requests. The requests may occur at any time and are stored in the stack. The loop labelled by $?x!xy$ is a symbolic representation of several loops: one per triple (q, x, y) with $q \in Q$, $x \in \Gamma$ and $y \in \Gamma$. Due to the symbolic loop, the set of increasing pairs of the $pPDA_{server}$ is equal to $Q \times \Gamma$ and there is always a transition increasing the height of the stack outgoing from any (q, a). Assume now that for any other transition, its weight does not depend on the size of the stack and that a transition $t = q \xrightarrow{?a!ab} q$ has weight $W_t(n) = c_t \times n$. Then $\mathcal{A}$ is increasing. The dependance on n means that due to congestion, the time to execute tasks of the server increases with the number of requests in the system and thus increase the probability of a new request that occurs at a constant rate. One is interested in computing the probability to reach (q_f, ε) from (q_0, ε) representing the probability that the server reaches an idle state having served all the incoming requests.

We establish that an increasing pPDA generates a divergent Markov chain.

Proposition 6. *Let* $\mathcal{A}$ *be an increasing pPDA. Then the Markov chain* $\mathcal{M}_\mathcal{A}$ *is divergent w.r.t. any* s_0 *and finite* A.

5 Conclusion and Perspectives

We have introduced the divergence property of Markov chains and designed two generic CRP-algorithms depending on the status of the reachability problem. Then we have studied the decidability of divergence for pPDA and for pPN for different kinds of weights and target sets. Finally, we have provided two useful classes of divergent models within pCS and pPDA.

In the future, we plan to study the model checking of polynomial pPDA (as a possible extension of [12]) and some heuristics to find functions f_0 and f_1.

References

1. Abdulla, P.A., Bertrand, N., Rabinovich, A.M., Schnoebelen, P.: Verification of probabilistic systems with faulty communication. Inf. Comput. **202**(2), 141–165 (2005)
2. Abdulla, P.A., Henda, N.B., Mayr, R.: Decisive Markov chains. Log. Methods Comput. Sci. **3**(4) (2007)
3. Baier, C., Katoen, J.: Principles of Model Checking. MIT Press, Heidelberg (2008)
4. Brázdil, T., Chatterjee, K., Kucera, A., Novotný, P., Velan, D., Zuleger, F.: Efficient algorithms for asymptotic bounds on termination time in VASS. In: Dawar, A., Grädel, E. (eds.) Proceedings of the 33rd Annual ACM/IEEE Symposium on Logic in Computer Science, LICS 2018, pp. 185–194. ACM (2018)
5. Brázdil, T., Esparza, J., Kiefer, S., Kucera, A.: Analyzing probabilistic pushdown automata. Formal Methods Syst. Des. **43**(2), 124–163 (2013)
6. Brázdil, T., Esparza, J., Kucera, A.: Analysis and prediction of the long-run behavior of probabilistic sequential programs with recursion (extended abstract). In: Proceedings of the 46th Annual IEEE Symposium on Foundations of Computer Science (FOCS), pp. 521–530. IEEE Computer Society (2005)
7. Brázdil, T., Kiefer, S., Kučera, A.: Efficient analysis of probabilistic programs with an unbounded counter. In: Gopalakrishnan, G., Qadeer, S. (eds.) CAV 2011. LNCS, vol. 6806, pp. 208–224. Springer, Heidelberg (2011). https://doi.org/10.1007/978-3-642-22110-1_18
8. Brázdil, T., Kiefer, S., Kucera, A.: Efficient analysis of probabilistic programs with an unbounded counter. J. ACM **61**(6), 41:1–41:35 (2014)
9. Brázdil, T., Kiefer, S., Kucera, A., Novotný, P., Katoen, J.: Zero-reachability in probabilistic multi-counter automata. In: Joint Meeting of the Twenty-Third EACSL Annual Conference on Computer Science Logic (CSL) and the Twenty-Ninth Annual ACM/IEEE Symposium on Logic in Computer Science (LICS), CSL-LICS 2014, pp. 22:1–22:10. ACM (2014)
10. Esparza, J., Kucera, A., Mayr, R.: Model checking probabilistic pushdown automata. In: Proceedings of 19th IEEE Symposium on Logic in Computer Science (LICS), pp. 12–21. IEEE Computer Society (2004)
11. Esparza, J., Kucera, A., Mayr, R.: Quantitative analysis of probabilistic pushdown automata: expectations and variances. In: Proceedings of the 20th IEEE Symposium on Logic in Computer Science (LICS), pp. 117–126. IEEE Computer Society (2005)
12. Esparza, J., Kucera, A., Mayr, R.: Model checking probabilistic pushdown automata. Logical Methods Comput. Sci. **2**(1) (2006)
13. Finkel, A., Haddad, S., Ye, L.: About decisiveness of dynamic probabilistic models. CoRR abs/2305.19564 (2023). https://doi.org/10.48550/arXiv.2305.19564, to appear in CONCUR'23
14. Finkel, A., Haddad, S., Ye, L.: Introducing divergence for infinite probabilistic models. CoRR abs/2308.08842 (2023). https://doi.org/10.48550/arXiv.2308.08842

15. Iyer, P., Narasimha, M.: Probabilistic lossy channel systems. In: Bidoit, M., Dauchet, M. (eds.) CAAP 1997. LNCS, vol. 1214, pp. 667–681. Springer, Heidelberg (1997). https://doi.org/10.1007/BFb0030633
16. Kemeny, J., Snell, J., Knapp, A.: Denumerable Markov Chains, 2nd edn. Springer, Heidelberg (1976). https://doi.org/10.1007/978-1-4684-9455-6
17. Kucera, A., Esparza, J., Mayr, R.: Model checking probabilistic pushdown automata. Log. Methods Comput. Sci. **2**(1) (2006). https://doi.org/10.2168/LMCS-2(1:2)2006
18. Santos, E.S.: Probabilistic grammars and automata. Inf. Control. **21**(1), 27–47 (1972). https://doi.org/10.1016/S0019-9958(72)90026-5
19. Vardi, M.Y.: Automatic verification of probabilistic concurrent finite-state programs. In: Proceedings of the 26th Annual Symposium on Foundations of Computer Science, pp. 327–338. IEEE Computer Society (1985)

A Framework for the Competitive Analysis of Model Predictive Controllers

Stijn Bellis, Joachim Denil, Ramesh Krishnamurthy(✉), Tim Leys, Guillermo A. Pérez, and Ritam Raha

University of Antwerp – Flanders Make, Antwerp, Belgium
ramesh.krishnamurthy@uantwerpen.be

Abstract. This paper presents a framework for the competitive analysis of Model Predictive Controllers (MPC). Competitive analysis means evaluating the relative performance of the MPC as compared to other controllers. Concretely, we associate the MPC with a regret value which quantifies the maximal difference between its cost and the cost of any alternative controller from a given class. Then, the problem we tackle is that of determining whether the regret value is at most some given bound. Our contributions are both theoretical as well as practical: (1) We reduce the regret problem for controllers modeled as hybrid automata to the reachability problem for such automata. We propose a reachability-based framework to solve the regret problem. Concretely, (2) we propose a novel CEGAR-like algorithm to train a deep neural network (DNN) to clone the behavior of the MPC. Then, (3) we leverage existing reachability analysis tools capable of handling hybrid automata with DNNs to check bounds on the regret value of the controller.

Keywords: Competitive analysis · Hybrid automata

1 Introduction

An optimal control problem (OCP) deals with finding a function $u(t)$, called a *control law* that assigns values to control variables for every time step $t \in \mathbb{R}_{\geq 0}$. The control law should minimize a given cost function $J[x(\cdot), u(\cdot), t_0, t_f]$ evaluated for a time interval (t_0, t_f) and subject to the state-equation constraints $\dot{x}(t) = f[x(t), u(t), t]$. Model predictive controllers (MPC) solve such a control problem for a given f. This paper presents an approach for the competitive analysis of MPC. Competitive analysis, in this context, means evaluating the relative performance of the MPC as compared to other controllers. Referring to the OCP, our approach assumes that a control law $u(t)$ is given to us. Further, we associate to $u(t)$ a *regret value*, which quantifies the maximal difference between its cost and the cost of any alternative control law from a given class $\mathcal{C}$. Formally, the regret of $u(t)$ is: $Reg(u) := \sup_{c \in \mathcal{C}} \sup_{t_f \in \mathbb{R}_{\geq 0}} J[x(\cdot), u(\cdot), t_0, t_f] - J[x(\cdot), c(\cdot), t_0, t_f]$. If $Reg(u) < r$, then we say that the control law $u(t)$ is r-competitive.

O. Bournez et al. (Eds.): RP 2023, LNCS 14235, pp. 141–154, 2023.
https://doi.org/10.1007/978-3-031-45286-4_11

In this work, we first show that the r-competitivity problem for controllers modeled as hybrid automata is interreducible with the reachability problem for hybrid automata. It follows that the r-competitivity problem is undecidable. Fortunately, this also points to using approximate reachability analysis tools to realize approximate competitive analysis. Based on the latter, we propose a counterexample-guided abstraction refinement (CEGAR) framework that abstracts a given MPC using a deep neural network (DNN) trained to clone the behavior of the MPC. This abstraction allows us to use reachability analysis tools such as Verisig [13] to overapproximate the regret value of the abstracted controller. As usual with CEGAR approaches, the refinement step is the main challenge: If the regret is deemed too high (and Verisig finds a real example of this), then this might be due to our abstraction of the controller as a DNN, the overapproximation incurred by the reachability tool, or it might be a real problem with the MPC. In our proposal, when we cannot match the high-regret example to a behavior of the MPC, we use the output of the reachability analysis tool to augment the dataset used for training the DNN.

As a final contribution, we report on a prototype implementation of our CEGAR framework using Verisig. We have used this prototype to analyze MPC for two well-known control problems. While the approach is promising, we conclude that further tooling support is required for the full automation of the framework.

Related Work. Chen et al. 2022 [5] conducted a survey on recent advancements in verifying cyber-physical systems and identified as understudied the verification of control systems whose performance is measured using cost functions. Indeed, we did not find many works on the verification of controllers with respect to the cost functions used to obtain them from an OCP instance. Further, to the best of our knowledge, there have been no previous works on the formal analysis of regret in hybrid systems. A notable exception is the recent work of Muvvala et al. [16] who propose regret minimization as a less pessimistic objective for robots involved in collaborations (e.g., with humans), as opposed to a sole emphasis on worst-case optimization. However, their regret analysis focuses on a higher planning level, distinct from the hybrid-dynamics level of the system, making it closer to the work of Hunter et al. [12] rather than the present one.

Behavioral cloning, also known as imitation learning, is a topic of increasing interest within artificial intelligence (see, e.g. [3,17,18]). We do not claim to have a new behavioral cloning algorithm. Rather, we have integrated a data aggregation step into our CEGAR algorithm for the competitive analysis of hybrid automata. Interestingly, contrary to previous uses of DNNs as proxies for MPC [6,13], we have observed that a successful competitive analysis (i.e., the tool says the controller is r-competitive for a small enough r) suggests one can use the DNN instead of the MPC! Although this does not guarantee that the MPC itself is r-competitive, the DNN demonstrates competitiveness. Moreover, evaluating the DNN to compute the control law proves to be relatively efficient.

2 Hybrid Automata and Competitive Analysis

A *hybrid automaton* (HA, for short) is an extension of a finite-state automaton equipped with a finite set of real-valued variables. The values of the variables change *discretely* along transitions and they do so *continuously*, over time while staying in a state. Formally it is a tuple $(Q, I, T, \Sigma, X, \mathit{jump}, \mathit{flow}, \mathit{inv})$, where:

- Q is a finite set of states and $I \subseteq Q$ is the subset of initial states,
- Σ is a finite alphabet,
- $T \subseteq Q \times \Sigma \times Q$ is a set of transitions, and
- X is a finite set of real-valued variables. We write $V \subseteq \mathbb{R}^X$ to denote the set of all possible valuations of X.
- $\mathit{jump} : T \to \mathit{Op}$ maps transitions to a set of *guards* and *effects* on the values of the variables. That is, $\mathit{Op} \subseteq V^2$ and for a transition $\delta \in T$, $\mathit{jump}(\delta) = (\mathit{guard}, \mathit{effect})$ implies that δ is "enabled" if the current valuation is *guard* and *effect* is the valuation after the transition. Intuitively, *jump* denotes the discrete changes in the variables along transitions. Usually, the guards and effects are encoded as first-order predicates over the reals, e.g. $\mathit{jump}(\delta) = (x > 2, x + 4)$ denotes the set $\{(v, v') \in V^2 \mid v(x) > 2 \text{ and } v'(x) = v(x) + 4\}$.
- $\mathit{flow} : Q \to F$, with $F \subseteq \{f : \mathbb{R}_{>0} \to V\}$, maps each state $q \in Q$ to a set F of functions f_q that give the continuous change in the valuation of the variables while in state q. Usually, the functions f_q are encoded as systems of first-order differential equations, e.g. $\dot{x} = 5$ denotes functions[1] $f(t)(x) = 5t + c$, where $c \in \mathbb{R}_{>0}$ is the value of x at time $t = 0$.
- $\mathit{inv} : Q \to 2^V$ maps each state $q \in Q$ to an invariant that constrains the possible valuations of the variables in q. Similar to *jump*, *inv* is usually encoded as first-order predicates over the reals.

Configurations and Runs. A configuration is a pair (q, v) where $q \in Q$ and $v \in V$ is a valuation of the variables in X. A configuration (q, v) is *valid* if $v \in \mathit{inv}(q)$. Let (q, v) and (q', v') be two valid configurations. We say (q', v') is a *discrete successor* of (q, v) if $\delta = (q, a, q') \in T$ for some $a \in \Sigma$ and $(v, v') \in \mathit{jump}(\delta)$. Similarly, (q', v') is a *continuous successor* of (q, v) if $q = q'$ and there exist $t_0, t_1 \in \mathbb{R}_{>0}$ and $f_q \in \mathit{flow}(q)$ such that $f_q(t_0) = v$, $f_q(t_1) = v'$ and for all $t_0 \le t \le t_1$, $f_q(t) \in \mathit{inv}(q)$.

A *run* ρ is a sequence of configurations $(q_0, v_0)(q_1, v_1) \dots (q_n, v_n)$ such that $q_0 \in I$, v_0 assigns 0 to all variables and, for all $0 \le i < n$, (q_{i+1}, v_{i+1}) is a discrete or continuous successor of (q_i, v_i). The REACH decision problem asks, for a given hybrid automaton A and configuration (q, v), whether there is a run of A whose last configuration is (q, v).

Parallel Composition. Let $A_i = (Q_i, I_i, T_i, \Sigma_i, X_i, \mathit{jump}_i, \mathit{flow}_i, \mathit{inv}_i)$ for $i = 1, 2$ be two HA. Then, $A = (Q, I, T, \Sigma, X, \mathit{jump}, \mathit{flow}, \mathit{inv})$ is the *parallel composition of* A_1 and A_2, written $A = A_1 \mid\mid A_2$, if and only if:

[1] Note that if X contains more variables than just x, this function is not unique.

- $Q = Q_1 \times Q_2$ and $I = I_1 \times I_2$,
- $\Sigma = \Sigma_1 \cup \Sigma_2$ and $X = X_1 \cup X_2$.
- The transition set T contains $(\langle q_1, q_2 \rangle, \sigma, \langle q_1', q_2' \rangle)$ if and only if there are $i, j \in \{1, 2\}$ such that $i \neq j$ and:
  - either $\sigma \in \Sigma_i \setminus \Sigma_j$, $(q_i, \sigma, q_i') \in T_i$, and $q_j = q_j'$;
  - or $\sigma \in \Sigma_i \cap \Sigma_j$, $(q_i, \sigma, q_i') \in T_i$, and $(q_j, \sigma, q_j') \in T_j$.
- The *jump* function is such that, for $\delta = (\langle q_1, q_2 \rangle, \sigma, \langle q_1', q_2' \rangle)$, we have that:
  - either $\sigma \in \Sigma_i \setminus \Sigma_j$ and $\mathit{jump}(\delta) = \mathit{jump}_i(\langle q_i, \sigma, q_i' \rangle)$ for some $i, j \in \{1, 2\}$ with $i \neq j$,
  - or $\sigma \in \Sigma_i \cap \Sigma_j$ and $\mathit{jump}(t) = \mathit{jump}_1(\langle q_1, \sigma, q_1' \rangle) \cap \mathit{jump}_2(\langle q_2, \sigma, q_2' \rangle)$.
- Finally, $\mathit{flow}(\langle q_1, q_2 \rangle) = \mathit{flow}_1(q_1) \cap \mathit{flow}_2(q_2)$, and
- $\mathit{inv}(\langle q_1, q_2 \rangle) = \mathit{inv}_1(q_1) \cap \mathit{inv}_2(q_2)$.

2.1 The Cost of Control

In this work, we use HA to model *hybrid systems* and *controllers*. In particular, we henceforth assume any HA $A = (Q, I, T, \Sigma, X, \mathit{jump}, \mathit{flow}, \mathit{inv})$ modelling a hybrid system has a designated *cost* variable $J \in X$. We make no such assumption for HA used to model controllers. Observe that from the definition of parallel composition, it follows that if A models a hybrid system, then $B = A \,||\, C$ also models a hybrid system—i.e. it has the cost variable J—for any HA C.

The following notation will be convenient: For a run $\rho = (q_0, v_0) \dots (q_n, v_n)$ we write J_ρ to denote the value $v_n(J)$. Further, we write $\rho \in A$, where ρ is a run of the hybrid automaton A. Now, the *maximal and minimal cost of a HA A* respectively are $\overline{J(A)} := \sup_{\rho \in A} J_\rho$ and, $\underline{J(A)} := \inf_{\rho \in A} J_\rho$.

2.2 Regret

Fix a hybrid-system HA $A = (Q, I, T, \Sigma, X, \mathit{jump}, \mathit{flow}, \mathit{inv})$. We define the *(worst-case) regret* $\mathit{Reg}(U)$ of a controller HA U as the maximal difference between the (maximal) cost incurred by the parallel composition of A and U—i.e. the controlled system—and the (minimal) cost incurred by an alternative controller HA from a set $\mathcal{C}$: $\mathit{Reg}(U) := \sup_{U' \in \mathcal{C}} (\overline{J(A \,||\, U)} - \underline{J(A \,||\, U')})$. The Regret problem asks, for given A, U, $\mathcal{C}$, and $r \in \mathbb{Q}$, whether $\mathit{Reg}(U) \geq r$.

3 Reachability and Competitive Analysis

In this section, we establish that the reachability and regret problems are interreducible. While this implies an exact algorithm for the competitive analysis of hybrid automata does not exist, it suggests the use of approximation algorithms for reachability as a means to realize an approximate analysis.

Theorem 1. *Let $\mathcal{C}$ be the set of all possible controllers. Then, the* Regret *problem reduces in polynomial time to the* Reach *problem.*

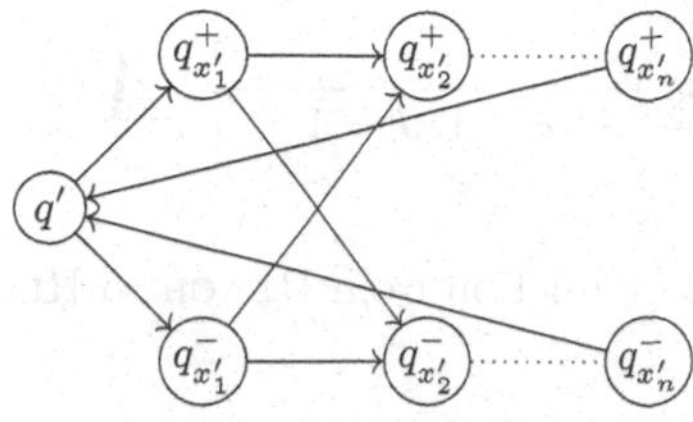

Fig. 1. Gadget for simulating any possible controller

Proof (of Theorem 1). Given a hybrid-system HA A, a controller U, a set of all possible controllers $\mathcal{C}$ and a regret bound $r \in \mathbb{Q}$, we will construct another HA $A' = (Q', I', T', \Sigma, X', \mathit{jump}', \mathit{flow}', \mathit{inv}')$ and a target configuration (q', v) of A' such that, (q', v) is reachable in A' if and only if $\mathit{Reg}(U) < r$ in $A \,||\, U$. Let us write $A = (Q, I, T, \Sigma, X, \mathit{jump}, \mathit{flow}, \mathit{inv})$ and note that $J \in X$ because A is a hybrid-system HA. We extend the automaton $A \,||\, U$ with a gadget to obtain A'. The idea is as follows: for every variable $y \in Y$ of $A \,||\, U$, we add a copy of it in the variable set X' of A' that simulates any possible choice of value for that variable by an alternative controller U'. The variable $J' \in X'$ calculates the cost of that alternative controller. Formally, $X' = Y \cup \{y' \mid y \in Y\}$.

To simulate any possible valuation of the variables, we introduce the gadget given in Fig. 1. For every variable x'_i such that x_i is a variable in U, the gadget contains two states $q^+_{x'_i}$ and $q^-_{x'_i}$. Then, $\mathit{flow}'(q^+_{x'_i})$ contains $\dot{x}'_i = 1$ and $\dot{x}'_j = 0$ for all $j \neq i$. Intuitively, this state allows us to positively update the value of x'_i to any arbitrary value. Similarly, $\mathit{flow}'(q^-_{x'_i})$ contains $\dot{x}'_i = -1$ and $\dot{x}'_j = 0$, $\forall j \neq i$, which allows it to negatively update the value of x'_i.

Now, we add a "sink" state q_{reach} and make it reachable from all the other states using transitions $\delta'_i \in T'$ such that $\mathit{jump}'(\delta'_i)$ contains guard of the form $J - J' \geq r$. Finally, from every state $q' \in Q'$, we add the option to go into its own copy of the gadget, set the values of the variables to any desired value and come back to the same state.

Note that if $(q_{\text{reach}}, \mathbf{0})$ is reachable in A', via a run $\rho \in A'$, then $J_\rho - J'_\rho \geq r$. As the gadget does not update the value of J and J', it is easy to see that $\mathit{Reg}(U) \geq r$. Now, if $(q_{\text{reach}}, \mathbf{0})$ is not reachable that means, $J_\rho - J'_\rho < r$ for all $\rho \in A'$. Now, as all possible controllers (in fact, all possible configurations of variables from U) can be simulated in A', it is easy to see that $\mathit{Reg}(U) < r$. □

Interestingly, the construction presented above does not preserve the property of being *initialized*. Intuitively, an initialized hybrid automaton is one that "resets" a variable x on transitions between states which have different flows for x. Alas, we do not know whether an alternative proof exists which does preserve the property of being initialized (and also being rectangular, a property which we do not formally define here). Such a reduction would imply the regret problem is decidable for rectangular and initialized hybrid automata.

We now proceed to stating and proving the converse reduction.

Theorem 2. *The* REACH *problem reduces in polynomial time to the* REGRET *problem.*

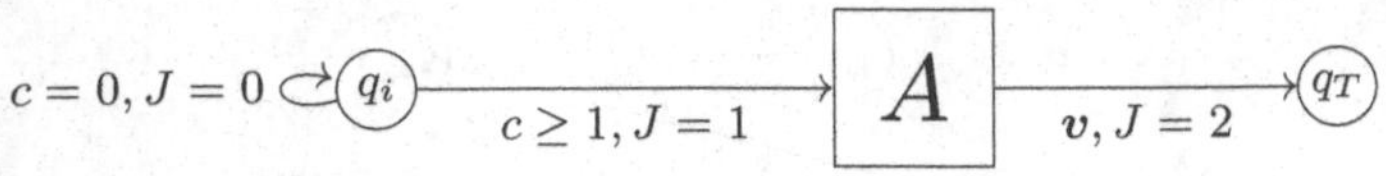

$c = 0, J = 0$ ↻ q_i → ($c \geq 1, J = 1$) → A → ($v, J = 2$) → q_T

Fig. 2. Reduction from REACH to REGRET

Proof (of Theorem 2). Given a HA A and a target configuration (q, v), we will construct a HA A' and a controller U such that $Reg(U) \geq 2$ with respect to $A' \,||\, U$ if and only if (q, v) is reachable in A. The reduction works for any set $\mathcal{C}$ of controllers that contains at least one controller that sets c to 0 all the time.

First, we add two states to A' so that $Q' = Q \cup \{q_i, q_T\}$. In A', q_i has a self-loop that can be taken if the value of c is 0 and the effect is that $J = 0$ (see Fig. 2). From q_i, we can also transition to the initial states of A if $c \geq 1$, and in doing so, we set J to 1. Finally, from the target state q in A, we can go to the new state q_T if the target valuation $\boldsymbol{v}$ is reached, and that changes the valuation of J to 2. The valuation of J does not change within A.

Note that the minimum cost incurred by a controller that constantly sets c to 0 in A' is 0, which is achieved by the run that loops on q_i. Now, if $(q, \boldsymbol{v})$ is reachable in A via run $\rho \in A$, then the maximum cost incurred by a controller that sets c to 1 occurs along a run $q_i \cdot \rho \cdot q_T$ and is 2, making $Reg(U) \geq 2$. On the other hand, if $(q, \boldsymbol{v})$ is not reachable in A, then the maximal value of J along any such run is 1, resulting in $Reg(U) < 2$. Our constructed controller U is such that it sets c to 1 all the time, and the above arguments give the desired result. □

Since the reachability problem is known to be undecidable for hybrid automata in general [10], it follows that our regret problem is also undecidable.

Corollary 1. *The* REGRET *problem is undecidable.*

4 CEGAR-Based Competitive Analysis

We present our CEGAR approach to realize approximate competitive analysis. To keep the discussion simple, we focus on continuous systems, specifically single-state hybrid automata. Since our goal is to approximate the regret of MPCs, we model controllers as hybrid automata that sample variable values at discrete-time intervals and determine control variable values using a deep neural network (DNN) trained to behave as the MPC. Concretely, our approach specializes the reduction in the proof of Theorem 1: We will work with a hybrid automaton $\mathcal{D}$ that abstracts the behavior of the controller using a DNN, and a hybrid automaton $\mathcal{N}$ that abstracts the behaviors of all alternative controllers. The overview of our framework is depicted in Fig. 3a.

4.1 Initial Abstraction and Analysis

Our proposed framework begins with the abstraction of the controller as a hybrid automaton $\mathcal{D}$ and the alternative controllers as $\mathcal{N}$. Each of these automata are

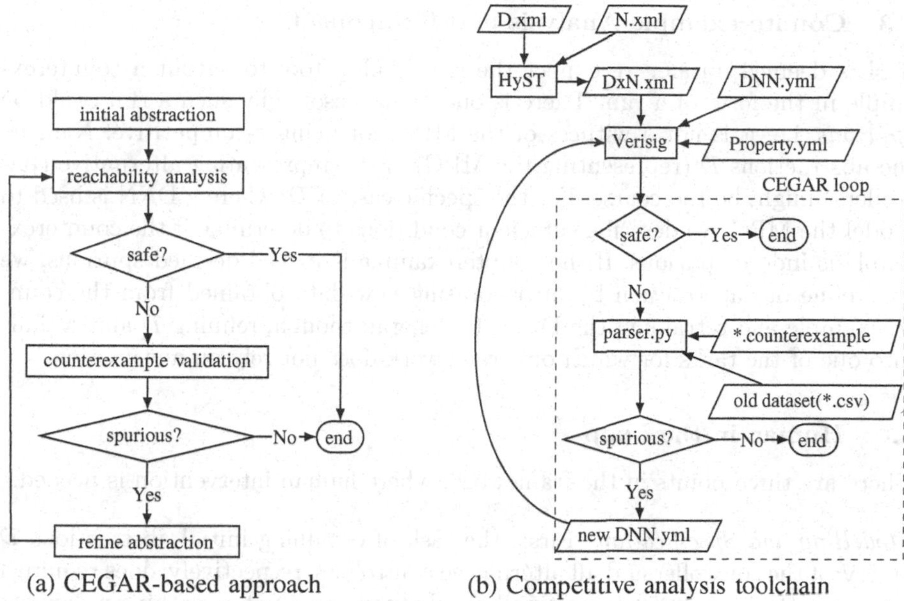

(a) CEGAR-based approach (b) Competitive analysis toolchain

Fig. 3. Flowchart depictions of our approach and our toolchain implementing it; We use ANSI/ISO standard flowchart symbols: the parallelogram blocks represent inputs/outputs, and the rectangular blocks represent processes or tools

assumed to have a cost variable, say $J_{\mathcal{D}}$ for $\mathcal{D}$ and $J_{\mathcal{N}}$ for $\mathcal{N}$. For a given value $r \in \mathbb{R}$, if we want to determine whether $\mathcal{D}$ is r-competitive then we add to $\mathcal{A} = \mathcal{D}||\mathcal{N}$ a new cost variable $J = J_{\mathcal{N}} - J_{\mathcal{D}}$. As is argued in Theorem 1, $\mathcal{D}$ should be r-competitive if and only if $\mathcal{A}$ can reach a configuration where the value of J is larger than r. Hence, we can apply any reachability set (overapproximation) tool to determine the feasibility of such a configuration.

4.2 Reachability Status

If the reachability tool finds that a configuration with $J \geq r$ is reachable in $\mathcal{A}$, we say it concludes $\mathcal{A}$ is *unsafe*. In that case, we will have to process the reachability witness. Otherwise, $\mathcal{A}$ is *safe*, and we can stop and conclude that $\mathcal{D}$ is r-competitive. Interestingly, $\mathcal{D}$ can now be used as an r-competitive replacement of the original controller! It is important to highlight that behavior cloning does not provide any guarantees regarding the relationship between the MPC and the DNN within $\mathcal{D}$. Consequently, even if we have evidence supporting the r-competitiveness of $\mathcal{D}$, we cannot infer the same for the MPC itself.

In the context of MPCs, this result is already quite useful. This is because MPCs have a non-trivial *latency* and memory usage before choosing a next valuation for the control variables (see, e.g. [11,14]). In our implementation described in the following section, $\mathcal{D}$ takes the form of a DNN. As DNNs can be evaluated rather efficiently, using the DNN instead of the original MPC is desirable.

4.3 Counterexample Analysis and Refinement

If $\mathcal{A}$ is deemed unsafe, we expect the reachability tool to output a counterexample in the form of a run. There is one main reason why such a run could be *spurious*, i.e. it is not a witness of the MPC not being r-competitive. Namely, the abstractions $\mathcal{D}$ (representing the MPC) or $\mathcal{N}$ (representing alternative controllers) might be too coarse. For the specific case of $\mathcal{D}$, where a DNN is used to model the MPC, we describe sufficient conditions to determine if the counterexample is indeed spurious. If the counterexample is indeed deemed spurious, we can refine our abstraction by incorporating new data obtained from the counterexample and retraining the DNN. In general, though, refining $\mathcal{D}$ and $\mathcal{N}$ falls into one of the tasks for which our framework does not rely on automation.

4.4 Human in the Loop

There are three points in the framework, where human intervention is needed.

Modelling and Specification. First, the task of obtaining initial abstractions $\mathcal{D}$ and $\mathcal{N}$ of the controller and all alternative controllers, respectively, does require a human in the loop. Indeed, crafting hybrid automata is not something we expect from every control engineer. In our prototype described in the next section, we mention partial support for obtaining $\mathcal{D}$ and $\mathcal{N}$ automatically when the MPC is given in the language of a particular OCP and optimization library.

Reachability Analysis. Second, reachability being an undecidable problem, most reachability analysis tools can not only output safe and unsafe as results. Additionally, they might output an "unknown" status. In this case, revisiting the abstractions $\mathcal{D}$ and $\mathcal{N}$, or even changing the options with which the tool is being used may require human intervention. In fact, we see this as an additional abstraction-refinement step which is considerably harder to automate since there is an absence of a counterexample to work with.

Abstraction Refinement. Finally, our framework does not say what to do if the counterexample being spurious is due to $\mathcal{N}$ being too coarse an approximation. This scenario can occur when $\mathcal{N}$ is purposefully modeled to discretize or approximate certain behaviors of alternative controllers to facilitate reachability analysis. However, for $\mathcal{D}$, we offer automation support by proposing the retraining of our DNN in the implementation. It might actually be needed to change the architecture of the DNN to obtain a better abstraction. This process can be automated, as increasing the number of layers is often sufficient according to the universal approximation theorem [4].

5 Implementation and Evaluation

We now present our implementation of the CEGAR-based competitive analysis method presented in the previous section, along with two case studies used for evaluation: the cart pendulum and an instance of motion planning.

5.1 Competitive Analysis Toolchain

Figure 3b gives a visual depiction of the toolchain in the form of a flowchart. Starting from the top, `D.xml, N.xml` are XML files encoding hybrid automata $\mathcal{D}$ and $\mathcal{N}$, respectively, in the SpaceEx modeling language [8]. The automaton $\mathcal{D}$ represents the controller, which could be a model predictive controller (MPC), and $\mathcal{N}$ represents a class of controllers that the MPC is compared against—see also Sect. 4.1. We use the HyST [2] translation tool for hybrid automata to generate the parallel composition $\mathcal{D} \,||\, \mathcal{N}$ (encoded in `DxN.xml`, again in the SpaceEx language). The composed automaton, along with the trained DNN and the property to be verified, are fed as inputs to Verisig. Verisig [13] is a tool that verifies the safety properties of closed-loop systems with neural network components. The tool takes a hybrid automaton, a trained neural network, and property specification files as inputs. It performs the reachability analysis and provides safety verification result. We then parse the output of Verisig to determine whether $\mathcal{D}$ is competitive enough (`parser.py`). If this is not the case, we realize a sound check to determine if the counterexample is spurious, in which case we use it to extend our dataset and further train the DNN.

5.2 Initial Abstraction and Training

Our toolchain is finetuned to work well for hybrid systems modeled in a tool called *Rockit* and MPCs obtained using the same tool. Rockit, which stands for Rapid Optimal Control Kit, is a tool designed to facilitate the rapid prototyping of optimal control problems, including iterative learning, model predictive control, system identification, and motion planning [9].

Our toolchain includes a utility that interfaces with the API of Rockit to automatically generate the hybrid automata $\mathcal{D}$ and $\mathcal{N}$ from a model of a control problem. While the use of Rockit is convenient, it is not required by our toolchain.

Based on a dataset (in our examples, we obtain it from Rockit), we train a DNN using *behavioral cloning*: we try to learn the behavior of an *expert* (in our case, the MPC) and replicate it. For this, we make use of the *Dagger algorithm* [18], which, after an initial round of training on the dataset from Rockit, will simulate traces using the DNN. The points that the neural network visits along these traces are then given to the expert, and the output of the expert is recorded. These new points and outputs are appended to the first dataset, and this new dataset is used to train a second DNN. This iterative process is done multiple times to make the DNN more robust. In all of our experiments, the TensorFlow framework [1] was used for the creation and training of the DNN.

5.3 Reachability Status

The regret property, encoded as a reachability property as is done in the proof of Theorem 1, is specified in the property file `Property.yml`, which also includes the initial states of $\mathcal{D} \,||\, \mathcal{N}$. Verisig provides three possible results: "safe" if no property violation is found, "unsafe" if there is a violation, and "unknown" if the property could not be verified, potentially due to a significant approximation error. In the latter two cases, a counterexample file (CE file) is generated.

5.4 Counterexample Analysis and Retraining

If the result is "unsafe", the next step is to compare the counterexample trajectory against the dataset generated from the controller code. If a matching trajectory is found, it indicates a real counterexample, meaning that this trajectory could potentially occur in the actual controller, and no further action is required. If a matching trajectory is not found, then it is a spurious counterexample that requires either retraining the DNN or fix(es) in $\mathcal{D}||\mathcal{N}$. Our toolchain automatically validates the counterexample by comparing the trajectories from Verisig and the controller as implemented in Rockit. To do so, since Rockit uses the floating-point representation of real numbers, we choose a decimal precision of $\epsilon = 10^{-3}$ for the comparison. In the case of a spurious counterexample that requires retraining the DNN, we update the existing dataset using Rockit to obtain additional labeled data based on the trajectory from the CE file.

The CE file from Verisig represents state variable values using interval arithmetic, while the controller dataset contains state variable values in $\mathbb{R}$ without intervals. To accommodate this difference, we choose to append to the dataset new entries: (a) the lower bounds of input intervals, (b) the upper bounds, and (c) a range[2] of intermediate input values within the intervals. For each of these, we also include the corresponding controller outputs. The generation of the updated dataset and the retraining of the DNN are performed automatically by our toolchain. A DNN trained on the new dataset is then fed to Verisig again along with `DxN.xml` and the `Property.yml`. This way, the CEGAR loop is repeated until one of the following conditions is true: (a) the counterexample is real, or (b) a maximum number of retraining iterations (determined by the user) is reached.

5.5 Experiments

In the sequel, we use our tool to analyze two control problems that have been implemented using the Rockit framework. The research questions we want to answer with the forthcoming empirical study are the following.

RQ1 Can we have a fully automated tool to perform the competitive analysis?
RQ2 Is the toolchain scalable? Why or why not?
RQ3 Does the approach help to improve confidence in (finite-horizon) competitivity of controllers?
RQ4 Does the approach help find bugs in controller design?

We now briefly introduce the two case studies, their dynamics, and how each of them are modeled so that our toolchain can be used to analyze them.

[2] Our toolchain splits each interval into n equally large segments and adds all points in the resulting lattice. In our experiments, we use $n = 4$.

Cart pendulum is a classic challenge in control theory and dynamics [7]. In it, an inverted pendulum is mounted on a cart that can move horizontally via an electronic servo system. The objective is to minimize a cost $J = F^2 + 100 * pos^2$, where F is the force applied to the cart and pos indicates the position of the cart. The values of F and pos are constrained within the range of $[-2, 2]$. The dynamics of the cart correspond to the physics of the system and depend on the mass of the cart and the pendulum and the length of the pendulum.

While the proof of Theorem 1 provides a sound way to model all alternative controllers in the form of $\mathcal{N}$, the construction combines continuous dynamics and non-determinism. Current hybrid automata tools do not handle non-trivial combinations of these two elements very well. Hence, we have opted to discretize the choice of control values for alternative controllers. Every time the DNN is asked for new control variable values in $\mathcal{D}$, the automaton $\mathcal{N}$ non-deterministically chooses new alternative values from a finite subset fixed by us a priori.

Motion planning involves computing a series of actions to move an object from one point to another while satisfying specific constraints [15]. In our case study, an MPC is used to plan the motion of an autonomous bicycle that is expected to move along a curved path on a 2D plane using a predefined set of waypoints. To prevent high-speed and skidding, the velocity (V) and the turning rate (δ, in radians) are constrained in the ranges $0 \leq V \leq 1$ and $-\pi/6 \leq \delta \leq \pi/6$. The objective is to minimize the sum of squared estimate of errors between the actual path taken by the bicycle and the reference path. Intuitively, the more the controller deviates from the reference path, the higher its cost.

Like in the cart pendulum case study, we discretize the alternative control variable valuations. A big difference is that the cost has both a *Mayer term* and a *Lagrangian* that depend on the location of the bicycle and the waypoints in an intricate way. In terms of modelling, this means that $\mathcal{D}$ and $\mathcal{N}$ have to "compute" closest waypoints relative to the current position of the bicycle.

Discussion. Towards an answer for **RQ1**, we can say that while our toolchain[3] somewhat automates our CEGAR, it still requires manual work (e.g. the initial training and choice of DNN architecture). Moreover, in the described case studies, we did not observe an MPC DNN that is labeled as competitive. This may be due to (over)approximations incurred by our framework and our use of Verisig. Despite this, we can answer **RQ4** positively as our toolchain allowed us to spot a bug hidden in the Rockit MPC solution for the cart pendulum. We observed in early experiments that the MPC was not competitive and short (run) examples of this were quickly found by Verisig. We then found that the objective function in Rockit was indeed not as intended by the developers.

The DNNs do show a trend towards copying the behavior of the MPC (see Fig. 4) even though we retrain a new DNN from scratch after each (spurious)

[3] All graphs and numbers can be reproduced using scripts from: https://doi.org/10.5281/zenodo.8255730.

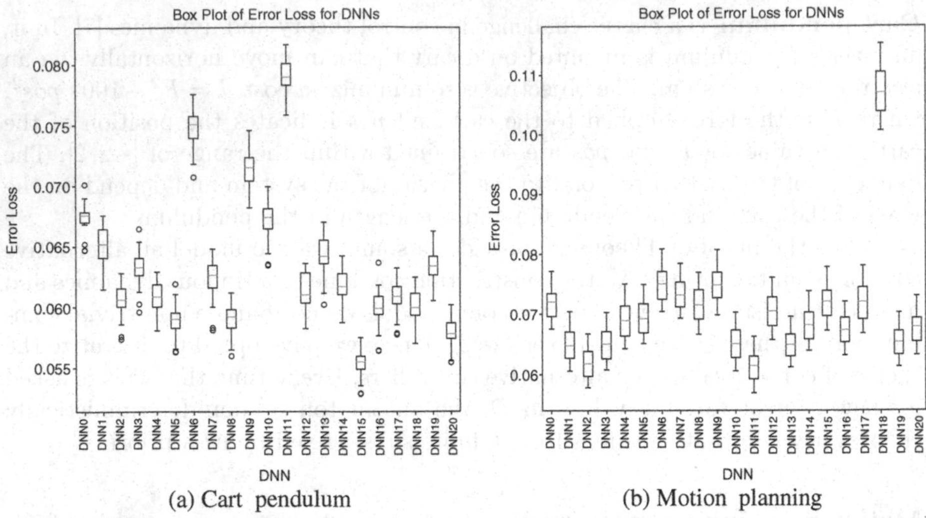

(a) Cart pendulum (b) Motion planning

Fig. 4. Boxplots showing the training losses of all DNNs against all test sets

counterexample obtained via Verisig and we (purposefully) randomize the choice of test and training set in each iteration. We do this to increase variability in the set of behaviors and the counterexamples used to extend the dataset. In the cart pendulum case study, we observe that in the iterations 2, 7, and 11, the number of discrete time steps during which the corresponding DNN can act while remaining competitive is larger than in the initial iteration. Hence, for **RQ3**, we conclude our toolchain can indeed help increase reliability in the DNN proxy being competitive, albeit only for a finite horizon. On the negative side, experiments for 20 iterations of retraining from spurious counterexamples take more than 90 min in both our case studies. This leads us to conclude that our toolchain does not yet scale as required for industrial-size case studies (**RQ2**).

6 Conclusion

Based on our theoretical developments to link the regret problem with the classical reachability problem, we proposed a CEGAR-based approach to realize the competitive analysis of MPCs via neural networks as proxies. We also presented an early proof-of-concept implementation of the approach. Now that we have a baseline, we strongly believe improvements in the form of algorithms and dedicated tools will allow us to improve our framework to the point where it scales for interesting classes of hybrid systems.

References

1. Abadi, M., et al.: TensorFlow: a system for large-scale machine learning. In: keeton, K., Roscoe, T. (eds.) 12th USENIX Symposium on Operating Systems Design and Implementation, OSDI 2016, Savannah, GA, USA, 2–4 November 2016, pp. 265–283. USENIX Association (2016). https://www.usenix.org/conference/osdi16/technical-sessions/presentation/abadi
2. Bak, S., Bogomolov, S., Johnson, T.T.: HYST: a source transformation and translation tool for hybrid automaton models. In: Proceedings of the 18th International Conference on Hybrid Systems: Computation and Control, pp. 128–133 (2015)
3. Bratko, I., Urbančič, T., Sammut, C.: Behavioural cloning: phenomena, results and problems. IFAC Proc. Vol. **28**(21), 143–149 (1995)
4. Chen, T., Chen, H.: Universal approximation to nonlinear operators by neural networks with arbitrary activation functions and its application to dynamical systems. IEEE Trans. Neural Netw. **6**(4), 911–917 (1995)
5. Chen, X., Sankaranarayanan, S.: Reachability analysis for cyber-physical systems: are we there yet? In: Deshmukh, J.V., Havelund, K., Perez, I. (eds.) NASA Formal Methods, NFM 2022. LNCS, vol. 13260. Springer, Cham (2022). https://doi.org/10.1007/978-3-031-06773-0_6
6. Clavière, A., Dutta, S., Sankaranarayanan, S.: Trajectory tracking control for robotic vehicles using counterexample guided training of neural networks. In: Benton, J., Lipovetzky, N., Onaindia, E., Smith, D.E., Srivastava, S. (eds.) Proceedings of the Twenty-Ninth International Conference on Automated Planning and Scheduling, ICAPS 2018, Berkeley, CA, USA, 11–15 July 2019, pp. 680–688. AAAI Press (2019). https://ojs.aaai.org/index.php/ICAPS/article/view/3555
7. Fantoni, I., Lozano, R., Lozano, R.: Non-linear Control for Underactuated Mechanical Systems. Springer, London (2002). https://doi.org/10.1007/978-1-4471-0177-2
8. Frehse, G., et al.: SpaceEx: scalable verification of hybrid systems. In: Gopalakrishnan, G., Qadeer, S. (eds.) CAV 2011. LNCS, vol. 6806, pp. 379–395. Springer, Heidelberg (2011). https://doi.org/10.1007/978-3-642-22110-1_30
9. Gillis, J., Vandewal, B., Pipeleers, G., Swevers, J.: Effortless modeling of optimal control problems with rockit. In: 39th Benelux Meeting on Systems and Control, Elspeet, The Netherlands, 10 March 2020–12 March 2020 (2020)
10. Henzinger, T.A., Kopke, P.W., Puri, A., Varaiya, P.: What's decidable about hybrid automata? J. Comput. Syst. Sci. **57**(1), 94–124 (1998). https://doi.org/10.1006/jcss.1998.1581
11. Hertneck, M., Köhler, J., Trimpe, S., Allgöwer, F.: Learning an approximate model predictive controller with guarantees. IEEE Control. Syst. Lett. **2**(3), 543–548 (2018). https://doi.org/10.1109/LCSYS.2018.2843682
12. Hunter, P., Pérez, G.A., Raskin, J.: Reactive synthesis without regret. Acta Informatica **54**(1), 3–39 (2017). https://doi.org/10.1007/s00236-016-0268-z
13. Ivanov, R., Carpenter, T., Weimer, J., Alur, R., Pappas, G., Lee, I.: Verisig 2.0: verification of neural network controllers using Taylor model preconditioning. In: Silva, A., Leino, K.R.M. (eds.) CAV 2021. LNCS, vol. 12759, pp. 249–262. Springer, Cham (2021). https://doi.org/10.1007/978-3-030-81685-8_11
14. Julian, K.D., Lopez, J., Brush, J.S., Owen, M.P., Kochenderfer, M.J.: Policy compression for aircraft collision avoidance systems. In: 2016 IEEE/AIAA 35th Digital Avionics Systems Conference (DASC), pp. 1–10. IEEE (2016)
15. LaValle, S.M.: Planning Algorithms. Cambridge University Press (2006)

16. Muvvala, K., Amorese, P., Lahijanian, M.: Let's collaborate: regret-based reactive synthesis for robotic manipulation. In: 2022 International Conference on Robotics and Automation, ICRA 2022, Philadelphia, PA, USA, 23–27 May 2022, pp. 4340–4346. IEEE (2022). https://doi.org/10.1109/ICRA46639.2022.9812298
17. Ross, S., Bagnell, D.: Efficient reductions for imitation learning. In: Teh, Y.W., Titterington, D.M. (eds.) Proceedings of the Thirteenth International Conference on Artificial Intelligence and Statistics, AISTATS 2010. JMLR Proceedings, Chia Laguna Resort, Sardinia, Italy, 13–15 May 2010, vol. 9, pp. 661–668. JMLR.org (2010). http://proceedings.mlr.press/v9/ross10a.html
18. Ross, S., Gordon, G.J., Bagnell, D.: A reduction of imitation learning and structured prediction to no-regret online learning. In: Gordon, G.J., Dunson, D.B., Dudík, M. (eds.) Proceedings of the Fourteenth International Conference on Artificial Intelligence and Statistics, AISTATS 2011. JMLR Proceedings, Fort Lauderdale, USA, 11–13 April 2011, vol. 15, pp. 627–635. JMLR.org (2011). http://proceedings.mlr.press/v15/ross11a/ross11a.pdf

Matching Patterns with Variables Under Simon's Congruence

Pamela Fleischmann[1], Sungmin Kim[3], Tore Koß[2], Florin Manea[2], Dirk Nowotka[1], Stefan Siemer[2(✉)], and Max Wiedenhöft[1]

[1] Department of Computer Science, Kiel University, Kiel, Germany
{fpa,dn,maw}@informatik.uni-kiel.de
[2] Department of Computer Science, University of Göttingen, Göttingen, Germany
{tore.koss,florin.manea,stefan.siemer}@cs.uni-goettingen.de
[3] Department of Computer Science, Yonsei University, Seoul, Republic of Korea
rena_rio@yonsei.ac.kr

Abstract. We introduce and investigate a series of matching problems for patterns with variables under Simon's congruence. Our results provide a thorough picture of these problems' computational complexity.

1 Introduction

A *pattern with variables* is a string α consisting of *constant letters* (or *terminals*) from a finite alphabet $\Sigma = \{1, \ldots, \sigma\}$, of size $\sigma \geq 2$, and *variables* from a potentially infinite set $\mathcal{X}$, with $\Sigma \cap \mathcal{X} = \emptyset$. Such a pattern α is mapped by a function h, called *substitution*, to a word by substituting the variables occurring in α by strings of constants, i.e., strings over Σ. For example, the pattern $\alpha = xx\mathtt{abab}yy$ can be mapped to the string of constants aaaaababbb by the substitution h defined by $h(x) = \mathtt{aa}$, $h(y) = \mathtt{b}$. In this framework, $h(\alpha)$ denotes the word obtained by substituting every occurrence of a variable x in α by $h(x)$ and leaving all the constants unchanged. If a pattern α can be mapped to a string of constants w, we say that α matches w; the problem of deciding, given a pattern α with variables and a string of constants w, whether there exists a substitution which maps α to w is called the *(exact) matching problem*, Match.

Exact Matching Problem: Match(α, w)
Input: Pattern α, $|\alpha| = m$, word w, $|w| = n$.
Question: Is there a substitution h with $h(\alpha) = w$?

Match is a heavily studied problem, which appears frequently in various areas of theoretical computer science. Initially, this problem was considered in language theory (e.g., pattern languages [5]) or combinatorics on words (e.g., unavoidable patterns [52]), with connections to algorithmic learning theory (e.g., the theory of descriptive patterns for finite sets of words [5,16,63]), and has by now found interesting applications in string solving and the theory of word equations [51], stringology (e.g., generalised function matching [4]), the theory of extended regular expressions with backreferences [8,24,29,30]), or database theory (mainly in relation to document spanners [15,25,27,28,46,58,59]).

O. Bournez et al. (Eds.): RP 2023, LNCS 14235, pp. 155–170, 2023.
https://doi.org/10.1007/978-3-031-45286-4_12

`Match` is NP-complete in general [5], and a more detailed image of the parameterised complexity of the matching problem is revealed in [17–19,55,57,62] and the references therein. A series of classes of patterns, defined by structural restrictions, for which `Match` is in P were identified [13,17,55]; moreover, for most of these classes, `Match` is $W[1]$-hard [14] w.r.t. the structural parameters used to define the respective classes.

Recently, Gawrychowski et al. [35,36] studied `Match` in an approximate setting: given a pattern α, a word w, and a natural number ℓ, one has to decide if there exists a substitution h such that $D(h(\alpha), w) \leq \ell$, where D is either the Hamming [35] or the edit distance [36]. Their results offered, once more, a detailed understanding of the approached matching problems' complexity (in general, and for classes of patterns defined by structural restrictions). The problems discussed in [35,36] can be seen in a more general setting: given a pattern α and a word w, decide if there exists a substitution h such that $h(\alpha)$ is similar to w, w.r.t. some similarity measure (Hamming resp. edit distance in [35,36] or string equality for exact `Match`). Thus, it seems natural to also consider various other string-equivalence relations as similarity measures, such as (k-)abelian equivalence [41,42] or k-binomial equivalence [26,50,56]. Here, we consider an approximate variant of `Match` using Simon's congruence $\sim_k$ [65].

Matching under Simon's Congruence: `MatchSimon`(α, w, k)
Input: Pattern α, $|\alpha| = m$, word w, $|w| = n$, and number $k \in [n]$.
Question: Is there a substitution h with $h(\alpha) \sim_k w$?

Let us recall the definition of Simon's congruence. A string u is a *subsequence* of a string v if u results from v by deleting some letters of v. Subsequences are well studied in the area of combinatorics of words and combinatorial pattern matching, and are well-connected to other areas of computer science (e.g., the handbook [51] or the survey [48] and the references therein). Let $\mathbb{S}_k(v)$ be the set of all subsequences of a given string v up to length $k \in \mathbb{N}_0$. Two strings v and v' are k-Simon congruent iff $\mathbb{S}_k(v) = \mathbb{S}_k(v')$. The problem of testing whether two given strings are k-Simon congruent, for a given k, was introduced by Imre Simon in his PhD thesis [64] as a similarity measure for strings, and was intensely studied in the combinatorial pattern matching community (see [11,20,31,39,66,67] and the references therein), before being optimally solved in [6,33]. Another interesting extension of these results, discussed in [44], brings us closer to the focus of this paper. There, the authors present an efficient solution for the following problem: given two words w, u and a natural number k, decide whether there exists a factor of w which is k-Simon congruent to u; this is `MatchSimon` with the input pattern $\alpha = xuy$ for variables x, y. Thus, it seems natural to consider, in a general setting, the problem of checking whether one can map a given pattern α to a string which is similar to w w.r.t. $\sim_k$. Moreover, there is another way to look at this problem, which seems interesting to us: the input word w and the number k are a succinct representation of $\mathbb{S}_k(w)$. So, `MatchSimon`(α, w, k) asks whether we can assign the variables of α in such a way that we reach a word describing the target set of subsequences of length k, as well.

One of the congruence-classes of Σ^* w.r.t. $\sim_k$ received a lot of attention: the class of k-subsequence universal words, those words which contain all k-length words as subsequences. This class was first studied in [40,60], and further investigated in [1,2,6,12,21,22,47,61] in contexts related to and motivated by formal languages, automata theory, or combinatorics, where the notion of universality is central (see [7,9,34,37,49,53,54] for examples in this direction). The motivation of studying k-subsequence universal words is thoroughly discussed in [12]. Here, we consider the following problem:

Matching a Target Universality: MatchUniv(α, k)
Input: Pattern α, $|\alpha| = m$, and $k \in \mathbb{N}_0$.
Question: Is there a substitution h with $\iota(h(\alpha)) = k$?

In this problem, $\iota(w)$ (the universality index of w) is the largest integer ℓ for which w is ℓ-subsequence universal. Note that MatchUniv can be formulated in terms of MatchSimon: the answer to MatchUniv(α, k) is yes iff the answer to MatchSimon($\alpha, (1 \cdots \sigma)^k, k$) is yes and the answer to MatchSimon(α, $(1 \cdots \sigma)^{k+1}, k+1$) is no. However, there is an important difference: for MatchUniv we are not explicitly given the target word w, whose set of k-length subsequences we want to reach; instead, we are given the number k which represents the target set more compactly (using only $\log k$ bits).

In the problems introduced above, we attempt to match (or reach), starting with a pattern α, the set of subsequences defined by a given word w (given explicitly or implicitly). A well-studied extension of Match is the satisfiability problem for word equations, where we are given two patterns α and β and are interested in finding an assignment of the variables that maps both patterns to the same word (see, e.g., [51]). This problem is central both to combinatorics on words and to the applied area of string solving [3,38]. In this paper, we extend MatchSimon to the problem of solving word equations under $\sim_k$, defined as follows.

Word Equations under Simon's Congruence: WESimon(α, β, k)
Input: Patterns α, β, $|\alpha| = m$, $|\beta| = n$, and $k \in [m+n]$.
Question: Is there a substitution h with $h(\alpha) \sim_k h(\beta)$?

Besides introducing these natural problems, our paper presents a rather comprehensive picture of their computational complexity. We start with MatchUniv, the most particular of them and whose input is given in the most compact way. In Sect. 3 we show that MatchUniv is NP-complete, and also present a series of structurally restricted classes of patterns, for which it can be solved in polynomial time. In Sect. 4, we approach MatchSimon and show that it is also NP-complete; some other variants of this problem, both tractable and intractable, are also discussed. Finally, in Sect. 5, we discuss WESimon and its variants, and characterise their computational complexity. The paper ends with a section pointing to a series of future research directions.

2 Preliminaries

Let $\mathbb{N} = \{1, 2, \ldots\}$ be the set of natural numbers. Let $[n] = \{1, \ldots, n\}$ and $[m:n] = [n] \setminus [m-1]$, for $m, n \in \mathbb{N}, m < n$. $\mathbb{N}_0$ denotes $\mathbb{N} \cup \{0\}$.

For a finite set $\Sigma = [\sigma]$, called *alphabet*, Σ^* denotes the set of all words (or strings) over Σ, with ε denoting the empty word. For $w \in \Sigma^*$, $|w|$ denotes its length, while $|w|_{\mathtt{a}}$ denotes the number of occurrences of $\mathtt{a} \in \Sigma$ in w. Further, $\Sigma^{\leq k}$ (resp. Σ^k) denotes the set of all words over Σ up to (resp. of) length $k \in \mathbb{N}$. Let $w[i]$ denote the ith letter in the string w, and let $\mathtt{alph}(w) = \{\mathtt{a} \mid |w|_{\mathtt{a}} \geq 1\}$ denote the set of different letters in w. To access the first occurrence of a letter $\mathtt{a} \in \Sigma$ after a position $i \in [|w|]$ in a word $w \in \Sigma^*$, define the *X-ranker* as a mapping $\mathtt{X} : \Sigma^* \times ([|w|] \cup \{0, \infty\}) \times \Sigma \to [|w|] \cup \{\infty\}$ with $(w, i, \mathtt{a}) \mapsto \min(\{j \in [i+1 : |w|] \mid w[j] = \mathtt{a}\} \cup \{\infty\})$ (cf. [68]). Notice that a lookup table for all possible X-ranker evaluations for some given $w \in \Sigma^*$ can be computed in linear time in $|w|$, where each item can be accessed in constant time [6,20]. In the special case of $\mathtt{X}(w, 0, \mathtt{a})$, we call this occurrence of $\mathtt{a}$ the *signature letter* $\mathtt{a}$ of w, for all $\mathtt{a} \in \mathtt{alph}(w)$. A *permutation* γ of an alphabet Σ is a string in Σ^σ with $\mathtt{alph}(\gamma) = \Sigma$. A string u is a *subsequence* of a string w if there exists a strictly increasing integer sequence $0 < i_1 < i_2 < \ldots < i_{|u|} \leq |w|$ with $w[i_j] = u[j]$ for all $j \in [|u|]$. For a given $k \in \mathbb{N}_0$, we use $\mathbb{S}_k(w)$ as the set of all subsequences of w with length at most k. A subsequence u of w is called a *substring* of w if there exists a position i of w such that $u = w[i]w[i+1] \cdots w[i+|u|-1]$. We write $w[i:j]$ for $w[i]w[i+1] \cdots w[j]$ for $1 \leq i \leq j \leq |w|$. Substrings $w[1:j]$ (resp., $w[i:|w|]$) are called *prefixes* (resp., *suffixes*) of w.

Two words $w_1, w_2 \in \Sigma^*$ are called *Simon k-congruent* ($w_1 \sim_k w_2$) if $\mathbb{S}_k(w_1) = \mathbb{S}_k(w_2)$ [65]. A word $w \in \Sigma^*$ is called *k-subsequence universal* (or short k-universal) for some $k \in \mathbb{N}$ if $\mathbb{S}_k(w) = \Sigma^{\leq k}$; this means that $w \sim_k (1 \cdots \sigma)^k$. The largest $k \in \mathbb{N}_0$ such that w is k-universal is the *universality index* of w, denoted by $\iota(w)$. In [39], Hébrard introduced the following unique factorisation of words.

Definition 1. *The* arch factorisation *of a word $w \in \Sigma^*$ is defined by $w = \mathtt{arch}_1(w) \cdots \mathtt{arch}_k(w)\mathtt{rest}(w)$ for some $k \in \mathbb{N}_0$ such that there exists a sequence $(i_j)_{j \leq k}$ with $i_0 = 0$, $i_j = \max\{\mathtt{X}(w, i_{j-1}, \mathtt{a}) \mid \mathtt{a} \in \Sigma\}$ for all $j \geq 1$, $\mathtt{arch}_j(w) = w[i_{j-1}+1 : i_j]$ whenever $1 \leq i_j < \infty$, and $\mathtt{rest}(w) = w[i_j : |w|]$, if $i_{j+1} = \infty$.*

Clearly, the number of arches of $w \in \Sigma^*$ is exactly $\iota(w)$. Extending the notion of arch factorisation, we define the arches and rest of $w \in \Sigma^*$ for $\mathtt{a} \in \mathtt{alph}(w)$ (cf. the arch jumping functions introduced in [61]) as well as the universality index for the respective letter $\mathtt{a}$. That is, we perform the arch factorisation and obtain the universality index for the suffix of w that starts after the first occurrence of $\mathtt{a}$.

Definition 2. *Let $w \in \Sigma^*$, $\mathtt{a} \in \mathtt{alph}(w)$, and $j \in [\iota(w)]$. The arches of signature letters are defined by $\mathtt{arch}_{\mathtt{a},j}(w) = \mathtt{arch}_j(w[\mathtt{X}(w,0,\mathtt{a})+1 : |w|])$ and $\mathtt{rest}_{\mathtt{a}}(w) = \mathtt{rest}(w[\mathtt{X}(w,0,\mathtt{a})+1 : |w|])$. The* universality index of $\mathtt{a}$ *is*

$\iota_{\mathtt{a}}(w) = \iota(w[X(w,0,\mathtt{a})+1 : |w|])$. *The last index w.r.t.* w *of* $\mathtt{arch}_{\mathtt{a},j}(w)$ *is defined as* $\mathtt{archEnd}_{\mathtt{a},j}(w) = \mathtt{X}(w,0,\mathtt{a}) + \sum_{i=1}^{j} |\mathtt{arch}_{\mathtt{a},i}(w)|$.

Now, we are interested in the smallest substrings of w that allow the completion of rests of specific prefixes of w to full arches. Hence, we define *marginal sequences*, which are breadth-first orderings of σ parallel arch factorisations, each starting after a signature letter of the word.

Definition 3. *Let* $w \in \Sigma^*$ *and* γ *be a permutation of* Σ *such that* $\mathtt{X}(w,0,\gamma[i])$ *is increasing w.r.t.* $i \in [\sigma]$. *From the arches for signature letters, we define the* marginal sequence *of integers of* $w \in \Sigma^*$ *inductively by* $\mathtt{M}_0(w) = 0$, $\mathtt{M}_i(w) = \mathtt{X}(w,0,\gamma[i])$ *for all* $i \in [\sigma]$, *and* $\mathtt{M}_{i\sigma+j}(w) = \mathtt{archEnd}_{\gamma[j],i}(w)$ *for* $j \in [\sigma]$, $i \in [\iota_{\gamma[j]}(w)]$. *Let* $\mathtt{M}_\infty(w) = |w|$ *denote the last element of the sequence.*

The sequence is called *marginal* because, for $j \in [\sigma]$, $w[\mathtt{M}_{i\sigma+j-1}(w)+1 : \mathtt{M}_{i\sigma+j}(w)]$ is the smallest prefix p of $w[\mathtt{M}_{i\sigma+j-1}(w)+1 : |w|]$ such that $\iota_{\gamma[j]}(w[1 : \mathtt{M}_{i\sigma+j-1}(w)]p) = i$. Note that the marginal sequence $\mathtt{M}_i(w)$ is non-decreasing. In the following, we define a slight variation of the *subsequence universality signature* $\mathtt{s}(w)$ introduced in [61].

Definition 4. *1. For* $w \in \Sigma^*$, *the* subsequence universality signature $\mathtt{s}(w)$ *of* w *is defined as the 3-tuple* $(\gamma, \mathcal{K}, \mathcal{R})$ *with a permutation* γ *of* $\mathtt{alph}(w)$, *where* $\mathtt{X}(w,0,\gamma[i]) > \mathtt{X}(w,0,\gamma[j]) \Leftrightarrow i > j$ *(*γ *consists of the letters of* $\mathtt{alph}(w)$ *in order of their first appearance in* w*) and two arrays* $\mathcal{K}$ *and* $\mathcal{R}$ *of length* σ *with* $\mathcal{K}[i] = \iota_{\gamma[i]}(w)$ *and* $\mathcal{R}[i] = \mathtt{alph}(\mathtt{rest}_{\gamma[i]}(w))$ *for all* $i \in [|\mathtt{alph}(w)|]$. *For all* $i \in [\sigma] \setminus \mathtt{alph}(w)$, *we have* $\mathcal{R}[i] = \Sigma$ *and* $\mathcal{K}[i] = -\infty$.
2. Conversely, for a permutation γ' *of* Σ, *an integer array* $\mathcal{K}'$ *and an alphabet array* $\mathcal{R}'$ *both of length* σ, *we say that the tuple* $(\gamma', \mathcal{K}', \mathcal{R}')$ *is a* valid *signature if there exists a string* w *that satisfies* $\mathtt{s}(w) = (\gamma,' \mathcal{K}', \mathcal{R}')$.

Note that, for $k_i = \iota_{\gamma[i]}(w)$, we have $\mathcal{R}[i] = \mathtt{alph}(w[\mathtt{M}_{k_i\sigma+i}(w)+1 : \mathtt{M}_\infty(w)])$, since $\mathtt{rest}_{\gamma[i]}(w) = w[\mathtt{M}_{k_i\sigma+i}(w)+1 : \mathtt{M}_\infty(w)]$.

A central notion to this work is that of patterns with variables. From now on, we consider two alphabets: $\Sigma = [\sigma]$ is an alphabet of constants (or terminals), and $\mathcal{X}$ a (possibly infinite) alphabet of variables, with $\mathcal{X} \cap \Sigma = \emptyset$. A pattern α is a string from $(\mathcal{X} \cup \Sigma)^*$, i.e., a string containing both constants and variables. For a pattern α, $\mathtt{var}(\alpha) = \mathtt{alph}(\alpha) \cap \mathcal{X}$ denotes the set of *variables* in α, while $\mathtt{term}(\alpha) = \mathtt{alph}(\alpha) \cap \Sigma$ is the set of *constants* (*terminals*) in α.

Definition 5. *A* substitution $h : (\mathcal{X} \cup \Sigma)^* \to \Sigma^*$ *is a morphism that acts as the identity on* Σ *and maps each variable of* $\mathcal{X}$ *to a (potentially empty) string over* Σ. *That is,* $h(\mathtt{a}) = \mathtt{a}$ *for all* $\mathtt{a} \in \Sigma$ *and* $h(x) \in \Sigma^*$ *for all* $x \in \mathcal{X}$. *We say that pattern* α matches *string* w *over* Σ *under a binary relation* $\sim$ *if there exists a substitution* h *that satisfies* $h(\alpha) \sim w$.

In the above definition, if $\sim$ is the string equality $=$, we say that the pattern α *matches* the string w instead of saying that α *matches* w under $=$.

The problems addressed in this paper, introduced in Sect. 1, deal with matching patterns to words under Simon's congruence $\sim_k$. For these problems, the input consists of patterns, words, and a number k. In general, we assume that each letter of Σ appears at least once, in at least one of the input patterns or words. E.g., for input pattern α and word w we assume that $\Sigma = \mathtt{term}(\alpha) \cup \mathtt{alph}(w)$. Hence, σ is upper bounded by the total length of the input words and patterns. Similarly, the total number of variables occurring in the input patterns is upper bounded by the total length of these patterns. However, in this paper, although the number of variables is not restricted, we assume that σ is a constant, i.e., $\sigma \in O(1)$. Clearly, the complexity lower bounds proven in this setting for the analysed problems are stronger while the upper bounds are weaker than in the general case, when no restriction is placed on σ. Note, however, that $\sigma \in O(1)$ is not an unusual assumption, being used in, e.g., [20].

Part of the results reported here are of algorithmic nature. The computational model we use for these is the unit-cost Word RAM model with memory words of logarithmic size, briefly presented in the full version of this paper [23] or in, e.g., [10].

3 MatchUniv

In this section, we discuss the `MatchUniv` problem. In this problem, we are given a pattern α and a natural number $k \leq n$, and we want to check the existence of a substitution h with $\iota(h(\alpha)) = k$. Note that $\iota(h(\alpha)) = k$ means both that $h(\alpha)$ is k-universal and that it is not $(k+1)$-universal. A slightly relaxed version of the problem, where we would only ask for $h(\alpha)$ to be k-universal is trivial (and, therefore, not interesting): the answer, in that case, is always positive, as it is enough to map one of the variables of α to $(1 \cdots \sigma)^k$. The main result of this section is that `MatchUniv` is NP-complete, which we will show in the following. To show that `MatchUniv`(α, k) is NP-hard, we reduce `3CNFSAT` (3-satisfiability in conjunctive normal form) to `MatchUniv`(α, k). We provide several gadgets allowing us to encode a `3CNFSAT`-instance φ as an `MatchUniv`-instance (α, k). Finally, we show that we can find a substitution h for the instance (α, k), such that $\iota(h(\alpha)) = k$, iff φ is satisfiable. We begin by recalling `3CNFSAT`.

3-Satisfiability for formulas in conjunctive normal form, `3CNFSAT`.

Input: Clauses $\varphi := \{c_1, c_2, \ldots, c_m\}$, where $c_j = (y_j^1 \vee y_j^2 \vee y_j^3)$ for $1 \leq j \leq m$, and y_j^1, y_j^2, y_j^3 from a finite set of boolean variables $X := \{x_1, x_2, \ldots, x_n\}$ and their negations $\bar{X} := \{\bar{x}_1, \bar{x}_2, \ldots, \bar{x}_n\}$.

Question: Is there an assignment for X, which satisfies all clauses of φ?

It is well-known that `3CNFSAT` is NP-complete (see [32,43] for a proof). With this result at hand, we can prove the following lower bound. Full details and figures illustrating the gadgets are given in the full version [23].

Lemma 1. `MatchUniv` *is* NP-*hard.*

Proof. We reduce 3CNFSAT to MatchUniv(α, k). Let us consider an instance of 3CNFSAT: formula φ given by m clauses $\varphi := \{c_1, c_2, \ldots, c_m\}$ over n variables $X := \{x_1, x_2, \ldots, x_n\}$ (for simplicity in notation we define $N = n + m$). We map this 3CNFSAT instance to an instance (α, k) of MatchUniv(α, k) with $k = 5n + m + 2$, the alphabet $\Sigma := \{0, 1, \#, \$\}$ and the variable set $\mathcal{X} := \{z_1, z_2, \ldots, z_n, u_1, u_2, \ldots, u_n\}$. More precisely, we want to show that there exists a substitution h to replace all the variables in α with constant words, such that $\iota(h(\alpha)) = 5n + m + 2$, if and only if the boolean formula φ is satisfiable. Our construction can be performed in polynomial time with respect to N. To present this construction, we will go through its building blocks, the so-called gadgets.

Before defining these gadgets, we introduce a renaming function for Boolean-variables $\rho : X \cup \bar{X} \to \mathcal{X}$ with $\rho(x_i) = z_i$ and $\rho(\bar{x}_i) = u_i$. Also, a substitution h with $\iota(h(\alpha)) = 5n + m + 2$ is called valid in the following.

The Binarisation Gadgets. These gadgets ensure the image of valid substitutions of z_i and u_i to be strings over $\{0, 1\}$.

At first, we construct the gadget $\pi_\# = (z_1 z_2 \cdots z_n u_1 u_2 \cdots u_n 01\,\$)^{N^6}\,\#$. We observe that for all possible substitutions h, we have two cases for the universality of the image of this gadget. On the one hand, assume that any of the variables is substituted under h by a string that contains a #. Then, the universality index of the image of this gadget will be $\iota(h(\pi_\#)) \geq N^6 > k$, which is too big for a valid substitution. On the other hand, when all the variables are substituted under h by strings that do not contain #, this gadget is mapped to a string which consists of exactly one arch because there is only one # at its very end. Thus, under a valid substitution h, the images of the variables z_i and u_i do not contain #. Note also that, in the arch factorisation of such a string $(h(\pi_\#)$, where h is a valid substitution) we have one arch and no rest. The gadget $\pi_\$ = (z_1 z_2 \cdots z_n u_1 u_2 \cdots u_n 01\,\#)^{N^6}\,\$$ is constructed analogously. This enforces that under a valid substitution h, the images of the variables z_i and u_i do not contain \$. In conclusion, the gadgets $\pi_\#$ and $\pi_\$$ ensure that under a valid substitution h, the images of the variables z_i and u_i contain only 0 and 1, i.e., they are binary strings.

The Boolean Gadgets. We use the following gadgets to force the image of each z_i and u_i to be either in 0^* or 1^*. Intuitively, mapping a variable z_i (respectively, u_i) to a string of the form 0^+ corresponds to mapping x_i (respectively, $\bar{x}_i$) to the Boolean value false. Similarly, mapping one of these string-variables to a string from 1^+ means mapping the corresponding boolean variable to true. For now, these gadgets just enforce that the image of any string-variable does not contain both 0 and 1; other gadgets will enforce that they are not mapped to empty words. We construct the gadget π_i^z (respectively π_i^u) for every string-variable z_i (respectively, u_i). More precisely, for all $i \in [n]$, we define two gadgets $\pi_i^z = (z_i\,\$\,\#)^{N^6}\,1001\,\$\,\#$ and $\pi_i^u = (u_i\,\$\,\#)^{N^6}\,1001\,\$\,\#$.

We now analyse the possible images of $\pi_i^z = (z_i\,\$\,\#)^{N^6}\,1001\,\$\,\#$ under various substitutions h. There are several ways in which z_i can be mapped to a string by h. Firstly, if the image of z_i contains both 0 and 1, then for the universality index of the image of $\pi^{\mathtt{z}}_{\mathtt{i}}$ under the respective substitution is $\iota(h(\pi^{\mathtt{z}}_{\mathtt{i}})) \geq N^6 > k$; such a substitution cannot be valid. Secondly, if the image of $\mathtt{z_i}$ is a string from $\mathtt{0}^* \cup \mathtt{1}^*$, then the universality of this gadget is exactly $\iota(h(\pi^{\mathtt{z}}_{\mathtt{i}})) = 2$. Similarly to the binarisation gadgets, in the arch factorisation of a string $h(\pi^{\mathtt{z}}_{\mathtt{i}})$, where h is a valid substitution, we have exactly two arches (and no rest). A similar analysis can be performed for the gadgets $\pi^{\mathtt{u}}_{\mathtt{i}} = (\mathtt{u_i}\,\$\,\#)^{N^6}\,\mathtt{1001}\,\$\,\#$. In conclusion, the gadgets $\pi^{\mathtt{z}}_{\mathtt{i}}$ and $\pi^{\mathtt{u}}_{\mathtt{i}}$ enforce that under a valid substitution h, the image of the variables $\mathtt{z_i}$ and $\mathtt{u_i}$ contains either only 0s or only 1s (or is empty).

The Complementation Gadgets. The role of these gadgets is to enforce the property that $\mathtt{z_i}$ and $\mathtt{u_i}$ are not both in $\mathtt{0}^+$ or not both in $\mathtt{1}^+$, for all $i \in [n]$. We construct the gadget $\xi_{\mathtt{i}} = \$\,\mathtt{z_i}\mathtt{u_i}\,\#$, for every $i \in [n]$. Let us now analyse the image of these gadgets under a valid substitution ($\pi_{\#}$ and $\pi_{\$}$ are mapped to exactly one arch each, and $\pi^{\mathtt{z}}_{\mathtt{i}}$ and $\pi^{\mathtt{u}}_{\mathtt{i}}$ are mapped to exactly two arches each). In this case, we observe that $\xi_{\mathtt{i}}$ is mapped to exactly one complete arch ending on the rightmost symbol # if and only if the image of one of the variables $\mathtt{z_i}$ and $\mathtt{u_i}$ has at least one 0 and the image of the other one has at least one 1. Further, let us consider the concatenation of two consecutive such gadgets $\xi_{\mathtt{i}}\xi_{\mathtt{i+1}}$ and assume that both $\mathtt{z_i}$ and $\mathtt{u_i}$ are mapped to strings over the same letter or at least one of them is mapped to the empty word. In that case, the first arch must close to the right of the \$ letter in ξ_{i+1}, hence $\xi_i\xi_{i+1}$ could not contain two arches. Thus, the concatenation of the gadgets $\xi_1 \cdots \xi_n$ is mapped to a string which has exactly n arches if and only if each gadget ξ_i is mapped to exactly one arch, which holds if and only if the image of one of the variables $\mathtt{z_i}$ and $\mathtt{u_i}$ has at least one 0 and the image of the other one has at least one 1. When assembling together all the gadgets, we will ensure that, in a valid substitution, this property holds: $\mathtt{z_i}$ and $\mathtt{u_i}$ are mapped to repetitions of different letters.

The Clause Gadgets. Let $c_j = (y_j^1 \vee y_j^2 \vee y_j^3)$ be a clause, with $y_j^1, y_j^2, y_j^3 \in X \cup \bar{X}$. We construct the gadget $\delta_{\mathtt{j}}$ for every clause c_j as $\$\,\mathtt{0}\rho(\mathtt{y}^1_{\mathtt{j}})\rho(\mathtt{y}^2_{\mathtt{j}})\rho(\mathtt{y}^3_{\mathtt{j}})\,\#$. Now, by all of the properties discussed for the previous gadgets, we can analyse the possible number of arches contained in the image of this gadget under a valid substitution. Firstly, note that if at least one of the variables $\rho(\mathtt{y}^1_{\mathtt{j}}), \rho(\mathtt{y}^2_{\mathtt{j}}), \rho(\mathtt{y}^3_{\mathtt{j}})$ is mapped to a string containing at least one 1, then this gadget will contain exactly one arch ending on its rightmost symbol #. Now consider the concatenation of two consecutive such gadgets $\delta_j\delta_{j+1}$, and assume that all the variables in δ_j are substituted by only 0s. In this case, the first arch must end to the right of the \$ symbol in δ_{j+1}, hence the string to which $\delta_{\mathtt{j}}\delta_{\mathtt{j+1}}$ is mapped could not contain two arches. The same argument holds if we look at the concatenation of the last complementation gadget and the first clause gadget, e.g. $\xi_n\delta_1$. Thus, the concatenation of the gadgets $\delta_1 \cdots \delta_m$ is mapped to a string which has exactly m arches if and only if each gadget δ_i is mapped to exactly one arch. This holds if and only if at least one of the string-variables occurring in δ_i is mapped to a string of 1s. When assembling together all the gadgets, we will ensure that at least one of the variables occurring in each gadget δ_i, for all $i \in [m]$, is mapped to a string of 1s in a valid substitution.

Final Assemblage. We finish the construction of the pattern α by concatenating all the gadgets. That is, $\alpha = \pi_{\#}\pi_{\$}\pi_1^{\mathtt{z}}\pi_1^{\mathtt{u}}\pi_2^{\mathtt{z}}\pi_2^{\mathtt{u}}\cdots\pi_{\mathtt{n}}^{\mathtt{z}}\pi_{\mathtt{n}}^{\mathtt{u}}\xi_1\xi_2\cdots\xi_{\mathtt{n}}\delta_1\delta_2\cdots\delta_{\mathtt{m}}$.

The Correctness of the Reduction. We can show that there exists a substitution h of the string variables of α with $\iota(h(\alpha)) = 5n + m + 2$ if and only if we can find an assignment for all Boolean-variables occurring in φ that satisfy all clauses $c_j \in \varphi$. If there is a satisfying assignment for Boolean-variables of φ, then we can give a canonical substitution h with $h(\rho(x_i)) = 1$ and $h(\rho(\bar{x}_i)) = 0$, if x_i true, and $h(\rho(x_i)) = 0$ and $h(\rho(\bar{x}_i)) = 1$, if x_i false. Conversely, by performing a left to right arch factorisation of $h(\alpha)$, while taking into account the intuition given for each gadget, shows that if a substitution h is valid, then we can define a satisfying assignment of φ by setting exactly those x_i to be true for which $h(\rho(x_i)) = 1^+$. Full details are given in the full version [23]. This concludes the sketch of our proof, and shows that $\mathtt{MatchUniv}(\alpha, k)$ is NP-hard. □

In the following we show that $\mathtt{MatchUniv}(\alpha, k)$ is in NP. One natural approach is to guess the images of the variables occurring in the input pattern α under a substitution h and check whether or not $\iota(h(\alpha))$ is indeed k. However, it is difficult to bound the size of the images of the variables of α under h in terms of the size of α and $\log k$ (the size of our input), since the strings we look for may be exponentially long. For example, consider the pattern $\alpha = X_1$: the length of the shortest k-universal string is $k\sigma$ [6], which is already exponential in $\log k$. Therefore, we consider guessing only the subsequence universality signatures for the image of each variable under the substitution. We show that it is sufficient to guess $|\mathtt{var}(\alpha)|$ subsequence universality signatures, one for each variable, instead of the actual images of the variables under a substitution h using the following proposition by Schnoebelen and Veron [61].

Proposition 1 ([61]). *For $u, v \in \Sigma^*$, we can compute $\mathtt{s}(uv)$, given the subsequence universality signatures $\mathtt{s}(u) = (\gamma_u, \mathcal{K}_u, \mathcal{R}_u)$ and $\mathtt{s}(v) = (\gamma_v, \mathcal{K}_v, \mathcal{R}_v)$ of each string, in time polynomial in $|\mathtt{alph}(uv)|$ and $\log t$, where t is the maximum element of $\mathcal{K}_u$ and $\mathcal{K}_v$.*

Once we have guessed the subsequence universality signatures of all variables in $\mathtt{var}(\alpha)$ under substitution h, we can compute $\iota(h(\alpha))$ in the following way. We first compute the subsequence universality signature of the maximal prefix of α that does not contain any variables. We then incrementally compute the subsequence universality signature of prefixes of the image of α. Let $\alpha = \alpha_1\alpha_2$, where we already have $\mathtt{s}(h(\alpha_1))$ from induction. If $\alpha_2[1]$ is a variable, we compute $\mathtt{s}(h(\alpha_1\alpha_2[1]))$ from $\mathtt{s}(h(\alpha_1))$ and the guessed subsequence universality signature for variable $\alpha_2[1]$, using Proposition 1. Otherwise, we take the maximal prefix w of α_2 that does not consist of any variables. We first compute $\mathtt{s}(w)$ and then compute $\mathtt{s}(h(\alpha_1 w))$ using Proposition 1. Once we have $\mathtt{s}(h(\alpha)) = (\gamma, \mathcal{K}, \mathcal{R})$, we compute $\iota(h(\alpha)) = \mathcal{K}[\sigma] + 1$. Note that the whole process can be done in a polynomial number of steps in $|\alpha|$, $\log k$, and σ due to Proposition 1, provided that the signatures are of polynomial size.

Thus, we now measure the encoding size of a subsequence universality signature and, as such, the overall size of the certificate for $\mathtt{MatchUniv}$ that we guess. We can use $\sigma!$ bits to encode a permutation γ of a subset of Σ. An integer between 1 and $\sigma - 1$ requires $\log \sigma$ bits. Naively, $\mathcal{R}$ requires $(2^\sigma)^\sigma$ bits because there can be 2^σ choices for each item. Finally, in the framework of our problem, note that $\mathcal{K}[1] - \mathcal{K}[|\gamma|] \leq 1$ by Schnoebelen and Veron [61], and that the values of $\mathcal{K}[i]$ are non-increasing in i. Therefore, we can encode $\mathcal{K}$ as a tuple (l, k') where $k' = \max\{\mathcal{K}[i] \mid 1 \leq i \leq |\gamma|\} \leq k$ and $l = |\{i \in [|\gamma|] \mid \mathcal{K}[i] = k'\}|$. This encoding scheme requires at most $\log \sigma + \log k$ bits. Summing up, the overall space required to encode a certificate that consists of $|\mathtt{var}(\alpha)|$ subsequence universality signatures takes at most $(1 + \sigma! + (2^\sigma)^\sigma + \log \sigma + \log k)|\mathtt{var}(\alpha)|$ bits. This is polynomial in the size of the input and the number of variables, because we assume a constant-sized alphabet, i.e. $\sigma \in O(1)$.

It remains to design a deterministic polynomial algorithm that tests the validity of the guessed subsequence universality signature. Assume that we have guessed the 3-tuple $(\gamma, \mathcal{K}, \mathcal{R})$. We claim that there are only constantly many strings we need to check to decide whether or not $(\gamma, \mathcal{K}, \mathcal{R})$ is a valid subsequence universality signature - allowing us a brute-force approach. Lemma 2 allows us to "pump down" strings with universality index greater than $(2^\sigma)^\sigma$, which is a constant. The proof and a figure illustrating it are given in the full version [23].

Lemma 2. *The tuple $(\gamma, \mathcal{K}_1, \mathcal{R})$ is a valid subsequence universality signature iff there exists $w \in \Sigma^*$ with $\iota(w) \leq (2^\sigma)^\sigma$, $\mathtt{s}(w) = (\gamma, \mathcal{K}_2, \mathcal{R})$, and $\mathcal{K}_1[t] - \mathcal{K}_2[t] = c \in \mathbb{N}_0$ for all $t \in [|\gamma|]$.*

Lemma 2 limits the search space for the candidate string corresponding to a tuple $(\gamma, \mathcal{K}, \mathcal{R})$ by mapping valid subsequence universality signatures to subsequence universality signatures for strings with universality index at most $(2^\sigma)^\sigma$. Therefore, we need to investigate those strings where there are up to $\sigma \cdot (1 + (2^\sigma)^\sigma) + 1$ terms in its marginal sequence. The following lemma bounds the length of the substring between two consecutive marginal sequence terms in such a string. Its proof can be found in the full version [23]. The conclusion of this line of thought follows then, in Corollary 1.

Lemma 3. *For a given string w, let $w = uvx$ where $v = w[\mathtt{M}_i(w)+1 : \mathtt{M}_{i+1}(w)] \neq \varepsilon$, and $u = w[1 : \mathtt{M}_i(w)]$, and $x = w[\mathtt{M}_{i+1}(w) + 1 : |w|]$ for some integer $i \geq 1$. For a permutation v' of $\mathtt{alph}(v)$ that ends with $v[|v|]$, we have $\mathtt{s}(uvx) = \mathtt{s}(uv'x)$.*

Corollary 1. *The tuple $(\gamma, \mathcal{K}, \mathcal{R})$ is a valid subsequence universality signature if and only if there exists a string w of length at most $\sigma \cdot (\sigma \cdot (1 + (2^\sigma)^\sigma) + 1)$ and a constant $c \in \mathbb{N}_0$ that satisfies $\mathtt{s}(w) = (\gamma, \mathcal{K} - c, \mathcal{R})$.*

We can now show the following result.

Lemma 4. *$\mathtt{MatchUniv}(\alpha, k)$ is in* NP.

Proof. Follows from Proposition 1 and Corollary 1. Firstly, for a guessed sequence of universality signatures $(\gamma_x, \mathcal{K}_x, \mathcal{R}_x)$, for $x \in \mathtt{var}(\alpha)$, we check their validity. For that, we enumerate all strings of length up to the constant $\sigma \cdot (\sigma \cdot (1 + (2^\sigma)^\sigma) + 1)$ over Σ and see if there exist strings w_x such that $\mathtt{s}(w_x) = (\gamma_x, \mathcal{K}_x - c_x, \mathcal{R}_x)$

for some constant $c_x \leq k$. Since σ is constant, this takes polynomial time. We then use Proposition 1 to check if the guessed signatures lead to an assignment h of the variables such that $\iota(h(\alpha)) = k$, as already explained. Since we have a polynomial size bound on the certificate and a deterministic verifier that runs in polynomial time, we obtain that $\texttt{MatchUniv}(\alpha, k)$ is in NP. □

Based on Lemmas 1 and 4, the following theorem follows.

Theorem 1. `MatchUniv` *is* NP*-complete.*

Further, we describe two classes of patterns, defined by structural restrictions on the input patterns, for which `MatchUniv` can be solved in polynomial time. The proof of Proposition 2 can be found in the full version [23].

Proposition 2. *a)* $\texttt{MatchUniv}(\alpha, k)$ *is in* P *when there exists a variable that occurs only once in* α. *As such,* $\texttt{MatchUniv}(\alpha, k)$ *is in* P *for the heavily studied class of regular patterns (see, e.g., [17] and the references therein), where each variable occurs only once. b)* $\texttt{MatchUniv}(\alpha, k)$ *is in* P *when* $|\texttt{var}(\alpha)|$ *is constant.*

4 MatchSimon

Further, we discuss the `MatchSimon` problem. For space reasons, the proofs of the results from this Section can be found in the full version [23].

In the case of `MatchSimon`, we are given a pattern α, a word w, and a natural number $k \leq n$, and we want to check the existence of a substitution h with $h(\alpha) \sim_k w$. The first result is immediate: `MatchSimon` is NP-hard, because $\texttt{MatchSimon}(\alpha, w, |w|)$ is equivalent to $\texttt{Match}(\alpha, w)$, and `Match` is NP-complete. To understand why this results followed much easier than the corresponding lower bound for `MatchUniv`, we note that in `MatchSimon` we only ask for $h(\alpha) \sim_k w$ and allow for $h(\alpha) \sim_{k+1} w$, while in `MatchUniv` $h(\alpha)$ has to be k-universal but not $(k+1)$-universal. So, in a sense, `MatchSimon` is not strict, while `MatchUniv` is strict. So, we can naturally consider the following problem.

Matching under Strict Simon's Congruence: $\texttt{MatchStrictSimon}(\alpha, w, k)$
Input: Pattern α, $|\alpha| = m$, word w, $|w| = n$, and $k \in [n]$.
Question: Is there a substitution h with $h(\alpha) \sim_k w$ and $h(\alpha) \not\sim_{k+1} w$?

Adapting the reduction from Lemma 1, we can show that `MatchStrictSimon` is NP-hard.

Following [45], we can also show an NP-upper bound: it is enough to consider as candidates for the images of the variables under the substitution h only strings of length $O((k+1)^\sigma)$; longer strings can be replaced with shorter, $\sim_k$-congruent ones, which have the same impact on the sets $\mathbb{S}_k(h(\alpha))$. The following holds.

Theorem 2. `MatchSimon` *and* `MatchStrictSimon` *are* NP*-complete.*

Finally, note that `MatchSimon` and `MatchStrictSimon` are in P when the input pattern is regular.

Proposition 3. *If* α *is a regular pattern, then both problems* $\texttt{MatchSimon}(\alpha, w, k)$ *and* $\texttt{MatchStrictSimon}(\alpha, w, k)$ *are in* P.

5 WESimon

In this section, we address the WESimon problem, where we are given two patterns α and β, and a natural number $k \leq n$, and we want to check the existence of a substitution h with $h(\alpha) \sim_k h(\beta)$. The first result is immediate: this problem is NP-hard because MatchSimon, which is a particular case of WESimon, is NP-hard.

To show that the problem is in NP, we need a more detailed analysis. If $k \leq |\alpha| + |\beta|$, the same proof as for the NP-membership of MatchSimon works: it is enough to look for substitutions of the variables with the image of each variable having length at most k^σ, and this is polynomial in the size of the input. If $k > |\alpha| + |\beta|$, and $\beta = w$ contains no variable, then this is an input for MatchSimon with k greater than the length of the input word w, and we have seen previously how this can be decided. Finally, if both α and β contain variables, then the problem is trivial, irrespective of k: the answer to any input is positive, as we simply have to map all variables to $(1 \cdots \sigma)^k$ and obtain two $\sim_k$-congruent words. Therefore, we have the following result.

Theorem 3. WESimon *is* NP*-complete.*

To avoid the trivial cases arising in the above analysis for WESimon, we can also consider a stricter variant of this problem:

Word Equations under Strict Simon's Congruence: WEStrictSimon(α, β, k)
Input: Patterns α, β, $|\alpha| = m$, $\beta = n$, and $k \in [m+n]$.
Question: Is there a substitution h with $h(\alpha) \sim_k h(\beta)$ and $h(\alpha) \not\sim_{k+1} h(\beta)$?

Differently from WESimon, we can show that this problem is NP-hard, even in the case when both sides of the pattern contain variables.

Lemma 5. WEStrictSimon *is* NP*-hard, even if both patterns contain variables.*

The proof of Lemma 5 can be found in the full version [23]. Regarding the membership in NP: if k is upper bounded by a polynomial function in $|\alpha| + |\beta|$ (or, alternatively, if k is given in unary representation), then the fact that WEStrictSimon is in NP follows as in the case of MatchStrictSimon. The case when k is not upper bounded by a polynomial in $|\alpha| + |\beta|$ remains open. We can show the following theorem.

Theorem 4. WEStrictSimon *is* NP*-complete, for* $k \leq |\alpha| + |\beta|$.

6 Conclusions

In this paper, we have considered the problem of matching patterns with variables under Simon's congruence. More precisely, we have considered three main problems MatchUniv, MatchSimon, and WESimon and we have given a rather comprehensive image of their computational complexity. These problems are NP-complete, in general, but have interesting particular cases which are in P. Interestingly, our NP or P algorithms work in (non-deterministic) polynomial time only in the case of constant input alphabet (their complexity being, in fact, exponential in the size σ of the input alphabet). It seems very interesting to characterize the parameterised complexity of these problems w.r.t. the parameter σ. In the light of Proposition 2, another interesting parameter to be considered in such a parameterised complexity analysis would be the number of variables. We conjecture that the problems are $W[1]$-hard with respect to both these parameters.

References

1. Adamson, D.: Ranking and unranking k-subsequence universal words. In: Frid, A., Mercaş, R. (eds.) Combinatorics on Words, WORDS 2023. LNCS, vol. 13899, pp. 47–59. Springer, Cham (2023). https://doi.org/10.1007/978-3-031-33180-0_4
2. Adamson, D., Kosche, M., Koß, T., Manea, F., Siemer, S.: Longest common subsequence with gap constraints. In: Frid, A., Mercaş, R. (eds.) Combinatorics on Words. WORDS 2023. Lecture Notes in Computer Science, vol. 13899, pp. 60–76. Springer, Cham (2023). https://doi.org/10.1007/978-3-031-33180-0_5
3. Amadini, R.: A survey on string constraint solving. ACM Comput. Surv. (CSUR) **55**(1), 1–38 (2021)
4. Amir, A., Nor, I.: Generalized function matching. J. Discrete Algorithms **5**, 514–523 (2007)
5. Angluin, D.: Finding patterns common to a set of strings. J. Comput. Syst. Sci. **21**(1), 46–62 (1980)
6. Barker, L., Fleischmann, P., Harwardt, K., Manea, F., Nowotka, D.: Scattered factor-universality of words. In: Jonoska, N., Savchuk, D. (eds.) DLT 2020. LNCS, vol. 12086, pp. 14–28. Springer, Cham (2020). https://doi.org/10.1007/978-3-030-48516-0_2
7. de Bruijn, N.G.: A combinatorial problem. Koninklijke Nederlandse Akademie v. Wetenschappen **49**, 758–764 (1946)
8. Câmpeanu, C., Salomaa, K., Yu, S.: A formal study of practical regular expressions. Int. J. Found. Comput. Sci. **14**, 1007–1018 (2003)
9. Chen, H.Z.Q., Kitaev, S., Mütze, T., Sun, B.Y.: On universal partial words. Electron. Notes Discrete Math. **61**, 231–237 (2017)
10. Crochemore, M., Hancart, C., Lecroq, T.: Algorithms on Strings. Cambridge University Press (2007)
11. Crochemore, M., Melichar, B., Tronícek, Z.: Directed acyclic subsequence graph - overview. J. Discrete Algorithms **1**(3–4), 255–280 (2003)
12. Day, J., Fleischmann, P., Kosche, M., Koß, T., Manea, F., Siemer, S.: The edit distance to k-subsequence universality. In: STACS, vol. 187, pp. 25:1–25:19 (2021)
13. Day, J.D., Fleischmann, P., Manea, F., Nowotka, D.: Local patterns. In: Proceedings of the 37th IARCS Annual Conference on Foundations of Software Technology and Theoretical Computer Science, FSTTCS 2017. LIPIcs, vol. 93, pp. 24:1–24:14 (2017)
14. Downey, R.G., Fellows, M.R.: Parameterized Complexity. Monographs in Computer Science. Springer, New York (1999). https://doi.org/10.1007/978-1-4612-0515-9
15. Fagin, R., Kimelfeld, B., Reiss, F., Vansummeren, S.: Document spanners: a formal approach to information extraction. J. ACM **62**(2), 12:1–12:51 (2015)

16. Fernau, H., Manea, F., Mercas, R., Schmid, M.L.: Revisiting Shinohara's algorithm for computing descriptive patterns. Theor. Comput. Sci. **733**, 44–54 (2018)
17. Fernau, H., Manea, F., Mercas, R., Schmid, M.L.: Pattern matching with variables: efficient algorithms and complexity results. ACM Trans. Comput. Theor. **12**(1), 6:1–6:37 (2020)
18. Fernau, H., Schmid, M.L.: Pattern matching with variables: a multivariate complexity analysis. Inf. Comput. **242**, 287–305 (2015)
19. Fernau, H., Schmid, M.L., Villanger, Y.: On the parameterised complexity of string morphism problems. Theor. Comput. Syst. **59**(1), 24–51 (2016)
20. Fleischer, L., Kufleitner, M.: Testing Simon's congruence. In: Potapov, I., Spirakis, P.G., Worrell, J. (eds.) MFCS 2018. LIPIcs, vol. 117, pp. 62:1–62:13. Schloss Dagstuhl - Leibniz-Zentrum für Informatik (2018)
21. Fleischmann, P., Germann, S., Nowotka, D.: Scattered factor universality-the power of the remainder. preprint arXiv:2104.09063 (published at RuFiDim) (2021)
22. Fleischmann, P., Höfer, J., Huch, A., Nowotka, D.: α-β-factorization and the binary case of Simon's congruence (2023)
23. Fleischmann, P., et al.: Matching patterns with variables under Simon's congruence (2023)
24. Freydenberger, D.D.: Extended regular expressions: succinctness and decidability. Theor. Comput. Syst. **53**, 159–193 (2013)
25. Freydenberger, D.D.: A logic for document spanners. Theor. Comput. Syst. **63**(7), 1679–1754 (2019)
26. Freydenberger, D.D., Gawrychowski, P., Karhumäki, J., Manea, F., Rytter, W.: Testing k-binomial equivalence. CoRR abs/1509.00622 (2015)
27. Freydenberger, D.D., Holldack, M.: Document spanners: from expressive power to decision problems. Theor. Comput. Syst. **62**(4), 854–898 (2018)
28. Freydenberger, D.D., Peterfreund, L.: The theory of concatenation over finite models. In: Bansal, N., Merelli, E., Worrell, J. (eds.) ICALP 2021, Proceedings. LIPIcs, vol. 198, pp. 130:1–130:17. Schloss Dagstuhl - Leibniz-Zentrum für Informatik (2021)
29. Freydenberger, D.D., Schmid, M.L.: Deterministic regular expressions with back-references. J. Comput. Syst. Sci. **105**, 1–39 (2019)
30. Friedl, J.E.F.: Mastering Regular Expressions, 3rd edn. O'Reilly, Sebastopol, CA (2006)
31. Garel, E.: Minimal separators of two words. In: Apostolico, A., Crochemore, M., Galil, Z., Manber, U. (eds.) CPM 1993. LNCS, vol. 684, pp. 35–53. Springer, Heidelberg (1993). https://doi.org/10.1007/BFb0029795
32. Garey, M.R., Johnson, D.S.: Computers and Intractability: A Guide to the Theory of NP-Completeness. W. H. Freeman & Co., New York, NY, USA (1979)
33. Gawrychowski, P., Kosche, M., Koß, T., Manea, F., Siemer, S.: Efficiently testing Simon's congruence. In: STACS 2021. LIPIcs, vol. 187, pp. 34:1–34:18. Schloss Dagstuhl - Leibniz-Zentrum für Informatik (2021)
34. Gawrychowski, P., Lange, M., Rampersad, N., Shallit, J.O., Szykula, M.: Existential length universality. In: Proceedings of the STACS 2020. LIPIcs, vol. 154, pp. 16:1–16:14 (2020)
35. Gawrychowski, P., Manea, F., Siemer, S.: Matching patterns with variables under Hamming distance. In: 46th International Symposium on Mathematical Foundations of Computer Science, MFCS 2021. LIPIcs, vol. 202, pp. 48:1–48:24 (2021)

36. Gawrychowski, P., Manea, F., Siemer, S.: Matching patterns with variables under edit distance. In: Arroyuelo, D., Poblete, B. (eds.) String Processing and Information Retrieval, SPIRE 2022. LNCS, vol. 13617, pp. 275–289. Springer, Cham (2022). https://doi.org/10.1007/978-3-031-20643-6_20
37. Goeckner, B., et al.: Universal partial words over non-binary alphabets. Theor. Comput. Sci. **713**, 56–65 (2018)
38. Hague, M.: Strings at MOSCA. ACM SIGLOG News **6**(4), 4–22 (2019)
39. Hébrard, J.: An algorithm for distinguishing efficiently bit-strings by their subsequences. Theoret. Comput. Sci. **82**(1), 35–49 (1991)
40. Karandikar, P., Schnoebelen, P.: The height of piecewise-testable languages with applications in logical complexity. In: CSL (2016)
41. Karhumäki, J., Saarela, A., Zamboni, L.Q.: On a generalization of Abelian equivalence and complexity of infinite words. J. Comb. Theor. Ser. A **120**(8), 2189–2206 (2013)
42. Karhumäki, J., Saarela, A., Zamboni, L.Q.: Variations of the Morse-Hedlund theorem for k-abelian equivalence. Acta Cybern. **23**(1), 175–189 (2017)
43. Karp, R.M.: Reducibility among combinatorial problems. In: Miller, R.E., Thatcher, J.W. (eds.) Proceedings of a Symposium on the Complexity of Computer Computations. The IBM Research Symposia Series, pp. 85–103. Plenum Press, New York (1972). https://doi.org/10.1007/978-1-4684-2001-2_9
44. Kim, S., Ko, S., Han, Y.: Simon's congruence pattern matching. In: Bae, S.W., Park, H. (eds.) 33rd International Symposium on Algorithms and Computation, ISAAC 2022, 19–21 December 2022, Seoul, Korea. LIPIcs, vol. 248, pp. 60:1–60:17. Schloss Dagstuhl - Leibniz-Zentrum für Informatik (2022)
45. Kim, S., Han, Y.S., Ko, S.K., Salomaa, K.: On Simon's congruence closure of a string. In: Han, Y.S., Vaszil, G. (eds.) DCFS 2022. LNCS, vol. 13439, pp. 127–141. Springer, Cham (2022). https://doi.org/10.1007/978-3-031-13257-5_10
46. Kleest-Meißner, S., Sattler, R., Schmid, M.L., Schweikardt, N., Weidlich, M.: Discovering event queries from traces: laying foundations for subsequence-queries with wildcards and gap-size constraints. In: 25th International Conference on Database Theory, ICDT 2022. LIPIcs, vol. 220, pp. 18:1–18:21 (2022)
47. Kosche, M., Koß, T., Manea, F., Siemer, S.: Absent subsequences in words. In: Bell, P.C., Totzke, P., Potapov, I. (eds.) RP 2021. LNCS, vol. 13035, pp. 115–131. Springer, Cham (2021). https://doi.org/10.1007/978-3-030-89716-1_8
48. Kosche, M., Koß, T., Manea, F., Siemer, S.: Combinatorial algorithms for subsequence matching: a survey. In: Bordihn, H., Horváth, G., Vaszil, G. (eds.) NCMA, vol. 367, pp. 11–27 (2022)
49. Krötzsch, M., Masopust, T., Thomazo, M.: Complexity of universality and related problems for partially ordered NFAs. Inf. Comput. **255**, 177–192 (2017)
50. Lejeune, M., Rigo, M., Rosenfeld, M.: The binomial equivalence classes of finite words. Int. J. Algebra Comput. **30**(07), 1375–1397 (2020)
51. Lothaire, M.: Combinatorics on Words. Cambridge University Press (1997)
52. Lothaire, M.: Algebraic Combinatorics on Words. Cambridge University Press (2002)
53. Martin, M.H.: A problem in arrangements. Bull. Am. Math. Soc. **40**(12), 859–864 (1934)
54. Rampersad, N., Shallit, J., Xu, Z.: The computational complexity of universality problems for prefixes, suffixes, factors, and subwords of regular languages. Fundam. Inf. **116**(1–4), 223–236 (2012)
55. Reidenbach, D., Schmid, M.L.: Patterns with bounded treewidth. Inf. Comput. **239**, 87–99 (2014). https://doi.org/10.1016/j.ic.2014.08.010

56. Rigo, M., Salimov, P.: Another generalization of abelian equivalence: binomial complexity of infinite words. Theor. Comput. Sci. **601**, 47–57 (2015)
57. Schmid, M.L.: A note on the complexity of matching patterns with variables. Inf. Process. Lett. **113**(19), 729–733 (2013)
58. Schmid, M.L., Schweikardt, N.: A purely regular approach to non-regular core spanners. In: Proceedings of the 24th International Conference on Database Theory, ICDT 2021. LIPIcs, vol. 186, pp. 4:1–4:19 (2021)
59. Schmid, M.L., Schweikardt, N.: Document spanners - a brief overview of concepts, results, and recent developments. In: International Conference on Management of Data, PODS 2022, pp. 139–150. ACM (2022)
60. Schnoebelen, P., Karandikar, P.: The height of piecewise-testable languages and the complexity of the logic of subwords. Logical Meth. Comput. Sci. **15**, 6:1–6:27 (2019)
61. Schnoebelen, P., Veron, J.: On arch factorization and subword universality for words and compressed words. In: Frid, A., Mercaş, R. (eds.) Combinatorics on Words, WORDS 2023. LNCS, vol. 13899, pp. 274–287. Springer, Cham. https://doi.org/10.1007/978-3-031-33180-0_21
62. Jantke, K.P.: Polynomial time inference of general pattern languages. In: Fontet, M., Mehlhorn, K. (eds.) STACS 1984. LNCS, vol. 166, pp. 314–325. Springer, Heidelberg (1984). https://doi.org/10.1007/3-540-12920-0_29
63. Shinohara, T., Arikawa, S.: Pattern inference. In: Jantke, K.P., Lange, S. (eds.) Algorithmic Learning for Knowledge-Based Systems. LNCS, vol. 961, pp. 259–291. Springer, Heidelberg (1995). https://doi.org/10.1007/3-540-60217-8_13
64. Simon, I.: Hierarchies of events with dot-depth one. Ph.D. thesis, University of Waterloo (1972)
65. Simon, I.: Piecewise testable events. In: Brakhage, H. (ed.) GI-Fachtagung 1975. LNCS, vol. 33, pp. 214–222. Springer, Heidelberg (1975). https://doi.org/10.1007/3-540-07407-4_23
66. Simon, I.: Words distinguished by their subwords (extended abstract). In: Proceedings of the WORDS 2003, vol. 27, pp. 6–13. TUCS General Publication (2003)
67. Troniĉek, Z.: Common subsequence automaton. In: Champarnaud, J.-M., Maurel, D. (eds.) CIAA 2002. LNCS, vol. 2608, pp. 270–275. Springer, Heidelberg (2003). https://doi.org/10.1007/3-540-44977-9_28
68. Weis, P., Immerman, N.: Structure theorem and strict alternation hierarchy for FO^2 on words. Log. Meth. Comput. Sci. **5**(3), 1–23 (2009)

HyperMonitor: A Python Prototype for Hyper Predictive Runtime Verification

Angelo Ferrando and Giorgio Delzanno(✉)

DIBRIS, Università degli Studi di Genova, Genoa, Italy
{angelo.ferrando,giorgio.delzanno}@unige.it

Abstract. We present HyperMonitor a Python prototype of a novel runtime verification method specifically designed for predicting race conditions in multithread programs. Our procedure is based on the combination of Inductive Process Mining, Petri Net Tranformations, and verification algorithms. More specifically, given a trace log, the Hyper Predictive Runtime Verifier (HPRV) procedure first exploits Inductive Process Mining to build a Petri Net that captures all traces in the log, and then applies semantic-driven transformations to increase the number of concurrent threads without re-executing the program. In this paper, we present the key ideas of our approach, details on the HyperMonitor implementation and discuss some preliminary results obtained on classical examples of concurrent C programs with semaphors.

Keywords: Runtime Verification · Process Mining · Petri Nets · Verification · Concurrent Programs

1 Introduction

In this paper, we present the main features and implementation details of the HyperMonitor prototype tool, a Python implementation of the Hyper Predictive Runtime Verification procedure (HPRV) designed for predicting race conditions in concurrent programs. The HPRV procedure is based on a novel combination of runtime verification, process mining, and Petri Nets verification algorithms. The main idea underlying our approach is to automatically infer potential race conditions by combining process mining and verification algorithms starting from a trace log obtained by executing a concurrent program. In our setting a trace is defined as a sequence of observed events (e.g. lock acquire and release operations) inferred from executions of concurrent programs (e.g. instrumented concurrent C programs).

The HPRV procedure is based on the following steps. We first apply inductive Process Mining (PM) to a given trace log in order to generate a Petri Net that represents at least every execution contained in the trace log. The resulting Petri Net takes into consideration possible reorderings of events of different threads. We then apply semantic-driven transformations to the computer Petri Net in ordert to consider executions with a larger number of threads w.r.t. those observed in the log. More specifically, starting from a model inferred

O. Bournez et al. (Eds.): RP 2023, LNCS 14235, pp. 171–182, 2023.
https://doi.org/10.1007/978-3-031-45286-4_13

from execution logs of K threads, we apply a series of Petri Net transformations to generate an over-approximation of the executions of $K + N$ threads. Based on practical considerations adopted in incomplete verification methods such as context-bounded model checking [13], we assume here that the most common race conditions can be detected with small values for N, the number of additional threads injected in the model. To illustrate, our procedure can be applied to a single threaded execution of a C program to predict possible race-conditions with two or more threads without re-executing the program. This way, we can increase the confidence level in legacy code that must be ported to a multithreaded scenario.

The Petri Net transformation is driven by the semantics of the observed operations. More specifically, we first assign a semantics to special operations on shared locks adding new places and transition to the Petri Net generated through PM. We then replicate parts of the Petri Net in order to model an over-approximation of the possible interleaving of multiple threads.

We can then apply Petri Nets verification algorithms, e.g., reachability and coverability, to the resulting model in order to predict potential race conditions or prove the considered trace set free from errors. For instance, if a system has been executed with 2 threads, our approach does not only check for violations considering such threads, but it also checks for larger numbers of threads (like for 4, 6, and so on). In this way, it is possible to expose wrong behaviours which would have not been observed by simply executing the system multiple times. This way, we try to combine techniques coming from Runtime Verification (RV) with those based on Petri Nets verification as shown in Fig. 1.

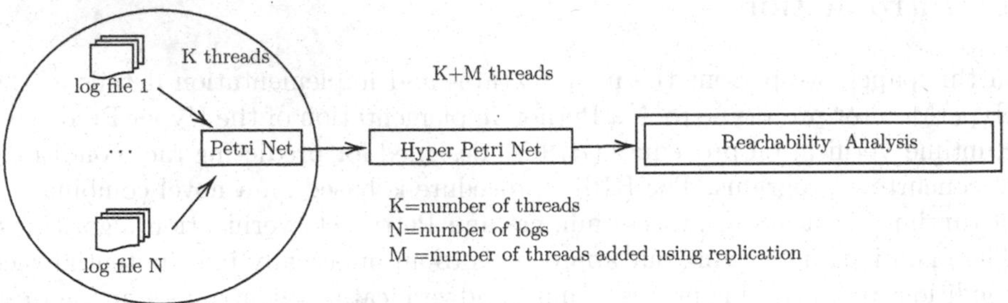

Fig. 1. General Scheme

2 Related Work

RV [2] focuses on checking the runtime behaviour of software/hardware systems. With respect to other formal verification techniques, such as Model Checking [4] and Theorem Provers [11], RV is considered more dynamic and lightweight. This is mainly due to its being completely focused on checking how the system behaves, while the latter is currently being executed (*i.e.*, running). In Predictive Runtime Verification (PRV), monitors do not only consider the observed system executions, but try to predict other possible behaviours as well. We can find at least two different kinds of PRV. The first use of prediction is to anticipate possible future behaviour.

In this setting, the main reason for applying such prediction is when the system under analysis is a safety-critical one. That is, a system where a failure can be costly (in terms of money, or human lives). In such systems, a monitor cannot reduce itself to report an error only when such error is observed in the system's execution. But, the monitor has to try to anticipate its presence and report it as soon as possible. Thus, in safety-critical scenarios, the prediction used at the monitor level is mainly focused on anticipating possible future events, in order to let the monitor anticipate its own verdict. Examples of this kind of prediction in RV can be found in [12,16]. The second kind of prediction is focused instead on the prediction of possible alternatives to the observed system's execution. The prediction in these cases is used to predict other possible executions, that have not been observed, but are nonetheless valid. Examples of this kind of prediction in RV can be found in [7–9,14,15].

To the best of our knowledge, no work exploiting PM to perform the prediction in RV has ever been studied before, except for [6]; however, in such a case, the model extracted through PM is not used to predict possible threads interleaving, but to predict future system's events.

3 Preliminaries

A Petri Net is a bipartite graph containing places and transitions, interconnected by directed arcs. Places (or states) can contain tokens. This is described through the marking function, which denotes how many tokens are present in each place. Tokens can move between states through transitions (or more precisely through arcs and transitions), and by doing so, they define the behaviours of the system. A Petri Net is a tuple $N = (P, T, F, W, M)$ where P and T are disjoint finite sets of places and transitions, respectively, $F \subseteq (P \times T) \cup (T \times P)$ is a set of (directed) arcs (or flow relations), $W : F \to \mathbb{N}$ is the arc weight mapping (where $W(f) > 0$ for all $f \in F$), $M : P \to \mathbb{N}$ is a marking function that assigns to each place a number of tokens (the current state of the net). M_0 is the initial marking that represents the initial distribution of tokens in the net. Let $\langle P, T, F, W, M\rangle$ be a Petri Net. We associate with it the transition system $\langle S, \Sigma, \Delta, I, AP, l\rangle$, where:

- $S = \{m \mid m : P \to \mathbb{N}\}$, $I = M_0$, $\Sigma = T$
- $\Delta = \{(m, t, m') \mid \forall_{p \in P}.m(p) \geq W(p, t) \wedge m'(p) = m(p) - W(p, t) + W(t, p)\}$
- $AP = P$, $l(m) = \{p \in P \mid m(p) > 0\}$

When $(m, t, m') \in \Delta$, we say that t is enabled in m and that its firing produces the successor marking m'. In this paper we only consider Petri nets in which the arc weight is always equal to one, i.e., each arcs removes/adds a single token. When drawing a Petri Net in the paper, we omit arc weights of 1. Also, we denote tokens on a place by black circles. Let m be a marking of a Petri Net $\langle P, T, F, W, M\rangle$. The set of markings reachable from m (the reachability set of m, written *reach*(m)), is the smallest set of markings such that: $m \in reach(m)$, and if (m', t, m'') for some $t \in T$, $m' \in reach(m)$, then $m'' \in reach(m)$. The set of reachable markings $reach(\langle P, T, F, W, M\rangle)$ is defined to be $reach(M)$. Let $\langle P, T, F, W, M\rangle$ be a Petri Net with associated transition system $\langle S, \Sigma, \Delta, I, AP, l\rangle$. The reachability graph is the rooted, directed graph $G = \langle S', \Delta', M\rangle$, where S' and Δ' are the restrictions of S and Δ to $reach(M)$. Note that, the reachability graph can be constructed in iterative fashion, starting with the initial marking and then adding, step for step, all reachable markings. Given a Petri Net $\langle P, T, F, W, M\rangle$ and a marking m. The reachability problem is to determine whether $m \in reach(M)$. A Petri Net $\langle P, T, F, W, M\rangle$ is said to be b-bounded for some $b \in \mathbb{N}$ if all markings in M have at most b tokens in all places, *i.e.*, $\forall_{p \in P}.M(p) = b$. A *b-bounded* Petri Net has at most $(b+1)^{|P|}$ reachable markings; for 1-*bounded* nets, the limit is $2^{|P|}$. The reachability problem for 1-*bounded* Petri Nets is known to be PSPACE [3].

4 An Overview of the HPRV Procedure

In our approach, we assume a system has been instrumented and log files have been generated through its execution, a common approach in RV [5]. In our setting, the events that can be generated through an execution of instrumented concurrent program comprise: (i) *read*(Thr, R) (resp. *write*(Thr, R)), the event denoting that thread T performs a read (resp. write) instruction on the shared resource R; (ii) *acquire*(Thr, L) (resp. *release*(Thr, L)), the event denoting that thread Thr acquires (resp. releases) lock L. Since we only focus on these events, the other instructions can be discarded, largely reducing the number of events that need to be observed.

4.1 Inductive Process Mining

Once the log files corresponding to (possibly multiple) runtime executions of the system are extracted, a model denoting the system's behaviour can be synthesised. We consider here models extracted by PM specified as a Petri Net [1]. To generate such Petri Net, we exploit the Inductive Miner algorithm [10]. This algorithm generates a 1-*bounded* Petri Net describing all traces of events passed as input. The resulting Petri Net is exhaustive and possibly incomplete. Exhaustive means that the Petri Net generated through PM recognises all traces given in input to the Inductive Miner algorithm, *i.e.*, there exists a sequence of transitions in the Petri Net that corresponds to the sequence of events of every trace in the log. In other words, the Petri Net captures all observed behaviours. Note that, the other way around is not necessarily true. Indeed, the Petri Net generated by the Inductive Miner algorithm can be seen as an over-approximation of the system. This is caused by the fact that in the process of generating the Petri Net, unobserved relations amongst events may be added. For instance, let us assume that one observed trace contains the sequence of events a b c (where a, b and c are any possible observable events in the system). While another trace contains $b\ d$. Then, the Petri Net would recognise both traces $a\ b\ c$, and $b\ d$; but also the never observed trace $a\ b\ d$.

The Petri Net produced via Inductive Miner is not complete in that it may not consider aspects that have never been observed in any trace belonging to the log files. This notion is closely related to branch coverage in software testing. Therefore, as in RV, we do not have a complete understanding over the system, because we can only evaluate what we can observe from the considered executions. This is a main difference w.r.t. other techniques, such as Model Checking [4], where it is assumed a perfect knowledge over the system behaviour.

Once we obtained the Petri Net describing the system's behaviour, we would like to use it to check for the presence/absence of data races.

4.2 Semantic-Driven Model Transformations

The Petri Net we can obtain through PM is not ready to answer verification questions. The main reason is that it does not assign any semantic to the events used for its generation. For instance, an acquire operation will be represented as a Petri Net transition without any synchronization with other transitions (e.g. a release operation).

To overcome this issue, we need an additional step, we called Petri Net enhancement, where we recognise the semantics of the observed events, and modify the Petri Net, consequently. In particular, the event that we need to consider is the one referring to the act of acquiring/releasing a lock; since this is the event that may determine a mutual exclusion access over shared resources.

The semantics of a lock is that of allowing only one (or more in case of semaphores) thread in a certain critical section (where one or more shared resources can be accessed). To replicate the same behaviour in the Petri Net, we need to enhance the transitions involving the acquiring/releasing of locks. The Enhancement algorithm in Fig. 1 takes in input a Petri Net $N = \langle P, T, F, W, M \rangle$ and returns a new Petri net $N' = \langle P', T, F', W, M' \rangle$ defined as follows. At line

Algorithm 1. Enhance(input $N = \langle P, T, F, W, M \rangle$, output $N' = \langle P', T, F', M' \rangle$)

1: $P' = P, F' = F, M' = M$
2: $Locks = GetLocks(T)$ ▷ Get all used locks
3: **for** $l \in Locks$ **do**
4:     $P' = P' \cup \{p_l\}$ ▷ Add lock place
5:     $M' = M' \cup \{p_l \to 1\}$ ▷ Add marking for lock place
6:     **for** $\langle p, acquire(l) \rangle \in F$ **do**
7:         $F' = F' \cup \{\langle p_l, acquire(l) \rangle\}$ ▷ Add arc from p_l to acquire transition
8:     **for** $\langle release(l), p \rangle \in F$ **do**
9:         $F' = F' \cup \{\langle release(l), p_l \rangle\}$ ▷ Add arc from release transition to p_l
return $N' = \langle P', T, F', W, M' \rangle$

1, the new set P' (resp. F' and M') which will contain the places (resp. arcs and markings) of the enhanced Petri Net is initialised. At line 2, the locks used by the system are extracted[1]. Then, the algorithm iterates over the set of locks so extracted (lines 3–9). For each lock l, an additional place (p_l) is created and added to the set P' of places (line 4). Since p_l is a place with a token, we also need to update the marking function consequently (line 5). That is the state which will be used to enforce synchronisations amongst traces using l. After that, for each transition corresponding to the acquisition of l (lines 6–7), a new arc is added in F' from p_l to the transition. In the opposite direction, the same thing is done, but for each transition corresponding to the release of l (lines 8–9), where a new arc is added in F' from the transition to p_l.

4.3 Hyper Projection over Additional Threads

Now that we have obtained the enhanced version of the Petri Net describing the system under analysis, we can verify the absence of data races. However, before the verification phase, there is another interesting post-processing modification to be applied. Such a modification is also one of the reasons for which this approach has been named "Hyper". Specifically, we can modify the Petri Net to increase the number of threads involved in the execution.

First of all, let us remark that the Petri Net we obtain through PM only considers as many threads as the ones that have been observed at runtime (*i.e.*, which have been reported in the trace log).

Algorithm 2. Project $(\langle P, T, F, W, M\rangle, N)$

$Locks = GetLocks(T)$ ▷ Get all used locks
$P' = P, T' = T, F' = F, M' = M$ ▷ Initialisation
for $i \in \{2, \dots, N\}$ **do**
  $P^i = \{p^i_j \mid p_j \in P\}$ ▷ Clone places
  $T^i = \{t^i_j \mid t_j \in T\}$ ▷ Clone transitions
  $F^i = \{\langle p^i_j, t^i_j\rangle \mid \langle p_j, t_j\rangle \in F\} \cup \{\langle t^i_j, p^i_j\rangle \mid \langle t_j, p_j\rangle \in F\}$ ▷ Clone arcs
  $M^i = \{p^i_j \rightarrow M(p_j) \mid p_j \in P\}$ ▷ Clone markings
  $P' = P' \cup (P^i \setminus \{p^i_l \in P^i \mid l \in Locks\})$ ▷ Update places but locks
  $T' = T' \cup T^i$ ▷ Update transitions
  $M' = M' \cup (M^i \setminus \{p^i_l \rightarrow \_ \mid p^i_l \in P^i \wedge l \in Locks\})$ ▷ Update markings but locks
  for $\langle p^i_j, t^i_j\rangle \in F^i$ **do**
    if $j \in Locks$ **then**
      $F' = F' \cup \{\langle p_j, t^i_j\rangle\}$ ▷ Replace lock place's arc
    else
      $F' = F' \cup \{\langle p^i_j, t^i_j\rangle\}$ ▷ Update arcs with new arc
  for $\langle t^i_j, p^i_j\rangle \in F^i$ **do**
    if $j \in Locks$ **then**
      $F' = F' \cup \{\langle t^i_j, p_j\rangle\}$ ▷ Replace lock place's arc
    else
      $F' = F' \cup \{\langle t^i_j, p^i_j\rangle\}$ ▷ Update arcs with new arc
return $\langle P', T', F', M'\rangle$

Algorithm 2 presents a projection function that can be used to increase the number of concurrent threads without re-executing the program. In more detail,

[1] Assuming each lock has a label to uniquely identify it through multiple traces, e.g., by considering the name of the variable used to store it, or similar.

the algorithm takes in input a Petri Net (in particular the Petri Net resulting from Algorithm 1), and the multiplier which specifies how many times to replicate the threads in the model. E.g., if $N = 2$, Algorithm 2 generates a Petri Net with twice the number of threads, and so on. At line 1, as in Algorithm 1, we extracts the lock labels (to identify the locking places added in the previous step). Then, in lines 3–20, the algorithms iterates over N. At each iteration, a clone of the Petri Net is created (lines 4–7). Such clone is exactly as the current Petri Net, except for the names of places, and transitions. For instance, if we have $P = \{p_1, p_2, p_3, p_l\}$, we would obtain $P^i = \{p_1^i, p_2^i, p_3^i, p_l^i\}$, with i from 1 to N. The same transformation is applied to transitions. While arcs and markings are simply carried out on the newly created places and transitions. After the clone Petri Net's components are created, the algorithms proceeds with the initialisation of the Hyper Petri Net's components (lines 8–10). In there, the newly created places and transitions are added to the places and transitions of the Petri Net. Then, the algorithm iterates on the arcs generated for the current clone Petri Net (F^i). For each arc, if it refers to a lock place (the ones added in Algorithm 1), a modified version of the arc is added to F', where p_j^i is replaced with p_j. This is needed to keep track of locks amongst different replications of the Petri Net. Otherwise, which means the arc does not refer to a lock place, the arc is added to F' unchanged. Note that, the iteration over the arcs is performed for the arcs from places to transitions (lines 11–15), and for the arcs from transitions to places (lines 16–20).

Once the repetitions have been done for the number of times required, we find in $\langle P', T', F', W, M' \rangle$ the resulting Petri Net. Note that, the loop of lines 3–20 iterates starting from 2; this means that if Algorithm 2 is called with $N = 1$, it returns directly the input Petri Net (which makes sense since $N = 1$ means no replication is needed).

The Algorithm 2 terminates in polynomial time w.r.t. N and the size of the Petri Net.

The entire Petri Net transformation algorithm can be refined in order to generate not only clones of event traces observed in a real execution trace but, for instance, to insert new places denoting locks reserved to a given set of event traces in order to distinguish shared and private (i.e. visibile within the scope of a given procedure) locks.

4.4 Verification

In our prototype, we focus on reachability properties for the Petri Net generated via the combination of inductive process mining and Algorithms 1 and 2. For instance, we can check whether bad states that represent data races can be reached in the execution of the computed Petri Net.

One possible way to solve the reachability problem is by generating the reachability graph. Then, to perform the formal verification step, it is enough to check whether a place containing a data race is (or not) reachable in such a graph.

Enhancement and projection both preserve the boundedness of the Petri Net. In fact, the enhancement step only adds locking places for synchronisation, while the projection step only replicates the Petri Net to simulate more threads. Both steps do not change the maximum number of tokens that a place may contain. Thus the final Petri Net is still a 1-*bounded* Petri Net.

As a natural extension we can also consider coverability problems (for bounded nets) to specify bad states using constraints on place subsets of places and applying parameterized verification procedures (e.g. covering graph, etc.).

Several other approaches can be applied to this kind of decision problems ranging from symbolic state exploration via model checking to static analysis via the marking equation or constraint solving in order to reduce the complexity of the verification phase as required in traditional runtime verification procedures.

The HyperMonitor Algorithm 3 summarizes the monitor resulting from combining the different steps described in this section. Each line in Algorithm 3

Algorithm 3. HyperMonitor $(Logs, N)$

1: $PetriNet = modelExtraction(Logs)$ ▷ Inductive Miner Algorithm [10]
2: $EnhancedPetriNet = Enhance(PetriNet)$ ▷ Algorithm 1
3: $HyperPetriNet = Project(EnhancedPetriNet, N)$ ▷ Algorithm 2
return $IsDataRaceNotReachable(HyperPetriNet)$

corresponds to one specific step. At line 1, the Inductive Miner algorithm is used to extract a Petri Net from log files given in input. Such log files have been obtained by executing the system under analysis. Note that, since we care about all possible behaviours observed by running the system, differently from standard RV techniques, we do not focus on a single execution at a time, but we consider all executions that have been observed (all the ones in the past executions as well). After the Petri Net has been generated, the algorithm continues with the enhancement of the latter by applying Algorithm 1. The resulting Enhanced Petri Net is then replicated w.r.t. the parameter N. In this way, the Enhanced Petri Net is projected on N dimensions (*i.e.*, it is replicated N times). This is obtained by applying Algorithm 2. The resulting Hyper Petri Net is then analysed and checked against the presence of data races. This last step is obtained by performing a reachability analysis over the Hyper Petri Net, where the algorithm looks whether states denoting data races may be reached. Algorithm 3 returns the answer given to the reachability problem, *i.e.*, in case at least one data race place is reachable in the Hyper Petri Net, Algorithm 3 returns false, while it returns true otherwise (so when no data race place can be reached whatsoever). Since inductive miner, transformations and projections require polynomial time w.r.t. the size of the Petri Net and the number N of replications, when a complete reachability algorithm is used in the last step of Algorithm 3, the whole procedure may require exponential time w.r.t. the size of the input model.

4.5 HPRV Implementation

The HPRV prototype implementation is publicly available as a GitHub repository[2]. Our procedure operates on trace logs given in XES (eXtensible Event Stream) format, an XML-based standard format for process mining tools. XES trace files are generated during the execution of an instrumented concurrent program. For implementing the PM step, we used the PM4Py[3] Python library. PM4Py is an open source PM platform written in Python, and is developed by the PM group of Fraunhofer Institute for Applied Information Technology[4]. PM4Py, amongst various algorithms, supports the Inductive Miner algorithm and generates a Petri Net starting from log files.

The Petri Net model transformations have been fully implemented by the authors in Python. The verification step has been implemented by extracting the reachability graph of the Petri Net derived from previous steps, and by searching for states containing data races in the latter.

4.6 Case-Studies

To experimentally evaluate our verification method, we have defined a set of instrumented C programs to capture the most common patterns (including typical errors) that we may find in concurrent programming when using libraries such as the pthread library. For instance, we consider the basic pattern with a single thread creation and an iterative body functions shown below.

```
#define INT2VOIDP(i) (void*)(uintptr_t)(i)
#define VOIDP2INT(i) (uintptr_t)(void*)(i)
int c = 0;

void* fnC1(void* a)
{
  int tmp; int k = VOIDP2INT(a);
  char thread_log_filename[20]; FILE* thread_log;

  thread_log = fopen ("log1", "a");

  for(int i=0;i<k;i++) {
      c=c+10;
      printEvent(thread_log,"Thread1","read_shared_data(c)");
      printEvent(thread_log,"Thread1","write_shared_data(c)");
  }

  fprintf(thread_log, "%s\n", "</trace>\n");
  fclose(thread_log);
  return(0);
}

int main(int argc, char* argv[])
{
    int rt1; pthread_t t1; int N=2;
    if (argc>1) N=atoi(argv[1]);
    void* val = INT2VOIDP(N);
    rt1=pthread_create(&t1, NULL, fnC1, val);
    pthread_join(t1, NULL);
    return 0;
}
```

The printEvent procedure is used to generate log items. Every read and write access is traced using a sequence of "read/write_shared_data(c)" labels. A single execution of the program appends the corresponding series of event items to the

---

[2] https://github.com/AngeloFerrando/HyperPredictiveRuntimeVerification.

[3] https://pm4py.fit.fraunhofer.de/.

[4] http://fit.fraunhofer.de/process-mining.

log file. Then, an auxiliary program is used to generated the header information needed to generate the XES format needed in the remaining part of the pipeline. We consider several other examples of single threaded programs parametric on input parameters such as the number of threads in the original C program (the number m of iterations in each thread, the number of distinct shared variables used in the different threads), and the following types of patterns for each thread:

P1 : $(\texttt{read(c) write(c)})^m$
P2 : create_lock(L) $(\texttt{lock(L) read(c) write(c) unlock(L)})^m$
P3 : create_lock(L) lock(L) read(c_1) write(c_1) unlock(L) ... lock(L) read(c_p) write(c_p) unlock(L)
P4 : create_lock(l) $(\texttt{lock(l) read(c) write(c) unlock(l)})^m$
P5 : create_lock(l) lock(l) read(c_1) write(c_1) unlock(l) ... lock(l) read(c_p) write(c_p) unlock(l)

where L (resp. l) denotes a shared (resp. private to the thread) lock.

Table 1. Experimental results for patterns $P1$–$P5$, where OTS = Original Transition System, CTS = Cloned Transition System): ✓ = the original (resp. version with cloned threads) C program is thread safe.

Pattern	#threads	#iterations	#shared variables	OTS?	Time[s]	#clones	CTS?	Time[s]
P1	1	10	1	✓	0.05–0.001	2	✗	0.07–0.003
	1	50	1	✓	0.08–0.001	2	✗	0.11–0.003
	1	100	1	✓	0.12–0.001	2	✗	0.13–0.003
	2	10	1	✗	0.25–0.004	2	✗	0.46–0.06
	2	50	1	✗	1.41–0.006	2	✗	2.49–0.05
	2	100	1	✗	3.95–0.009	2	✗	5.30–0.072
P2	1	10	1	✓	0.09–0.0009	2	✓	0.16–0.006
	1	50	1	✓	0.18–0.0008	2	✓	0.21–0.005
	1	100	1	✓	0.38–0.0009	2	✓	0.49–0.006
	2	10	1	✓	0.17–0.006	2	✓	0.24–0.43
	2	50	1	✓	0.35–0.05	2	✓	0.52–0.42
	2	100	1	✓	0.71–0.005	2	✓	0.93–0.51
P3	1	10	10	✓	0.28–0.002	2	✓	0.36–0.02
	1	50	50	✓	2.45–0.009	2	✓	3.58–0.20
	1	100	100	✓	17.74–0.03	2	✓	19.66–0.56
	2	10	10	✓	0.37–0.02	2	✓	1.01–1.97
	2	50	50	✓	5.17–0.16	2	✓	7.09–16.78
	2	100	100	✓	36.63–0.59	2	✓	39.02–101.64
P4	1	10	1	✓	0.09–0.001	2	✗	0.14–0.01
	1	50	1	✓	0.17–0.001	2	✗	0.23–0.008
	1	100	1	✓	0.37–0.001	2	✗	0.52–0.01
	2	10	1	✗	0.15–0.007	2	✗	0.19–1.34
	2	50	1	✗	0.34–0.008	2	✗	0.47–1.77
	2	100	1	✗	0.69–0.009	2	✗	0.81–2.03
P5	1	10	10	✓	0.19–0.002	2	✗	0.37–0.06
	1	50	50	✓	2.15–0.01	2	✗	3.21–1.14
	1	100	100	✓	18.33–0.03	2	✗	21.17–6.88
	2	10	10	✗	0.15–0.009	2	✗	0.26–2.09
	2	50	50	✗	0.33–0.05	2	✗	0.47–2.02
	2	100	100	✗	0.71–0.008	2	✗	0.87–2.03

Table 1 reports experiments we carried out on the previously listed patterns. Each pattern is tested w.r.t. various parameters. Amongst them we may find the number of threads used, the number of iterations per thread (the m parameter used before), whether the program or its replicated version (with double the number of thread) are thread safe. We name OTS (Original Transition System) and CTS (Cloned Transition System) the columns reporting such verification results. For the CTS version, we also report the multiplier used in the thread replication (the N parameter in Algorithm 2, which in Table 1 is set to 2, *i.e.*, the threads are duplicated through the projection).

Moreover, in each experiment, the execution time is reported. The first component is the time required to extract the Petri Net and perform its enhancement. While the second component is the verification time (*i.e.*, the time required to perform the reachability analysis). Each C program has been executed 10 times to generate the log file used by the PM algorithm to synthesise the Petri Net. The obtained results are consistent with the considered C programs with respect to the safety violations detected by our procedure. Note that, the time execution is always less than 1 s, except for the more stressed experiments. Even though these are initial results the seem promising considering that we applied an exact reachability procedure.

5 Conclusions

In this paper, we have presented the main features of HyperMonitor, a prototype tool written in Python for predicting race conditions on (instrumented) concurrent C code. The tool is based on a pipeline that combines inductive process mining with *ad hoc* Petri net transformations and verification procedures used for different types of reachability problems. The considered examples are available on Github and consists of instrumented concurrent C programs (pthreads library) that generate event traces in XES format when executed. Future directions include further development of the tool, above all w.r.t. the logs extraction and the analysis which can be performed on the Petri Net. For instance, other synchronisation mechanisms can be introduced, like barriers. Furthermore, approximated algorithms could be applied in order to lower the complexity of the verification step.

References

1. Augusto, A., et al.: Automated discovery of process models from event logs: review and benchmark. IEEE Trans. Knowl. Data Eng. **31**(4), 686–705 (2019)
2. Bartocci, E., Falcone, Y., Francalanza, A., Reger, G.: Introduction to runtime verification. In: Bartocci, E., Falcone, Y. (eds.) Lectures on Runtime Verification. LNCS, vol. 10457, pp. 1–33. Springer, Cham (2018). https://doi.org/10.1007/978-3-319-75632-5_1
3. Cheng, A., Esparza, J., Palsberg, J.: Complexity results for 1-safe nets. In: Shyamasundar, R.K. (ed.) FSTTCS 1993. LNCS, vol. 761, pp. 326–337. Springer, Heidelberg (1993). https://doi.org/10.1007/3-540-57529-4_66

4. Clarke, E.M.: Model checking. In: Ramesh, S., Sivakumar, G. (eds.) FSTTCS 1997. LNCS, vol. 1346, pp. 54–56. Springer, Heidelberg (1997). https://doi.org/10.1007/BFb0058022
5. Falcone, Y., Havelund, K., Reger, G.: A tutorial on runtime verification. In: Broy, M., Peled, D.A., Kalus, G. (eds.) Engineering Dependable Software Systems. NATO Science for Peace and Security Series, D: Information and Communication Security, vol. 34, pp. 141–175. IOS Press (2013)
6. Ferrando, A., Delzanno, G.: Incrementally predictive runtime verification. In: Monica, S., Bergenti, F. (eds.) Proceedings of the 36th Italian Conference on Computational Logic. CEUR Workshop Proceedings, Parma, Italy, 7–9 September 2021, vol. 3002, pp. 92–106. CEUR-WS.org (2021)
7. Havelund, K.: Using runtime analysis to guide model checking of Java programs. In: Havelund, K., Penix, J., Visser, W. (eds.) SPIN 2000. LNCS, vol. 1885, pp. 245–264. Springer, Heidelberg (2000). https://doi.org/10.1007/10722468_15
8. Havelund, K., Rosu, G.: An overview of the runtime verification tool Java pathexplorer. Formal Meth. Syst. Des. **24**(2), 189–215 (2004)
9. Huang, J., Meredith, P.O., Rosu, G.: Maximal sound predictive race detection with control flow abstraction. In: O'Boyle, M.F.P., Pingali, K. (eds.) ACM SIGPLAN Conference on Programming Language Design and Implementation, PLDI 2014, Edinburgh, United Kingdom, 09–11 June 2014, pp. 337–348. ACM (2014). https://doi.org/10.1145/2594291.2594315
10. Leemans, S.J.J., Fahland, D., van der Aalst, W.M.P.: Discovering block-structured process models from event logs - a constructive approach. In: Colom, J.-M., Desel, J. (eds.) PETRI NETS 2013. LNCS, vol. 7927, pp. 311–329. Springer, Heidelberg (2013). https://doi.org/10.1007/978-3-642-38697-8_17
11. Loveland, D.W.: Automated Theorem Proving: A Logical Basis. Fundamental Studies in Computer Science, vol. 6. North-Holland (1978)
12. Pinisetty, S., Jéron, T., Tripakis, S., Falcone, Y., Marchand, H., Preoteasa, V.: Predictive runtime verification of timed properties. J. Syst. Softw. **132**, 353–365 (2017)
13. Qadeer, S., Rehof, J.: Context-bounded model checking of concurrent software. In: Halbwachs, N., Zuck, L.D. (eds.) TACAS 2005. LNCS, vol. 3440, pp. 93–107. Springer, Heidelberg (2005). https://doi.org/10.1007/978-3-540-31980-1_7
14. Savage, S., Burrows, M., Nelson, G., Sobalvarro, P., Anderson, T.E.: Eraser: a dynamic data race detector for multithreaded programs. ACM Trans. Comput. Syst. **15**(4), 391–411 (1997)
15. Şerbănuţă, T.F., Chen, F., Roşu, G.: Maximal causal models for sequentially consistent systems. In: Qadeer, S., Tasiran, S. (eds.) RV 2012. LNCS, vol. 7687, pp. 136–150. Springer, Heidelberg (2013). https://doi.org/10.1007/978-3-642-35632-2_16
16. Zhang, X., Leucker, M., Dong, W.: Runtime verification with predictive semantics. In: Goodloe, A.E., Person, S. (eds.) NFM 2012. LNCS, vol. 7226, pp. 418–432. Springer, Heidelberg (2012). https://doi.org/10.1007/978-3-642-28891-3_37

Generalized ARRIVAL Problem for Rotor Walks in Path Multigraphs

David Auger, Pierre Couchenay(✉), Loric Duhazé, and Kossi Roland Etse

DAVID Laboratory, UVSQ, Université Paris Saclay, 45 avenue des Etats-Unis, 78000 Versailles, France
Pierre.Coucheney@uvsq.fr

Abstract. Rotor walks are cellular automata that determine deterministic traversals of particles in a directed multigraph using simple local rules, yet they can generate complex behaviors. Furthermore, these trajectories exhibit statistical properties similar to random walks.

In this study, we investigate a generalized version of the reachability problem known as ARRIVAL in Path Multigraphs, which involves predicting the number of particles that will reach designated target vertices. We show that this problem is in NP and co-NP in the general case. However, we exhibit algebraic invariants for Path Multigraphs that allow us to solve the problem efficiently, even for an exponential configuration of particles. These invariants are based on harmonic functions and are connected to the decomposition of integers in rational bases (This paper is an extended abstract. For a more explanatory approach please refer to the full version [2]).

Keywords: Rotor walks · cellular automata · discrete harmonic function

1 Introduction

The *rotor routing*, or *rotor walk model*, has been studied under different names: *eulerian walkers* [15,16] and *patrolling algorithm* [17]. It shares many properties with a more algebraically focused model: *abelian sandpiles* [4,13]. General introductions to this cellular automaton can be found in [11] and [13].

Here is how a rotor walk works: in a directed graph, each vertex v with an outdegree of k has its outgoing arcs numbered from 1 to k. Initially, a particle is placed on a starting vertex, and the following process is repeated. On the initial vertex, the particle moves to the next vertex following arc 1. The same rule then applies on subsequent vertices. However, when a vertex is revisited, the particle changes its movement to the next arc, incrementing the number until the last arc is used. Then, the particle restarts from arc 1 if it visits this vertex again. The trajectories defined by this deterministic process has been shown to share statistical properties with random walks [14].

This simple rule defines the rotor routing, which exhibits many interesting properties. Particularly, if the graph is sufficiently connected, the particle will

O. Bournez et al. (Eds.): RP 2023, LNCS 14235, pp. 183–198, 2023.
https://doi.org/10.1007/978-3-031-45286-4_14

eventually reach certain target vertices known as sinks. The problem of determining, given a starting configuration (numbering) of arcs and an initial vertex, which sink will be reached first, is known as the ARRIVAL problem. It was defined in [5], along with a proof that the problem belongs to the complexity class NP ∩ co-NP. Although the problem is not known to be in P, [9] showed that it belongs to the smaller complexity class UP ∩ co-UP. Furthermore, a subexponential algorithm based on computing a Tarski fixed point was proposed in [10].

Despite these general bounds, little is known about efficiently solving the problem in specific graph classes, especially when extending it to the routing of multiple particles. In [1], we addressed the problem in multigraphs with a tree-like structure and provided a linear algorithm for solving it with a single particle. However, the recursive nature of the algorithm provided limited insights into the structure of rotor walks in the graph. We also examined the structure of rotor walks and the so-called sandpile group in the case of a simple directed path, where simple invariants can explain the behavior of rotor walks.

In this work, we focus specifically on a family of multigraphs that consist of directed paths with a fixed number of arcs going left and right on each vertex, with a sink located at both ends of the path. We present an efficient algorithm for solving the ARRIVAL problem in this general context (see Theorems 5 and 6, and the example that follows), considering a potentially exponential number of particles and antiparticles, a concept introduced in [11]. Our approach involves introducing algebraic invariants for rotor walks and chip-firing, enabling a complete description of the interplay between particle configurations and rotor configurations/walks. These invariants are derived from harmonic functions in graphs, which are functions invariant under chip-firing. Additionally, we introduce a related concept for rotor configurations called arcmonic functions, inspired by [12].

An essential tool for analyzing rotor routing in Path Multigraphs is the decomposition of integer values, which is closely associated with the AFS number system [8], where numbers are decomposed into rational bases. While we draw inspiration from these results, our approach focuses on proving precisely what is necessary, using our own methodology.

Additionally, we derive other outcomes, such as the cardinality of the Sandpile Group of Path Multigraphs or its cyclic structure. These results can also be derived from Kirchoff's Matrix-Tree Theorem or the notion of co-eulerian graphs [7]. Nevertheless, our results remain self-contained.

2 Mechanics and Tools for Rotor Routing in Multigraphs

Proofs of results that are not in the document can be found in the full version [2].

2.1 Multigraphs

A **directed multigraph** G is a tuple $G = (V, A, \text{head}, \text{tail})$ where V and A are respectively finite sets of *vertices* and *arcs*, and *head* and *tail* are maps from A

to V defining incidence between arcs and vertices. An arc with tail x and head y is said to be from x to y. Note that multigraphs can have multiple arcs with the same head and tail, as well as loops.

For a vertex $u \in V$, we denote by $A^+(u)$ the subset of arcs going out of u, i.e. $A^+(u) = \{a \in A \mid \text{tail}(a) = u\}$ and $\deg^+(u) = |A^+(u)|$ is the outdegree of u. We denote by V_0 the set of vertices with positive outdegree and S_0 vertices with zero outdegree, i.e. **sinks**. A directed multigraph is **stopping** if for every vertex u, there is a directed path from u to a sink. **In this whole paper, we suppose that G is a stopping multigraph.**

In the second part of this work, we consider the **Path multigraph** $P_n^{x,y}$ on $n+2$ vertices which is a multigraph $G = (V_0 \cup S_0, A, \text{head}, \text{tail})$ such that:

- $V_0 = \{u_1, u_2, ..., u_n\}$ and $S_0 = \{u_0, u_{n+1}\}$;
- for $k \in [\![1, n]\!]$, we have $\deg^+(u_k) = x + y$ with x arcs from u_k to u_{k+1} and y arcs from u_k to u_{k-1}
- u_0 and u_{n+1} are considered as **sinks** with no outgoing arcs.

This graph is clearly stopping if $x+y \geq 1$. See Fig. 1 for a representation of $P_n^{2,3}$.

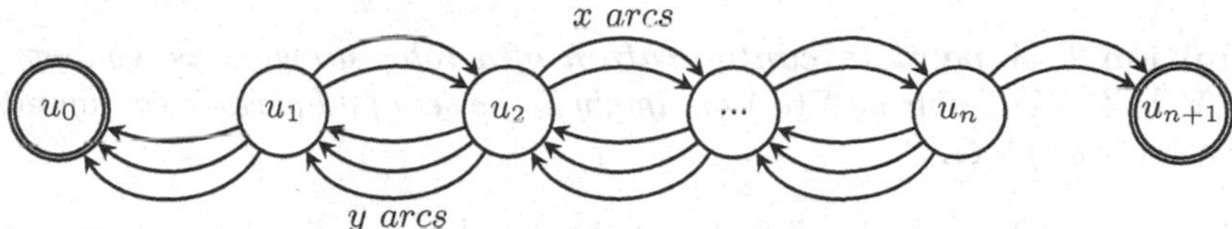

Fig. 1. The Path Multigraph $P_n^{2,3}$.

We consider the case $n \geq 1$, and $1 \leq x < y$ with x, y coprime.

2.2 Rotor Structure

If $u \in V_0$, a **rotor order** at u is an operator denoted by θ_u such that:

- $\theta_u : A^+(u) \to A^+(u)$;
- for all $a \in A^+(u)$, the orbit $\{a, \theta_u(a), \theta_u^2(a), ..., \theta_u^{\deg^+(u)-1}(a)\}$ of a under θ_u is equal to $A^+(u)$, where $\theta_u^k(a)$ is the composition of θ_u applied to arc a exactly k times.

A **rotor order** for G is then a map $\theta : A \to A$ such that the restriction θ_u of θ to $A^+(u)$ is a rotor order at u for every $u \in V_0$. Note that all θ_u as well as θ are one to one. If $C \subseteq V_0$, the composition of operators θ_u for all $u \in C$ does not depend on the order of composition since they act on disjoint sets $A^+(u)$; we denote by θ_C this operator and θ_C^{-1} is its inverse. Finally, we use the term **rotor graph** to denote a stopping multigraph together with a rotor order θ.

In $P_n^{x,y}$, we define a rotor order by simply considering all arcs going right before all arcs going left, cyclically (see Fig. 2). Formally, let a_i^k denote for $i \in [\![0, x-1]\!]$ the x arcs from u_k to u_{k+1} and for $i \in [\![x, x+y-1]\!]$ the y arcs from u_k to u_{k-1}; then we define $\theta(a_i^k) = a_j^k$ with $j = i + 1 \mod x + y$.

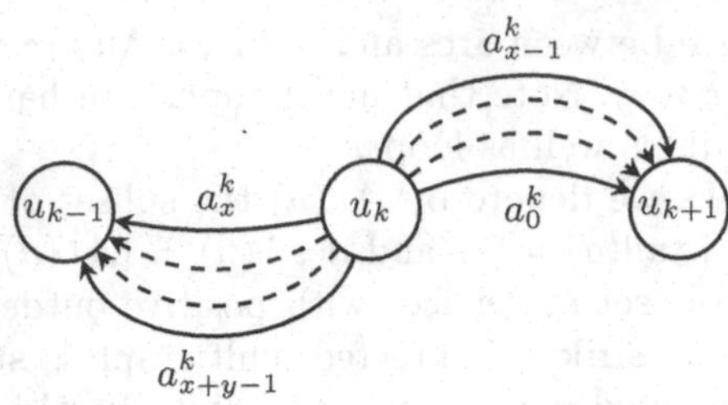

Fig. 2. Rotor order at a vertex u_k in the Path Multigraph $P_n^{x,y}$

2.3 Configurations

Definition 1. *A* ***rotor configuration*** *of a rotor graph G is a mapping ρ from V_0 to A such that $\rho(u) \in A^+(u)$ for all $u \in V_0$. We denote by $\mathcal{R}(G)$ or simply $\mathcal{R}$ the set of all rotor configurations of the rotor graph G.*

The **graph induced** by ρ on $G = (V, A, \text{head}, \text{tail})$ is $G(\rho) = (V, \rho(V_0), \text{head}, \text{tail})$ in which each vertex in V_0 has outdegree one.

Definition 2. *A* ***particle configuration*** *of a rotor graph G is a mapping σ from V to $\mathbb{Z}$. We denote by $\Sigma(G)$ or simply Σ the set of all particle configurations of the rotor graph G.*

The set $\Sigma(G)$ can be identified with $\mathbb{Z}^V$ and has a natural structure of additive abelian group. If $u \in V$, we identify u with the element of $\Sigma(G)$ which is one on vertex u and zero elsewhere. Thus we can write, e.g. $\sigma + 3u$ to denote the configuration obtained from $\sigma \in \Sigma$ by adding 3 to $\sigma(u)$.

If $\sigma(u) \geq 0$, we interpret it as a number of particles on vertex u, whereas if $\sigma(u) \leq 0$ it can be interpreted as antiparticles, or simply a debt of particles. The **degree** of a particle configuration σ is defined by $\deg(\sigma) = \sum_{u \in V} \sigma(u)$.

Finally, a **rotor-particle** configuration is an element of $\mathcal{R}(G) \times \Sigma(G)$.

2.4 Rotor Routing

Definition 3. *Let G be a rotor graph, we define operators indexed by vertices $u \in V_0$ on $\mathcal{R}(G) \times \Sigma(G)$:*

- *$move_u^+ : \mathcal{R}(G) \times \Sigma(G) \to \mathcal{R}(G) \times \Sigma(G)$ is defined by*

$$move_u^+(\rho, \sigma) = (\rho, \sigma + head(\rho(u)) - u);$$

- *$turn_u^+ : \mathcal{R}(G) \times \Sigma(G) \to \mathcal{R}(G) \times \Sigma(G)$ is defined by $turn_u^+(\rho, \sigma) = (\theta_u \circ \rho, \sigma)$.*

Note that $\theta_u \circ \rho$ is the rotor configuration equal to ρ on all vertices except in u where θ has updated the arc. Applying move_u^+ to (ρ, σ) can be interpreted as moving a particle from u to the head of arc $\rho(u)$, whereas applying turn_u^+ updates the rotor configuration at u. It is easy to see that these operators are bijective on $\mathcal{R}(G) \times \Sigma(G)$, and we denote by move_u^- and turn_u^- their inverses.

We now define the routing operators by $\text{routing}_u^+ = \text{turn}_u^+ \circ \text{move}_u^+$, and its inverse is obviously $\text{routing}_u^- = \text{move}_u^- \circ \text{turn}_u^-$. Routing a rotor-particle configuration (ρ, σ) consists in applying a series of routing^+ and routing^- operators. Since they act on different vertices and disjoint sets of arcs, the following result is straightforward.

Lemma 1. *The family of operators* routing_u^+ *and* routing_u^- *for all* $u \in V_0$ *commute.*

Since the order in which routing operators are applied does not matter, we define a **routing vector** as a map from V_0 to $\mathbb{Z}$. We define routing^r as the operator obtained by composing all elements of the family $\{(\text{routing}_u^+)^{r(u)}\}_{u \in V_0}$, in any order, where the exponent $r(u)$ stands for composition of the operator or its inverse with itself, depending on the sign of $r(u)$. We shall use the term *routing* when we apply any operator routing^r as well.

2.5 Legal Routing and ARRIVAL

Applying routing_u^+ to $(\rho, \sigma) \in \mathcal{R} \times \Sigma$ is said to be a **legal routing** if $\sigma(u) > 0$. A sequence of configurations obtained by successive legal routings $(\rho_0, \sigma_0)(\rho_1, \sigma_1) \cdots (\rho_k, \sigma_k)$ is **maximal** if for all $u \in V_0$ we have $\sigma_k(u) \leq 0$, i.e. no other legal routing can be applied.

The classic version of the commutativity result for rotor routing is the following:

Proposition 1 ([13]). *For all* $(\rho, \sigma) \in \mathcal{R} \times \Sigma$ *with* $\sigma \geq 0$*, there is a unique* (ρ', σ') *with* $\sigma'(u) = 0$ *for all* $u \in V_0$*, such that all maximal legal routings from* (ρ, σ) *end in* (ρ', σ')*. Furthermore, all legal routings can be continued in such a maximal legal routing.*

For such a maximal legal routing, we shall say that (ρ, σ) is **fully routed** to sinks, and write $(\rho', \sigma') = \text{routing}_L^\infty(\rho, \sigma)$ where the L stands for legal.

The original ARRIVAL problem consists in the following decision problem: *if* $(\rho, \sigma) \in \mathcal{R} \times \Sigma$ *with* $\sigma \geq 0$ *and* $\deg(\sigma) = 1$*, if* $(\rho', \sigma') = \text{routing}_L^\infty(\rho, \sigma)$*, for a given sink* $s \in S_0$*, does* $\sigma'(s) = 1$ *?*

This problem is known to be in NP and co-NP, but the best algorithm known to this date (see [10]) has complexity $2^{O(\sqrt{|V|}\log|V|)}$ in the case of a switch graph (where $\deg^+(u) \leq 2$ for every vertex u). We shall now generalize this problem to any number of positive and negative particles, and remove the legality assumption. In the next two subsections, we define equivalence classes for rotor configurations and particle configurations respectively which will be the basis for defining algebraic invariants for rotor walks.

2.6 Equivalence Classes of Rotors

Definition 4. *Two rotor-particle configurations* (ρ, σ) *and* (ρ', σ') *are said to be equivalent, which we denote by* $(\rho, \sigma) \sim (\rho', \sigma')$*, if there is a routing vector* r *such that* $\text{routing}^r(\rho, \sigma) = (\rho', \sigma')$*.*

It is easy to see that this defines an equivalence relation on $\mathcal{R} \times \Sigma$.

Definition 5. *Two rotor configurations ρ, ρ' are said to be equivalent, which we denote by $\rho \sim \rho'$, if there is $\sigma \in \Sigma$ such that $(\rho, \sigma) \sim (\rho', \sigma)$.*

In this case, the relation is true for any $\sigma \in \Sigma$, and it defines an equivalence relation on $\mathcal{R}$.

Cycle Pushes. Suppose that $\rho \in \mathcal{R}$ and let C be a directed circuit in $G(\rho)$. The **positive cycle push** of C in ρ transforms ρ into $\theta_C \circ \rho$. Similarly, if C is a directed circuit in $G(\theta^{-1} \circ \rho)$, the **negative cycle push** transforms ρ into $\theta_C^{-1} \circ \rho$. A **sequence of cycle pushes** is a finite or infinite sequence of rotor configurations (ρ_i) such that each ρ_{i+1} is obtained from ρ_i by a positive or negative cycle push.

Note that if C is a directed circuit in $G(\rho)$, for any $\sigma \in \Sigma$, we can obtain $(\theta_C \circ \rho, \sigma)$ by applying routing^{r_C} to (ρ, σ), and if C is a circuit in $G(\theta^{-1} \circ \rho)$, then $(\theta_C^{-1} \circ \rho, \sigma)$ is equal to routing$^{-r_C}(\rho, \sigma)$, where in both cases r_C is the routing vector consisting in routing once every vertex of C. In other words, a cycle push is a shortcut in the routing of a particle on the circuit.

Theorem 1. *Given two rotor configurations ρ and ρ', $\rho \sim \rho'$ if and only if ρ' can be obtained from ρ by a sequence of cycle pushes.*

Whenever rotor configurations are equivalent, they eventually route particles identically since positive and negative cycle pushes correspond to adding or removing closed circuits in trajectories. In particular, it is easy to see that it is always possible to route any (ρ, σ) to a (ρ', σ') such that $\sigma'(u) = 0$ for all $u \in V_0$. Let us denote by routing$^{\infty}(\rho, \sigma)$ the nonempty set of these configurations.

Theorem 2. *Let $(\rho_1, \sigma_1) \in$ routing$^{\infty}(\rho, \sigma)$. Then $(\rho_2, \sigma_2) \in$ routing$^{\infty}(\rho, \sigma)$ if and only if $\rho_1 \sim \rho_2$ and $\sigma_1 = \sigma_2$.*

Corollary 1. *If $\sigma \geq 0$, then if $(\rho', \sigma') =$ routing$_L^{\infty}(\rho, \sigma)$ and $(\rho_1, \sigma_1) \in$ routing$^{\infty}(\rho, \sigma)$, we have $\sigma_1 = \sigma'$ and $\rho' \sim \rho_1$.*

The GENERALIZED ARRIVAL problem is: *given any (σ, ρ), compute σ_1 for any* $(\rho_1, \sigma_1) \in$ routing$^{\infty}(\rho, \sigma)$.

Corollary 1 shows that this problem contains the original ARRIVAL problem. On the other hand, the decision version of GENERALIZED ARRIVAL belongs to NP and co-NP, a certificate being a routing vector r; one may compute efficiently the configuration routing$^{r}(\rho, \sigma)$ and check that we obtain 0 particles on V_0.

Acyclic Configurations. We say that $\rho \in \mathcal{R}$ is **acyclic** if $G(\rho)$ contains no directed cycles. It amounts to saying that the set of arcs $\rho(V_0)$ forms in G a directed forest, rooted in the sinks of G.

Proposition 2 ([11]). *Each equivalence class of rotor configurations contains exactly one acyclic configuration.*

2.7 Equivalence Classes of Particles

Definition 6. *Two particle configurations σ, σ' are said to be equivalent, which we denote by $\sigma \sim \sigma'$, if there is $\rho \in \mathcal{R}$ such that $(\rho, \sigma) \sim (\rho, \sigma')$.*

In this case, the relation is true for any $\rho \in \mathcal{R}$, and it defines an equivalence relation on Σ.

Define the Laplacian operator Δ as the linear operator from $\mathbb{Z}^{V_0}$ to Σ, defined for $u \in V_0$ by

$$\Delta(u) = \sum_{a \in A^+(u)} (\text{head}(a) - \text{tail}(a))$$

The vector $\Delta(u)$, when added to a particle configuration σ, corresponds to transferring a total of $\deg^+(u)$ particles from u to every outneighbour of u. The transformation from σ to $\sigma + \Delta(u)$ is called **firing** σ at u. This firing is legal if $\sigma(u) \geq \deg^+(u)$.

A **firing vector** is simply an element of $r \in \mathbb{Z}^{V_0}$, and we can fire simultaneously vertices according to this vector by

$$\sigma + \Delta(r) = \sigma + \sum_{u \in V_0} r(u)\Delta(u).$$

Proposition 3. *For any two particle configurations σ, σ' we have $\sigma \sim \sigma'$ if and only if there exists a firing vector r with $\sigma' = \sigma + \Delta(r)$.*

By analogy with maximal legal routings, define a maximal legal firing as a sequence of legal firings from σ to another particle configuration σ' such that finally σ' is **stable**, meaning that $\sigma'(u) < \deg^+(u)$ for all $u \in V_0$, i.e. no more legal firing are possible.

Proposition 4 ([3]). *If G is stopping, for all particle configurations σ there is a unique configuration σ' such that every maximal sequence of legal firings leads to σ', and every sequence of legal firings can be continued in such a maximal sequence (in particular, all legal sequences are finite).*

This stable configuration σ' is the **stabilization** of σ and denoted σ°.

2.8 Sandpile Group

We point out that the equivalence relation on particles defined in the previous section is not equivalent to the construction of the so-called Sandpile Group. In the case of a stopping rotor graph, the Sandpile Group is obtained from particle configurations equivalence classes by furthermore identifying configurations which have the same value on V_0. More precisely, define a relation $\sim_S$ by

$$\sigma \sim_S \sigma' \Leftrightarrow \exists \sigma_1, \sigma \sim \sigma_1 \text{ and } \forall u \in V_0, \sigma'(u) = \sigma_1(u).$$

Proposition 5 ([13]).

- *The quotient of Σ by $\sim_S$ has an additive structure inherited from Σ, and it is a finite abelian group called the Sandpile Group and denoted by $SP(G)$;*
- *the order of $SP(G)$ is the number of acyclic rotor configurations in G.*

3 Main Results for Path Multigraphs

In this part, we summarize our results, and the rest of the paper will introduce the tools used to prove them. From now on, we consider only graphs of the family $P_n^{x,y}$, and the letter G denotes such a graph.

3.1 Case $x = y = 1$

First, let us recall the results obtained about Path Graphs $P_n^{1,1}$ in [1] in order to understand how they compare to the case $P_n^{x,y}$ when $0 < x < y$ are coprime. Technically, these results were stated only for nonnegative particle configurations but they still hold in the general case.

In the case $x = y = 1$, define for any particle configuration σ and rotor configuration ρ

$$h(\sigma) = \sum_{i=0}^{n+1} i \cdot \sigma(u_i) \text{ and } g(\rho) = |i : \text{head}(\rho(u_i)) = u_{i-1}|$$

i.e. $g(\rho)$ is the number of arcs in $G(\rho)$ pointing to the left.

The next result completely solves GENERALIZED ARRIVAL in $P_n^{1,1}$ for any number of particles and antiparticles.

Theorem 3. *In the case $x = y = 1$, for all $(\rho, \sigma) \in \mathcal{R} \times \Sigma$, the number of particles on sink u_{n+1} in any configuration of routing$^\infty(\rho, \sigma)$ is equal to the unique $m \in \mathbb{Z}$ such that*

$$0 \leq g(\rho) - h(\sigma) + m(n+1) \leq n, \text{ i.e. } m = \lceil \frac{h(\sigma) - g(\rho)}{n+1} \rceil.$$

Additionally, we can describe the structure of the Sandpile Group of $P_n^{1,1}$ and its action on rotor configurations. Define $\bar{h}$ and $\bar{g}$ as h and g modulo $n+1$.

Theorem 4. *(i) The Sandpile Group $SP(P_n^{1,1})$ is cyclic of order $n+1$;*

(ii) the map $\bar{h} : \Sigma \to \mathbb{Z}/(n+1)\mathbb{Z}$ quotients by $\sim_S$ into an isomorphism between $SP(P_n^{1,1})$ and $\mathbb{Z}/(n+1)\mathbb{Z}$;

(iii) the map $\bar{g} : \mathcal{R} \to \mathbb{Z}/(n+1)\mathbb{Z}$ quotients into a bijection between rotor equivalence classes and $\mathbb{Z}/(n+1)\mathbb{Z}$;

(iv) the action of the sandpile group on rotor equivalence classes can be understood in the following way: let (ρ, σ) be a rotor-particle configuration and $(\rho', \sigma') \in$ routing$^\infty(\rho, \sigma)$. Then ρ' is in class $\bar{g}(\rho') = \bar{g}(\rho) - \bar{h}(\sigma)$.

As an example, consider the case $P_3^{1,1}$, which is depicted on Fig. 3, with the particle configuration σ equal to $(-8, 5, 10, -5, 12)$ from left to right and ρ as depicted. We see that ρ has 2 arcs going left so that $g(\rho) = 2$, while we have

$$h(\sigma) = -8 \cdot 0 + 5 \cdot 1 + 10 \cdot 2 - 5 \cdot 3 + 12 \cdot 4 = 58.$$

From Theorem 3, we deduce the final configuration σ' of the full routing of (ρ, σ) counts $m = 14$ particles ending on the right sink u_4 and $-8+5+10-5+12-14 = 0$ particles on u_0.

From Theorem 4, we deduce that any final rotor configuration ρ' in the routing will be such that $\bar{g}(\rho') = 2 - 58 = 0 \mod 4$, so that all its arcs will point right, hence ρ' is the acyclic configuration of this class.

Fig. 3. Rotor routing a particle configuration in $P_3^{1,1}$. The particle configuration is written in squares in vertices and the initial rotor configuration is given by full arcs while other arcs are dashed. Note that there is a unique rotor order in this graph.

3.2 Case $0 < x < y$ Coprime

We now state our results in the case this paper is concerned about. Compare this with Theorem 3. In both theorems, we use

$$F = \sum_{i=0}^{n} x^{n-i} y^i.$$

Theorem 5. *Suppose that $0 < x < y$ are coprime and consider the rotor multigraph $P_n^{x,y}$.*

(i) There exists a linear function $h : \Sigma \to \mathbb{Z}$ and a function $g : \mathcal{R} \to \mathbb{Z}$ such that, for all $(\rho, \sigma) \in \mathcal{R} \times \Sigma$, the number of particles on sink u_{n+1} in any configuration of routing$^\infty(\rho, \sigma)$ is equal to m if and only if

$$g(\rho) - h(\sigma) + mF \in g(\mathcal{R});$$

(ii) the set $g(\mathcal{R})$ is a finite set of nonnegative integers, and membership in $g(\mathcal{R})$ can be tested in linear time; moreover the unique integer m satisfying the previous condition can be found in time $O(n \log x)$, and it satisfies

$$m - \lceil \frac{h(\sigma) - g(\rho)}{F} \rceil \in [\![0, x-1]\!].$$

(iii) More generally, if (ρ, σ) and (ρ', σ') are rotor-particle configurations, then $(\rho, \sigma) \sim (\rho', \sigma')$ if and only if

$$g(\rho) - h(\sigma) = g(\rho') - h(\sigma') \text{ and } \deg(\sigma) = \deg(\sigma').$$

Note that, in the case $x = 1$, we have $m = \lceil \frac{h(\sigma)-g(\rho)}{F} \rceil$ as in the case $x = y = 1$ and no further algorithm is needed.

This is now the version of Theorem 4 in our present case. We define $\bar{h}$ and $\bar{g}$ as equal respectively to h and g modulo F.

Theorem 6. *Suppose that $0 < x < y$ are coprime and consider the rotor multigraph $P_n^{x,y}$.*

(i) The Sandpile Group of $P_n^{x,y}$ is cyclic of order F;
(ii) The map $\bar{h} : \Sigma \to \mathbb{Z}/F\mathbb{Z}$ quotients by $\sim_S$ into an isomorphism between $SP(P_n^{x,y})$ and $\mathbb{Z}/F\mathbb{Z}$;
(iii) The map $\bar{g}$ quotients by $\sim$ into a bijection between rotor equivalence classes and $\mathbb{Z}/F\mathbb{Z}$;
(iv) The action of the sandpile group on rotor equivalence classes can be understood in the following way: let (ρ, σ) be a rotor-particle configuration and $(\rho', \sigma') \in routing^\infty(\rho, \sigma)$. Then ρ' is in class $\bar{g}(\rho') = \bar{g}(\rho) - \bar{h}(\sigma)$.

As an example, we consider the Path Multigraph $P_3^{2,3}$. The graph is depicted on Fig. 4, together with harmonic values (values of h, inside vertices) and arcmonic values (values of g, on arcs).

Consider for instance the particle configuration $\sigma = (-8, 5, 13, -5, 12)$ from left to right such that

$$h(\sigma) = -8 \times 0 + 5 \times 8 + 13 \times 20 - 5 \times 38 + 12 \times 65 = 890,$$

and the rotor configuration $\rho = (a_1^1, a_1^2, a_1^3)$ such that $g(\rho) = 12 + 18 + 27 = 57$. We have $F = 65$, and $g(\mathcal{R}) = \{0, 8, 12, 16, 18, 20, 24, 26, 27, 28, 30, 32, 34, 35, 36, 38, 39, 40, 42, 43, 44, 45, 46, 47, 48, 50, 51, 52, 53, 54, 55, 56, 57, 58, 59, 60, 61, 62, 63, 64, 66, 67, 68, 69, 70, 71, 72, 74, 75, 76, 78, 79, 80, 82, 84, 86, 87, 88, 90, 94, 96, 98, 102, 106, 114\}$.

The only value v in $g(\mathcal{R})$ equal to $g(\rho) - h(\sigma) = -833 \mod 65$ is $12 = -833 + 13 * 65$. Since $\deg(\sigma) = 17$, in the end of the routing there are 13 particles on sink u_4 and 4 particles on sink u_0. The final rotor configuration ρ' satisfies

$$\bar{g}(\rho') = \bar{g}(\rho) - \bar{h}(\sigma) = -833 \mod 65 = 12 \mod 65$$

so $g(\rho') = 12$ by looking in $g(\mathcal{R})$ (in this case, this gives a unique possibility for ρ').

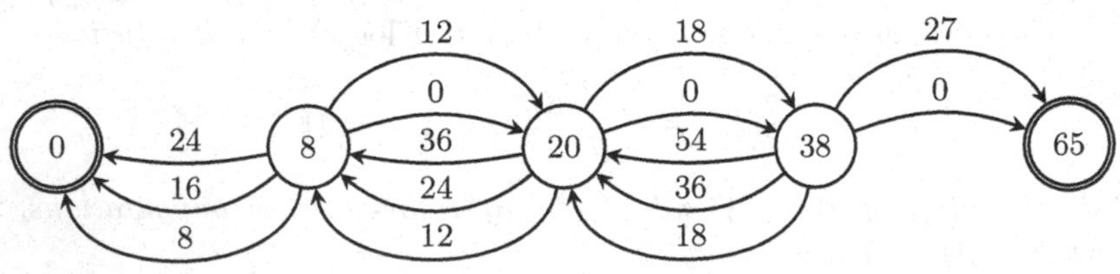

Fig. 4. Harmonic and arcmonic values on $P_3^{2,3}$. The values of h and g are given respectively in vertices and on arcs.

4 Harmonic and Arcmonic Functions in the Path

In the rest of the paper, we fix $n > 0$ and coprime integers x, y such that $0 < x < y$, and consider the Path Multigraph $P_n^{x,y}$ as defined in Subsect. 2.1.

4.1 Definition of h and g

First, we define the linear function $h : \Sigma \to \mathbb{Z}$ which will serve as an invariant for the firing operation and enable the characterization of particle equivalence classes. Initially, we define h on vertices and then extend it by linearity to Σ.

Lemma 2. *The linear function $h : \Sigma \to \mathbb{Z}$ defined by $h(u_0) = 0$ and*

$$h(u_k) = \sum_{i=0}^{k-1} x^{n-i} y^i$$

for $k \in [\![1, n+1]\!]$ is harmonic on G, i.e. for any $u \in V_0$ we have $h(\Delta(u)) = 0$.

Corollary 2. *For any particle configurations σ, σ', if $\sigma \sim \sigma'$ then $h(\sigma) = h(\sigma')$.*

It turns out that $h(u_k)$ is the number of acyclic configurations in $P_n^{x,y}$ that contain a directed path from u_k to u_{n+1}. In particular, $h(u_{n+1})$ is the number of rooted forests, which is also the number of particle equivalence classes and rotor equivalence classes [13].

We now define a similar function for rotor configurations, designed to be invariant on equivalence classes of rotors configurations. We introduce the term arcmonic for these functions that correspond to harmonic functions but on arcs.

Proposition 6. *The linear function $g : \mathbb{Z}^A \to \mathbb{Z}$, defined by*

$$g(a_j^k) = \sum_{i=0}^{j-1} (h(head(a_i^k)) - h(u_k))$$

for all $k \in [\![1, n]\!]$ and $j \in [\![0, x+y-1]\!]$ (in particular, $g(a_0^k) = 0$) is arcmonic, i.e. it satisfies for all directed circuits C in $G(\rho)$, $g(C) = g(\theta(C))$, where C is identified with the sum of arcs $\sum_{a \in C} a$.

By identifying a rotor configuration ρ with the formal sum of its arcs, we define

$$g(\rho) = \sum_{u \in V_0} g(\rho(v)).$$

Corollary 3. *If $\rho \in \mathcal{R}, \rho' \in \mathcal{R}$ are such that $\rho \sim \rho'$, then $g(\rho) = g(\rho')$.*

The exact values of g are given by:

Proposition 7. *For $j \in [\![0, x+y-1]\!]$ and $k \in [\![1, n]\!]$,*

$$g(a_j^k) = \begin{cases} jd_k & \text{if } j \in [\![0, x]\!] \\ (x+y-j)d_{k-1} & \text{if } j \in [\![x+1, x+y-1]\!] \end{cases}$$

where, for every $k \geq 0$, $d_k = x^{n-k}y^k$.

Remark that, for every $k \in [\![0, n]\!], d_k = h(u_{k+1}) - h(u_k)$. See Fig. 4 for an example of harmonic and arcmonic values on $P_3^{2,3}$. In this example, $d_0 = 8$, $d_1 = 12$, $d_2 = 18$, $d_3 = 27$, and $d_4 = \frac{81}{2}$.

Proposition 8. *If (ρ, σ) and (ρ', σ') are rotor-particle configurations, then if $(\rho, \sigma) \sim (\rho', \sigma')$ we have*

$$g(\rho) - h(\sigma) = g(\rho') - h(\sigma').$$

It turns out that Corollary 3 and Proposition 8 are equivalences.

4.2 Stable Decomposition of Arcmonic Values

In the light of Proposition 8, it becomes important to characterize which integers are of the form $g(\rho)$ for some $\rho \in \mathcal{R}$. If $\rho \in \mathcal{R}$, by Proposition 7, $g(\rho)$ can be decomposed as a sum

$$g(\rho) = \sum_{k=0}^{n} c_k d_k,$$

with $c_k \in [\![0, x+y-1]\!]$ for all $k \in [\![1, n]\!]$; recall that $d_k = x^{n-k}y^k$.

This decomposition is not unique since all equivalent rotor configurations share the same value.

Theorem 7. *Every integer $v \geq 0$ has unique decomposition of the form*

$$v = \sum_{k=0}^{n} c_k d_k + c_{n+1} d_{n+1}$$

with $c_k \in [\![0, y-1]\!]$ for $k \in [\![0, n]\!]$ and $c_{n+1} \in x\mathbb{Z}$. We call this decomposition the **stable decomposition** *of v and denote it by $c[v]$.*

Note that $d_{n+1} = \frac{y^{n+1}}{x}$. A special case is the case $x = 1$ where if $v < y^{n+1}$, the stable decomposition of v coincides with the decomposition of v in base y up to the n-th element.

Proof. We establish the uniqueness of this stable decomposition. The existence relies on the lemmas presented subsequently.

Suppose that v admits two stable decompositions $c^1 = (c_0^1, \ldots, c_{n+1}^1)$ and $c^2 = (c_0^2, \ldots, c_{n+1}^2)$. Recall that, for $i \in \{1, 2\}$, $c_{n+1}^i \in x\mathbb{Z}$. Then:

$$\sum_{k=0}^{n} c_k^1 d_k + c_{n+1}^1 d_{n+1} = \sum_{k=0}^{n} c_k^2 d_k + c_{n+1}^2 d_{n+1} \mod y$$

which amounts to

$$c_0^1 d_0 = c_0^2 d_0 \mod y.$$

Since $d_0 = x^n$ and y are coprime, and $0 \leq c_0^1, c_0^2 \leq y-1$, we obtain $c_0^1 = c_0^2$. Now, consider $v' = \frac{v - c_0^1 x^n}{y}$, then, for $i \in \{1,2\}$,

$$v' = \sum_{k=0}^{n-1} c_{k+1}^i x^{n-1-k} y^k + c_{n+1}^i \frac{y^n}{x}$$

and one can apply the same reasoning iteratively on v' to show that $c_1^1 = c_1^2$, $c_2^1 = c_2^2$, etc. And finally that $c^1 = c^2$. □

To prove the existence of the stable decomposition, we rely on another device named **Engel Machine** [6]. The Engel Machine $E_n^{x,y}$ is the Multigraph defined on the set $\{u_0, u_1, \cdots, u_n\} \cup \{u_{n+1}, s\}$, where every vertex u_i for $i \in [\![0,n]\!]$ has x arcs going to u_{i+1} and $y-x$ arcs going to s. Since we assumed $y > x$, then $y - x > 0$. Vertices s and u_{n+1} are sinks. We say that a particle configuration σ in $E_n^{x,y}$ is **nonnegative** if $\sigma(u_i) \geq 0$ for $i \in [\![0,n]\!]$ (whereas sinks may have a negative value). See Fig. 5 for an example.

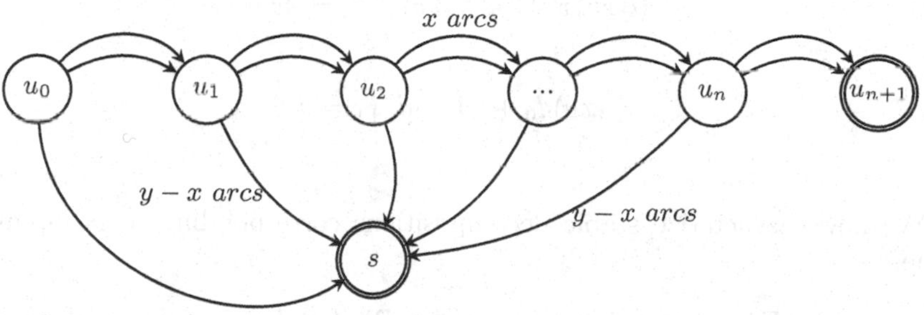

Fig. 5. The Engel Machine $E_n^{2,3}$.

We define a function h_E on the vertices of this graph that will turn out to be harmonic on $E_n^{x,y}$. This function is defined by

$$h_E(s) = 0 \text{ and } h_E(u_k) = d_k \text{ for k in } [\![0, n+1]\!]$$

and extend it to particle configurations by linearity.

We shall be mainly concerned with the h_E value of particle configurations in the Engel Machine. In order to keep notation simple, and since $h_E(s) = 0$, the value of configurations on s never matters and we identify particle configurations in $c \in \Sigma(E_n^{x,y})$ with words $c \in \mathbb{Z}^{n+2}$. In particular, for any $v \geq 0$, the notation $c[v]$ denotes the word corresponding to the stable decomposition of v, as well as a (stable) particle configuration (we can suppose that its value on s is always 0). Note that $h_E(c[v]) = v$ by construction. Conversely, remark that any nonnegative *stable* configuration c with $h_E(c) = v$ gives the unique *stable* decomposition of v.

Lemma 3. *The function h_E is harmonic on $E_n^{x,y}$.*

Proof. Consider the particle configuration c' obtained from c by firing vertex u_k, $k \in [\![0, n]\!]$. Then:

$$h_E(c') - h_E(c) = -yd_k + xd_{k+1} = 0.$$

□

In order to compute a stable decomposition for v, one simply has to find any configuration c with $h_E(c) = v$ and then stabilize c. The proof of the next lemma provides a method for computing such a configuration c. Together with Lemma 3, this completes the proof of Theorem 7.

Lemma 4. *For any $v \geq 0$, there exists a nonnegative configuration c in $E_n^{x,y}$ with $h_E(c) = v$.*

Proof. Since x^{n+1} and y^{n+1} are coprime, by Bezout's theorem there are integers α, β such that

$$\alpha x^{n+1} + \beta y^{n+1} = 1$$

and we can choose $\alpha \geq 0$. It follows that

$$(\alpha xv)x^{n+1} + (\beta xv)y^{n+1} = xv$$

and

$$(\alpha xv)d_0 + (\beta xv)d_{n+1} = v.$$

□

We now characterize stable decompositions corresponding to an arcmonic value.

Theorem 8. *For any $v \in \mathbb{Z}$, we have $v \in g(\mathcal{R})$ if and only if the regular expression*

$$e_d = [\![0, y-1]\!]^* \cdot 0 \cdot [\![1, x]\!]^* \cdot 0$$

matches $c[v]$.

Theorem 8 is the main tool for recognizing values m such that $g(\rho) - h(\sigma) + mF \in g(\mathcal{R})$ (see Theorem 5).

Open Problems and Future Works. In this paper, we addressed the generalized version of the ARRIVAL problem in the Path Multigraph $P_n^{x,y}$. Moreover, we investigated the Sandpile Group structure and its action on rotor configurations when x and y are coprime. However, when x and y are not coprime, we observed that the characterization of classes by harmonic and arcmonic functions becomes inadequate, necessitating the inclusion of more comprehensive algebraic invariants. We are currently working on a project that presents a theory of arcmonic and harmonic functions applicable to general graphs, which will be submitted soon to publication.

Moreover, it is worth considering other scenarios, such as variations in x and y across different vertices or changes in the rotor order. These cases pose interesting questions that require further investigation. We regard them as open problems that warrant additional research.

Acknowledgements. This work was supported by a public grant as part of the Investissement d'Avenir project, reference ANR-11-LABX-0056-LMH, LabEx LMH.

References

1. Auger, D., Coucheney, P., Duhazé, L.: Polynomial time algorithm for arrival on tree-like multigraphs. In: 47th International Symposium on Mathematical Foundations of Computer Science, MFCS 2022. Schloss Dagstuhl-Leibniz-Zentrum für Informatik (2022)
2. Auger, D., Coucheney, P., Duhazé, L., Etse, K.R.: Generalized arrival problem for rotor walks in path multigraphs. arXiv preprint arXiv:2307.01897 (2023)
3. Björner, A., Lovász, L.: Chip-firing games on directed graphs. J. Algebraic Combin. **1**, 305–328 (1992)
4. Björner, A., Lovász, L., Shor, P.W.: Chip-firing games on graphs. Eur. J. Comb. **12**(4), 283–291 (1991)
5. Dohrau, J., Gärtner, B., Kohler, M., Matoušek, J., Welzl, E.: ARRIVAL: a zero-player graph game in NP ∩ coNP. In: Loebl, M., Nešetřil, J., Thomas, R. (eds.) A Journey Through Discrete Mathematics, pp. 367–374. Springer, Cham (2017). https://doi.org/10.1007/978-3-319-44479-6_14
6. Engel, A.: The probabilistic abacus. Educ. Stud. Math. **6**, 1–22 (1975). https://doi.org/10.1007/BF00590021
7. Farrell, M., Levine, L.: Coeulerian graphs. Proc. Am. Math. Soc. **144**(7), 2847–2860 (2016)
8. Frougny, C., Klouda, K.: Rational base number systems for p-adic numbers. RAIRO-Theor. Inf. Appl.-Informatique Théorique et Applications **46**(1), 87–106 (2012)
9. Gärtner, B., Hansen, T.D., Hubácek, P., Král, K., Mosaad, H., Slívová, V.: ARRIVAL: next stop in CLS. In: 45th International Colloquium on Automata, Languages, and Programming, ICALP 2018. Schloss Dagstuhl-Leibniz-Zentrum fuer Informatik (2018)
10. Gärtner, B., Haslebacher, S., Hoang, H.P.: A subexponential algorithm for ARRIVAL. In: ICALP 2021, vol. 198, pp. 69:1–69:14 (2021)
11. Giacaglia, G.P., Levine, L., Propp, J., Zayas-Palmer, L.: Local-to-global principles for rotor walk. arXiv preprint arXiv:1107.4442 (2011)
12. Hoang, P.H.: On Two Combinatorial Reconfiguration Problems: Reachability and Hamiltonicity. Ph.D. thesis, ETH Zurich (2022)
13. Holroyd, A.E., Levine, L., Mészàos, K., Peres, Y., Propp, J., Wilson, D.B.: Chip-firing and rotor-routing on directed graphs. In: Sidoravicius, V., Vares, M.E. (eds.) In and Out of Equilibrium 2. Progress in Probability, vol. 60, pp. 331–364. Birkhäuer Basel (2008). https://doi.org/10.1007/978-3-7643-8786-0_17
14. Holroyd, A.E., Propp, J.: Rotor walks and Markov chains. In: Algorithmic Probability and Combinatorics, pp. 105–126 (2010)

15. Povolotsky, A., Priezzhev, V., Shcherbakov, R.: Dynamics of Eulerian walkers. Phys. Rev. E **58**(5), 5449 (1998)
16. Priezzhev, V.B., Dhar, D., Dhar, A., Krishnamurthy, S.: Eulerian walkers as a model of self-organized criticality. Phys. Rev. Lett. **77**(25), 5079 (1996)
17. Yanovski, V., Wagner, I.A., Bruckstein, A.M.: A distributed ant algorithm for protect efficiently patrolling a network. Algorithmica **37**(3), 165–186 (2003)

Author Index

O. Bournez et al. (Eds.): RP 2023, LNCS 14235, p. 199, 2023.
https://doi.org/10.1007/978-3-031-45286-4

Printed in the United States
by Baker & Taylor Publisher Services

Printed in the United States
by Baker & Taylor Publisher Services